CO$_2$ IN SEAWATER:
EQUILIBRIUM, KINETICS, ISOTOF

The front cover shows the effect of various biogeochemical processes on the carbonate chemistry in seawater (cf. Fig. 1.1.3). The x- and y-axes represent total dissolved inorganic carbon (DIC) and total alkalinity (TA), while the labeled diagonal lines indicate levels of constant dissolved carbon dioxide ($[CO_2]$). For example, $CaCO_3$ formation reduces DIC by one and TA by two units, respectively, therefore driving the system to higher $[CO_2]$ levels (cf. Broecker and Peng, 1989). Invasion of atmospheric CO_2 into the ocean increases DIC, while release of CO_2 to the atmosphere has the opposite effect. TA stays constant in these two cases.

Elsevier Oceanography Series
Series Editor: David Halpern (1993–)

Elsevier Oceanography Series, 65

CO$_2$ IN SEAWATER: EQUILIBRIUM, KINETICS, ISOTOPES

Richard E. Zeebe[1,2,]* **and Dieter Wolf-Gladrow**[1,]**

[1]*Alfred Wegener Institute for Polar and Marine Research*
PO Box 12 01 61
D-27515 Bremerhaven
Germany

[2]*Lamont-Doherty Earth Observatoryof Columbia University*
Rte 9W, Palisades
NY 10964-8000
USA

*Fax: +49 (471) 4831 1425; Email: rzeebe@awi-bremerhaven.de
**Fax: +49 (471) 4831 1425; Email: dwolf@awi-bremerhaven.de

ELSEVIER
Amsterdam - Boston - London - New York - Oxford - Paris - San Diego
San Francisco - Singapore - Sydney

ELSEVIER B.V.
Radarweg 29
P.O. Box 211, 1000 AE
Amsterdam, The Netherlands

ELSEVIER Inc.
525 B Street
Suite 1900, San Diego
CA 92101-4495, USA

ELSEVIER Ltd
The Boulevard
Langford Lane, Kidlington,
Oxford OX5 1GB, UK

ELSEVIER Ltd
84 Theobalds Road
London WC1X 8RR
UK

First edition 2001
Second impression 2003 (with corrections)
Third impression 2005 (with corrections)

British Library Cataloguing in Publication Data
A catalogue record is available from the British Library.

Library of Congress Cataloging in Publication Data
A catalog record is available from the Library of Congress.

ISBN (this volume) 0 444 50579 2 (hardbound)
                     0 444 50946 1 (paperback)
ISSN (series)   0422 9894

♾ The paper used in this publication meets the requirements of ANSI/NISO Z39.48-1992 (Permanence of Paper).

Transferred to digital printing in 2007.

Preface

Carbon dioxide is the most important greenhouse gas after water vapor in the atmosphere of the earth. More than 98% of the carbon of the atmosphere-ocean system is stored in the oceans as dissolved inorganic carbon. The carbon reservoirs of ocean and atmosphere may be pictured as a dog (the ocean) and its tail (the atmosphere). An understanding of the dynamics of the global carbon cycle and of changes of atmospheric CO_2 concentrations in the past and future therefore demands a comprehension of the role of the ocean in the carbon cycle. Analyses of air trapped in the ice sheets of Greenland and Antarctica have revealed that atmospheric carbon dioxide concentrations varied between glacial and interglacial times, with low values during glacials. These natural variations are most probably driven by oceanic processes. With the beginning of the industrial revolution, the anthropogenic influence on the global carbon cycle became increasingly important. The burning of coal, gas, and oil and the change in land use including deforestation resulted in an increase of atmospheric carbon dioxide that is comparable to the increase from the last glacial to preindustrial times. Wagging the tail provokes a response of the dog. Invasion of 'anthropogenic' carbon dioxide into the ocean has already led to an appreciable increase of the acidity of the surface ocean since the year 1800.

Comprehension of past and prediction of future changes of the marine carbon cycle requires an understanding of several questions of which two are of major importance. (1) Which processes are responsible for the variations of atmospheric carbon dioxide concentrations on glacial-interglacial time scales and (2) How does the ocean (including the biota) respond to anthropogenic perturbations and natural variations? Working on these questions indispensably demands an interdisciplinary approach in which scientists from different backgrounds join their abilities and efforts to benefit from each other.

The authors of this book have been working for several years in an interdisciplinary group which encompasses biologists, physicists, mathematicians, and geologists. In our everyday work we have experienced that the key for understanding critical processes of the marine carbon cycle is a sound knowledge of the seawater carbonate chemistry, including equilibrium and nonequilibrium properties as well as stable isotope fractionation. Unfortunately, it appears to be quite difficult for non-chemists to obtain a good knowledge of these subjects from, e.g., original contributions to chemical journals. Text books on chemical oceanography usually include an introduction to the equilibrium properties of the carbonate system. However, hitherto there is no coherent description of equilibrium *and* nonequilibrium properties *and* of stable isotope fractionation among the elements of the carbonate system in form of a comprehensible text. It is our intention to provide an overview and a synthesis of these subjects which should be useful for graduate students and researchers in various fields such as biogeochemistry, chemical oceanography, paleoceanography, marine biology, marine chemistry, marine geology, and others. In addition to the presentation of well known topics in the book, outcome of original research is included which has not been published previously (see, for instance, Sections 1.6, 2.3.5, 3.5.3, and Appendix B and C).

One of our main goals is to provide a quantitative description of the topics discussed in connection with the carbonate system. In this regard, the treatment given in the current book differs from many other presentations which are often of qualitative nature. We feel that our approach is a very useful one because it provides the reader with strong tools that can be used in her/his own studies and research. Naturally, this requires a little mathematics. We have tried to keep the mathematical level as low as possible in order to make the text accessible to a wide range of scientists from different disciplines. However, an adequate description of reaction kinetics, for example, inevitably requires the application of ordinary differential equations. Partial derivatives are used in the derivation of the Revelle factor. Diffusion is governed by a partial differential equation which simplifies, however, to an ordinary differential equation by restriction to stationary problems in one spatial dimension. The Schrödinger equation, which is a partial differential equation, is only briefly mentioned in the appendix. Whenever a mathematical or physical derivation is given in the text, a smaller font size is used to indicate that these sections or paragraphs may be skipped by the reader when studying the subject for the first time. Details and elaborate calculations are given in the appendix.

For the most part, we have attempted to explain the concepts of chemistry, biology, and physics used in the text. However, since the aspects discussed touch on many different branches of various disciplines, it was impossible to recapitulate all basics in detail. In these cases it might be useful for the reader to consult additional text books.

The outline of the book is as follows. The text begins with an introduction to the equilibrium properties of the carbonate system (Chapter 1, Equilibrium) in which basic concepts such as equilibrium constants, alkalinity, pH scales, and buffering are discussed. In addition, the application of these concepts is emphasized, including a discussion of the Revelle factor and future scenarios of atmospheric CO_2 concentrations. Chapter 2 (Kinetics) deals with the nonequilibrium properties of the seawater carbonate chemistry. Whereas principles of chemical kinetics are recapitulated, reaction rates and relaxation times of the carbonate system are considered in detail. Chapter 3 (Stable Isotope Fractionation) provides a general introduction to stable isotope fractionation and describes the partitioning of carbon, oxygen, and boron isotopes between the species of the carbonate system. The appendix contains formulas for the equilibrium constants of the carbonate system, mathematical expressions to calculate carbonate system parameters, answers to exercises and more. Numerical routines for the calculation of carbonate system parameters are available on our web-page: 'http://www.awi-bremerhaven.de/Carbon/co2book.html'.

Last but not least, a few comments on the exercises are added. Problems with one asterisk (*) should be very easy to solve (in a few minutes); those with two asterisks require more thinking or somewhat lengthy ('... after some algebra ...') calculations. Exercises with three asterisks are difficult. They are addressed to the reader who is interested in solving advanced problems and puzzles. Answers to exercises are given in Appendix D.

Acknowledgments. We are indebted to many people who helped us to write this book. Some of them spent a lot of time on discussing aspects of the carbonate chemistry with us, others provided comments on the manuscript or proof-read the manuscript. Many scientists have encouraged our work by the stimulating outcome of their own research. Unfortunately, it is impossible to name them all here. Being aware that any list must be incomplete, we still thank the following people for their support (in alphabetical order): D. C. E. Bakker, J. Bijma, A. G. Dickson, A. Engel, K. Fahl, N. G. Hemming, M. Hoppema, H. Jansen, M. M. Rutgers van der Loeff, F. Louanchi, J. D. Ortiz, U. Riebesell, A. Sanyal, H. J. Spero, M. H. C. Stoll, Ch. Völker, D. W. R. Wallace, and A. Wischmeyer. We thank H. J. W. de Baar for his suggestions to improve the manuscript. R. E. Z. is particularly grateful to Mira Djokić for her patience.

We gratefully acknowledge the reviewers of the present book who improved the manuscript by their constructive comments. They are: H. J. W. de Baar, J. M. Hayes, A. Körtzinger, A. Mucci, E. Usdowski, and R. Wanninkhof.

Contents

Chapter 1

Equilibrium

Next to nitrogen, oxygen and argon, carbon dioxide is the most abundant gas in the earth's atmosphere. Next to water vapor it is the most important greenhouse gas. In contrast to nitrogen and oxygen most carbon dioxide of the combined atmosphere - ocean system is dissolved in water (98%), because carbon dioxide is not simply dissolved in water as other gases, but it reacts with water and forms bicarbonate and carbonate ions.

Although the carbonate system in seawater comprises only a few components, essentially CO_2, HCO_3^-, CO_3^{2-}, H^+, OH^-, and may be described by equations derived from the law of mass action, its behavior in response to perturbations is in some cases not easily predictable by intuitive reasoning. A doubling of the CO_2 concentration in the atmosphere will not cause a doubling of the total dissolved inorganic carbon, DIC, at equilibrium, but results in an increase of only $\sim 10\%$. This unexpectedly low increase is due to the dissociation of carbon dioxide and the simultaneous change of pH, see Section 1.5 on Revelle factor. Another example is biological precipitation of calcium carbonate, which will remove inorganic carbon from the oceanic surface layer, but does not result in further uptake of atmospheric carbon dioxide. On the contrary, because of a change of alkalinity, carbon dioxide will outgas as a consequence of production of calcium carbonate!

The main goal of this chapter is to present the equilibrium aspects of the carbonate system. The basic equations derived from the law of mass action allow us to calculate the ratios between the different forms of dissolved inorganic carbon (Section 1.1). Chemical concepts such as alkalinity, pH, and fugacity will be discussed in some detail (Sections 1.2 – 1.4) in order to make the text understandable also for non-chemists. The Revelle factor, which is important for the determination of oceanic CO_2 uptake, is introduced in Section 1.5. The final section (Section 1.6) of this chapter contains several interesting problems that can be addressed with knowledge of the equilibrium properties of the carbonate system. In particular, the first two problems, $CaCO_3$ formation and Revelle factor, are of relevance to the global carbon cycle. Values of the equilibrium constants and their dependence on temperature, salinity, and pressure can be found in Appendix A. In Appendix B it is shown how to calculate all components of the carbonate system from any two given quantities, for instance, from $[CO_2]$ and pH. The appendices are meant for reference only.

Various aspects of equilibrium properties of the carbonate system have been discussed, for example, in reviews by Skirrow (1965, 1975), in DOE (1994), or in textbooks such as Drever (1982), Morel and Hering (1993), Millero (1996), Stumm and Morgan (1996), to name only a few.

1.1   The carbonate system

In the ocean, carbon dioxide exists in three different inorganic forms: as free carbon dioxide, $CO_2(aq)$ = aqueous carbon dioxide, as bicarbonate, HCO_3^-,[1] and as carbonate ion, CO_3^{2-} (see Figure 1.1.1). A fourth form is H_2CO_3 (true carbonic acid); the concentration of H_2CO_3 is, however, much smaller than that of $CO_2(aq)$ ($\lesssim 0.3\%$). The sum of the two electrically neutral forms, true carbonic acid, H_2CO_3, and aqueous carbon dioxide, $CO_2(aq)$, which are chemically not separable, is usually denoted by CO_2 or $H_2CO_3^*$. We will use the former notation in this book:

$$[CO_2] = [CO_2(aq)] + [H_2CO_3] \, , \tag{1.1.1}$$

where brackets represent total stoichiometric concentrations. Note that in the literature also the symbol CO_{2T} is used. In thermodynamic equilibrium with gaseous carbon dioxide ($CO_2(g)$):

$$CO_2(g) \overset{K_0}{\rightleftharpoons} CO_2 \, , \tag{1.1.2}$$

the concentration of CO_2 is given by Henry's law (see Section 1.5) with K_0 being the solubility coefficient of CO_2 in seawater. The carbonate species are related by the following equilibria:

$$CO_2(aq) + H_2O \rightleftharpoons H_2CO_3 \rightleftharpoons HCO_3^- + H^+ \rightleftharpoons CO_3^{2-} + 2H^+ \, . \tag{1.1.3}$$

---

[1]The prefix 'bi' in bicarbonate has the following origin: the carbonate ion, CO_3^{2-}, may bind a second single positive ion in addition to H^+.

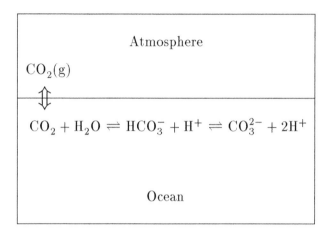

Figure 1.1.1: Schematic illustration of the carbonate system in the ocean. CO_2 is exchanged between atmosphere and ocean via equilibration of $CO_2(g)$ and dissolved CO_2. Dissolved CO_2 is part of the carbonate system in seawater that includes bicarbonate, HCO_3^-, and carbonate ion, CO_3^{2-}.

It is noted that equilibria are considered and not reaction pathways. Thus the hydroxylation $CO_2 + OH^- \rightleftharpoons HCO_3^-$, does not show up in Eq. (1.1.3). Using CO_2 instead of carbonic acid and aqueous carbon dioxide (Eq. (1.1.1)), the equilibria (Eq. (1.1.3)) simplify:

$$CO_2 + H_2O \overset{K_1}{\rightleftharpoons} HCO_3^- + H^+ \overset{K_2}{\rightleftharpoons} CO_3^{2-} + 2H^+ \qquad (1.1.4)$$

where K_1 and K_2 are equilibrium constants, often referred to as the first and second dissociation constants of carbonic acid, respectively. For the description of the carbonate system in seawater, stoichiometric equilibrium constants are used which are related to concentrations:

$$K_1^* = \frac{[HCO_3^-][H^+]}{[CO_2]} \qquad (1.1.5)$$

$$K_2^* = \frac{[CO_3^{2-}][H^+]}{[HCO_3^-]} . \qquad (1.1.6)$$

Stoichiometric equilibrium constants depend on temperature T, pressure P, and salinity S and are conventionally denoted by a star (refer to discussion in Section 1.1.3, 1.1.6, and Appendix A).

The sum of the dissolved forms CO_2, HCO_3^-, and CO_3^{2-}, is called total dissolved inorganic carbon, which we will denote by DIC or ΣCO_2:

$$DIC \equiv \Sigma CO_2 = [CO_2] + [HCO_3^-] + [CO_3^{2-}] . \qquad (1.1.7)$$

Note that in the literature also the symbols TCO_2 and C_T are used. A further essential quantity for the description of the carbonate system is the alkalinity, which is closely related to the charge balance in seawater. One might say that while DIC keeps track of the carbon, the alkalinity keeps track of the charges. The carbonate alkalinity, CA, is defined as[2]:

$$CA \;\; = \;\; [HCO_3^-] + 2[CO_3^{2-}] \,, \tag{1.1.8}$$

where the carbonate ion, CO_3^{2-}, is counted twice because it has a double negative charge. Note that the current treatment of alkalinity is a simplification and that the carbonate alkalinity is part of the total alkalinity, TA, which also includes boron compounds and more (see below):

$$\begin{aligned} TA \;\; = \;\; & [HCO_3^-] + 2[CO_3^{2-}] + [B(OH)_4^-] + [OH^-] - [H^+] \\ & + \text{minor components} \,. \end{aligned}$$

The concept of total alkalinity is extensively examined in Section 1.2.

The quantities introduced above are used for the quantitative description of the carbonate system in seawater. The two equilibrium conditions, Eqs. (1.1.5) and (1.1.6), the mass balance for total inorganic carbon, Eq. (1.1.7), and the charge balance, Eq. (1.1.8), constitute four equations with six unknown variables $[CO_2]$, $[HCO_3^-]$, $[CO_3^{2-}]$, $[H^+]$, DIC, and CA. As a result, when $2 = 6 - 4$ variables are known, the system is determined and all other components can be calculated. Theoretically, this goal could be achieved by measuring any two of the six quantities. In principle, however, only $[CO_2]$, $[H^+]$, DIC, and TA can be measured directly. This is the reason, for example, why for a quantitative description the dissolved boron species and other minor species have to be taken into account as they contribute to the total alkalinity. The procedure how to determine all components of the carbonate system from any two given quantities, including boron compounds, is demonstrated in Appendix B.

As an example, consider the case in which DIC and $[H^+]$ (i.e. pH) have been obtained by direct measurement. The concentrations of CO_2, HCO_3^-, and CO_3^{2-} and CA can then be expressed as functions of DIC and $[H^+]$:

$$[CO_2] \;\; = \;\; DIC \left/ \left(1 + \frac{K_1^*}{[H^+]} + \frac{K_1^* K_2^*}{[H^+]^2} \right) \right. \tag{1.1.9}$$

$$[HCO_3^-] \;\; = \;\; DIC \left/ \left(1 + \frac{[H^+]}{K_1^*} + \frac{K_2^*}{[H^+]} \right) \right. \tag{1.1.10}$$

$$[CO_3^{2-}] \;\; = \;\; DIC \left/ \left(1 + \frac{[H^+]}{K_2^*} + \frac{[H^+]^2}{K_1^* K_2^*} \right) \right. \tag{1.1.11}$$

$$CA \;\; = \;\; [HCO_3^-] + 2[CO_3^{2-}] \,. \tag{1.1.12}$$

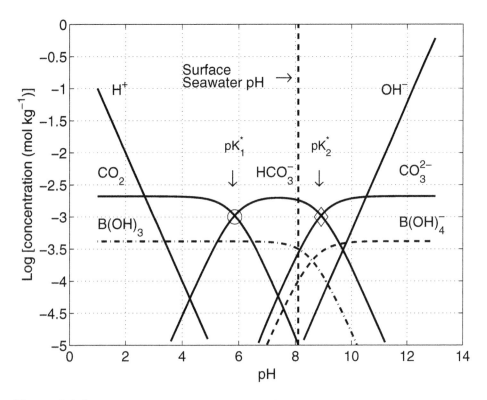

Figure 1.1.2: Carbonate system: Bjerrum plot (named after N. Bjerrum who invented the graphical representation of equilibrium relationships in 1914); DIC = 2.1 mmol kg^{-1}, $S = 35$, $T_c = 25°C$. The circle and the diamond indicate $pK_1^* = 5.86$ and $pK_2^* = 8.92$ of carbonic acid. The values of pK_B^* and pK_W^* used are 8.60 and 13.22, respectively (DOE, 1994). Note that in seawater, the relative proportions of CO_2, HCO_3^-, and CO_3^{2-} control the pH and not vice versa as this plot might suggest (see text).

Let us assume that a surface seawater sample in equilibrium with today's atmosphere at $pCO_2 = 365 \ \mu$atm has a pH of 8.1 and DIC = 2.1 mmol kg^{-1} at a salinity[3] $S = 35$ and $T_c = 25°C$. Using Eqs. (1.1.9)-(1.1.11), we calculate $[CO_2] = 10.4 \ \mu$mol kg^{-1}, $[HCO_3^-] = 1818 \ \mu$mol kg^{-1}, and $[CO_3^{2-}] = 272 \ \mu$mol kg^{-1}; the constants used for the calculations are summarized in DOE (1994), see Appendix A. In other words, the percentage of the dis-

---

[2]The alkalinity is expressed in units of mol kg^{-1}. Note that the unit eq kg^{-1} is also widely used in the literature.

[3]Note that no unit is assigned to the practical salinity S (for definition see e.g. Müller, 1999). Roughly, $S = 35$ corresponds to ~ 35 g salt per kg seawater. If not stated otherwise, quantities will be given for $S = 35$ and $T_c = 25°C$ in order to allow comparison with values in the chemical literature. The calculation of quantities at other temperatures and salinities will be left as an exercise to the reader.

solved species is $[CO_2] : [HCO_3^-] : [CO_3^{2-}] \simeq 0.5\% : 86.5\% : 13\%$. Thus, at typical seawater conditions, bicarbonate is the dominant species, followed by carbonate ion, whereas dissolved carbon dioxide is present only in small concentrations. This is illustrated in Figure 1.1.2 by the crossover between the concentration curves and the dashed vertical line at $pH = 8.1$. Also indicated in Figure 1.1.2 are the pH values at which the concentration of CO_2 equals the concentration of HCO_3^- and at which the concentration of HCO_3^- equals the concentration of CO_3^{2-}. These pH values correspond to pK_1^* and pK_2^*, the pK values of the first and second dissociation constants of carbonic acid, respectively (cf. box on pK values).

---

pK-values. Mathematically, the pK value of an equilibrium constant, K, is simply the negative common logarithm of K:

$$pK := -\log_{10}(K) .$$

This is in analogy to the pH value which is the negative common logarithm of $[H^+]$. Chemically, the pK value has an interesting interpretation. Consider, for example, the first acidity constant of carbonic acid:

$$K_1^* = \frac{[H^+][HCO_3^-]}{[CO_2]} .$$

Let us assume that in a given solution the concentration of CO_2 is equal to the concentration of HCO_3^-. It follows that $[H^+]$ is equal to K_1^*:

$$[H^+] = K_1^* \qquad \text{at } [CO_2] = [HCO_3^-]$$

and thus (taking the negative common logarithm) that the pH of the solution is equal to pK_1^*, i.e.:

$$pH = pK_1^* \qquad \text{at } [CO_2] = [HCO_3^-] .$$

This feature is graphically indicated by the circle in Figure 1.1.2 where the curves of CO_2 and HCO_3^- intersect. Consequently, $[CO_2]$ is larger than $[HCO_3^-]$ for pH values below pK_1^* (and vice versa). A similar reasoning holds for pK_2^*: $[HCO_3^-]$ is equal to $[CO_3^{2-}]$ at $pH = pK_2^*$.

Because the Bjerrum plot (Figure 1.1.2) shows the concentrations of the carbonate species as a function of pH, one might be tempted to believe that the pH is controlling the concentrations and relative proportions of the carbonate species in the ocean. However, the reverse is true: the carbonate system is the natural buffer for the seawater pH. For instance, if strong acid is added to seawater, HCO_3^- and CO_3^{2-} ions are transformed to CO_2 and the pH will remain between 8 and 6. Only after almost 3 mmol H^+ per

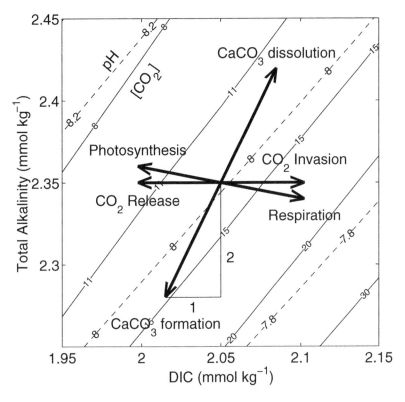

Figure 1.1.3: Effect of various processes on DIC and TA (arrows). Solid and dashed lines indicate levels of constant dissolved CO_2 (in μmol kg^{-1}) and pH, respectively, as a function of DIC and TA. $CaCO_3$ formation, for example, reduces DIC by one and TA by two units, therefore driving the system to higher CO_2 levels and lower pH. Invasion of atmospheric CO_2 into the ocean increases DIC, while release of CO_2 to the atmosphere has the opposite effect. TA stays constant in these two cases.

kg seawater have been added, the pH will drop appreciably (see, however, Sillén (1961, 1967); Holland (1984) for further reading).

Many processes affecting the carbonate system in the ocean are best described by considering the change of DIC and TA that is associated with them (Figure 1.1.3). For example, the invasion of anthropogenic CO_2 leads to an increase of DIC but does not change TA because the charge balance is not affected (see Section 1.5). The formation of $CaCO_3$ decreases both DIC and TA. For each mole of $CaCO_3$ precipitated, one mole of carbon and one mole of double positively charged Ca^{2+} ions are taken up which leads to a decrease of DIC and TA in a ratio of 1:2 (Section 1.6.1). As a result, the system shifts to higher CO_2 levels and lower pH (Figure 1.1.3). Photosynthesis reduces DIC and slightly increases TA because in addition to inorganic carbon, nutrients are taken up (Section 1.2.7). As a result of these various processes, the carbonate species and pH adjust according to the equilibrium conditions given by the set of equations (1.1.5)-(1.1.8) that has to be obeyed simultaneously. In summary, it is very useful to keep in mind that the pH and the concentrations of the carbonate species in the ocean are governed by the distribution of DIC and TA in many cases. This fact and the various processes depicted in Figure 1.1.3 will be elaborated in subsequent chapters.

Water equilibrium

The carbonate system includes water, H_2O, and its dissociation products H^+ and OH^-:

$$H_2O \quad \overset{K_W}{\rightleftharpoons} \quad H^+ + OH^-$$

where K_W is the dissociation constant, or ion product, of water. The stoichiometric equilibrium constant is defined as:

$$K_w^* = [H^+][OH^-] . \tag{1.1.13}$$

It is important to note that the symbol 'H$^+$' represents hydrate complexes associated with H_3O^+ and $H_9O_4^+$ rather than the concentration of free hydrogen ions. Free hydrogen ions do not exist in any significant amount in aqueous solutions. It is, however, convenient to refer to $[H^+]$ as the hydrogen ion concentration. This subject and the different pH scales which are used to determine the hydrogen ion concentration in aqueous solutions are discussed in more detail in Section 1.3.

Boric acid-borate equilibrium

For quantitative calculations of the carbonate system, boric acid, $B(OH)_3$, and borate, $B(OH)_4^-$, and some other minor species have to be taken into account as well. This is because those minor species contribute to the total alkalinity (TA) from which, in combination with DIC, carbon system parameters are frequently determined. The boric acid - borate equilibrium can be written as:

$$B(OH)_3 + H_2O \quad \overset{K_B}{\rightleftharpoons} \quad B(OH)_4^- + H^+$$

where K_B is the dissociation constant of boric acid. The stoichiometric equilibrium constant is defined as:

$$K_B^* = \frac{[B(OH)_4^-][H^+]}{[B(OH)_3]} . \tag{1.1.14}$$

The total boron concentration B_T is given by

$$B_T = [B(OH)_4^-] + [B(OH)_3] . \tag{1.1.15}$$

1.1.1 Effect of temperature, salinity, and pressure

As mentioned before, temperature, salinity, and pressure influence the values of the dissociation constants. It can be derived from thermodynamics that the equilibrium constant is related to the standard free energy of the reaction. Varying the temperature or the pressure of the system results in a change of this energy and thus of the thermodynamic equilibrium constant. Using the laws of thermodynamics, expressions for the temperature and pressure effects on equilibrium constants can be derived. The details of the calculations for temperature and pressure will not be discussed here - we refer the reader to e.g. Millero (1979) and Millero (1982).

From what has been said so far, it is comprehensible that the equilibrium constants (K^*'s) depend on temperature and pressure. The effect of salinity, however, is not *a priori* comprehensible since the K^*'s should not depend on e.g. the composition of the solution. The reason for the salinity dependence of the dissociation constants which are used for the description of the carbonate system (so-called 'stoichiometric constants'), is that they are not the 'true thermodynamic constants'. Thermodynamic constants are expressed in terms of ion activities whereas stoichiometric constants are expressed in terms of concentrations - this subject is discussed in more detail in Section 1.1.3.

Figure 1.1.4 shows the dependence of pK_1^* and pK_2^* on temperature, salinity, and pressure. As is obvious, shifts in the pK^* values lead to shifts in the relative proportions of CO_2, HCO_3^-, and CO_3^{2-} at a given pH. With respect to the reference values of the pK^*'s at $T_c = 25°C$, $S = 35$, and $P = 1$ atm, a decrease of temperature or salinity ($T_c = 0°C$ or $S = 0$) results in an increase of the pK^* values (cf. Table 1.1.1). Consequently, when comparing e.g. seawater at $S = 35$ and fresh water at $S = 0$ at the same pH and temperature, the relative proportion of CO_3^{2-} ions, with respect to $[CO_2]$ and $[HCO_3^-]$, will be appreciably higher in seawater than in fresh water.

Another important example of the dependence of the carbonate system parameters on e.g. T and P are the different chemical properties of water masses within the surface and the deep ocean. Let us consider a water parcel which is cooled from 25°C to 0°C, and is then sinking from the surface ocean ($S = 35$, and $P = 1$ atm) into the deep ocean ($S = 35$, and $P = 300$ atm, i.e., depth ~ 3 km). The conserved quantities, when only temperature and pressure are changed, are DIC and TA. Note that this example considers a hypothetical 'abiotic' case because biological processes such as primary

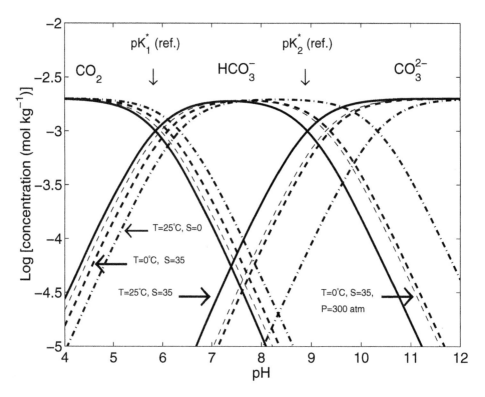

Figure 1.1.4: Illustration of the effect of temperature, pressure, and salinity on pK_1^* and pK_2^*. The reference case is $T_c = 25°C$, $S = 35$, and $P = 1$ atm (solid line). See Table 1.1.1 for values. Note that DIC = 2 mmol kg^{-1} in all cases.

production and calcification which do change DIC and TA in the real ocean are not considered. As is obvious from Figure 1.1.4, the drop in temperature from 25° to 0°C causes a large increase in the pK^* values which is only slightly modified by the pressure change. At constant DIC = 2 mmol kg^{-1}, and TA = 2.44 mmol kg^{-1}, the carbonate ion concentration of the sinking water parcel would drop from 316 to 286 μmol kg^{-1}. In reality much larger differences in [CO$_3^{2-}$] between surface and deep ocean are observed, i.e. ~ 300 μmol kg^{-1} in the surface vs. ~100 μmol kg^{-1} in the deep ocean. These differences are mainly due to biological processes which produce vertical gradients in DIC and TA.

In summary, decreasing T, S, or P results in an increase of pK_1^* and pK_2^* of carbonic acid. The same is true for the pK^*'s of the dissociation constants of water, pK_W^*, and boric acid, pK_B^*, cf. Table 1.1.1.

Table 1.1.1: Influence of salinity, temperature, and pressure on pK_1^*, pK_2^*, pK_W^*, and pK_B^*.

$T_c(°C)$	S	P (atm)	pK_1^*	pK_2^*	pK_W^*	pK_B^*
25	35	1	5.86^a	8.92^a	13.22^a	8.60^a
25	0	1	6.35^b	10.33^b	14.00^c	9.24^d
0	35	1	6.11^a	9.38^a	14.31^a	8.91^a
0	35	300	5.96^e	9.29^e	14.16^e	8.75^e

[a] DOE (1994) - for a discussion of seawater pK^*'s used in this book, see Section 1.1.6.
[b] Usdowski (1982).
[c] Stumm and Morgan (1996).
[d] Hershey et al. (1986).
[e] Pressure correction from Millero (1995).

Exercise 1.1 (*)

What is the dominant carbonate species at typical seawater pH of 8.2 ($\sim$ surface ocean) and 7.8 ($\sim$ deep ocean)? See Appendix D for answers.

Exercise 1.2 (**)

Using Eqs. (1.1.9)-(1.1.11) and the values of pK_1^* and pK_2^* at $T_c = 25°C$ and $S = 35$ given in Table 1.1.1, calculate the concentrations of CO_2, HCO_3^-, and CO_3^{2-} at pH 8.2 and DIC $= 2$ mmol kg^{-1}.

1.1.2 Ionic strength and activity coefficient

The major difference between fresh water and seawater is the total concentration and the relative proportion of ions dissolved in the solution. In concentrated solutions, the ions interact with each other and thus do not exhibit their full potential to react chemically with other chemical compounds. Before getting into the details of this subject, let us briefly summarize the content of the current section with the following example.

Imagine a volume of water on its journey from the spring of a river to the ocean. As the solution changes from fresh water to seawater, i.e. the number of ions in solution increases, the ion activity decreases due to (a) long-range electrostatic interactions and (b) ion pairing and complex formation. Because the activity of different chemical species with different charges are affected in a different way, the ratio of their activities changes. Considering the example of pK_2^* (see Figure 1.1.4), the activity of CO_3^{2-} is decreasing more strongly than, e.g. the activity of HCO_3^-. Consequently, the pK_2^* of carbonic acid in seawater is smaller (K_2^* is greater) than its corresponding value in fresh water. This explains why seawater has a higher CO_3^{2-} concentration (relative to the concentrations of CO_2 and HCO_3^-) than fresh water at the same pH.

Ionic strength

The quantity used to characterize aqueous solutions that contain different concentrations of ions is the ionic strength. The ionic strength of the medium, I, is defined as:

$$I = \frac{1}{2} \Sigma\, c_i\, z_i^2 \,, \tag{1.1.16}$$

where c_i is the concentration and z_i the charge of ion i in solution. The sum runs over all ions present in the medium which gives e.g. for a pure NaCl solution:

$$I = \frac{1}{2} \left([Cl^-] \times 1 + [Na^+] \times 1 \right) .$$

Although NaCl is the major salt component of seawater, the properties of seawater and pure NaCl solutions are quite different. Seawater is a complex solution of electrolyte mixtures of unlike charge types where many more ions such as Mg^{2+} and SO_4^{2-} have to be considered (for chemical composition of seawater, see Table A.12.3, Appendix A). The ionic strength of seawater is approximately 0.7:

$$I = \frac{1}{2} \left([Cl^-] \times 1 + [Na^+] \times 1 + [Mg^{2+}] \times 4 + [SO_4^{2-}] \times 4 + ... \right) \approx 0.7 \,, \tag{1.1.17}$$

which corresponds to a seawater salinity of $S \sim 35$. The ionic strength of seawater may be calculated from salinity by (DOE, 1994):

$$I = \frac{19.924\ S}{1000 - 1.005\ S} \,.$$

As is obvious from Figure 1.1.5, the ionic strength is a fairly linear function of S within the range $30 < S < 40$ and might be approximated by $I \simeq 0.02\ S$.

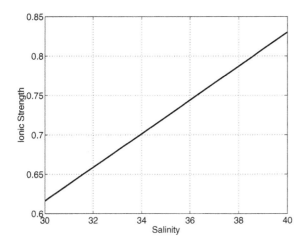

Figure 1.1.5: Relationship between salinity and ionic strength in seawater.

Activity coefficient

Central to the description of the chemical behavior of an ion dissolved in fresh water and in seawater is its activity. This concept might be illustrated as follows (note that the illustration is not a rigorous analogy). Consider two automobiles, one of them driving through New York City during rush hour, the other one driving through Death Valley at night. While the car in New York City is hindered by the interaction with all other vehicles in the streets (corresponding to the interaction with a large number of ions present in solution), the car in Death Valley is free to move (the solution is highly diluted). One might say that the 'activity' of the car in the New York City traffic is low while the 'activity' of the car in Death Valley is high. The chemical quantity which expresses such a behavior of ions in solution is indeed called activity.

The activity of a chemical species A, denoted by $\{A\}$, is related to the concentration of this species, $[A]$, by the activity coefficient γ_A:

$$\{A\} = \gamma_A [A] \ .$$

Ideally, the activity coefficient is 1.0, referring to the infinite dilution activity for which the concentrations of all solutes approach zero. It is to note that this definition refers to a reference state in pure water and that pH scales in seawater, for example, are based on seawater as the reference state (cf. Section 1.3 and Wedborg et al. (1999)). In dilute solutions of simple electrolytes, deviations from ideal behavior are caused by long-range electrostatic interactions. For those interactions approximations can be derived to describe the dependence of activity coefficients on the ionic strength (see below). In seawater the situation is more complicated because of (a) the higher ionic strength of seawater and (b) because various different ions with different charges are present. This leads to ion pairing and complex formation in electrolyte mixtures of unlike charge types such as seawater. For example, negatively charged carbonate ions (CO_3^{2-}) are associated with positively charged Mg^{2+} or Na^+ ions, forming ion pairs such as $NaCO_3^-$ or $MgCO_3^0$. The most important ion pair equilibria in seawater are (Skirrow, 1975):

$$Ca^{2+} + CO_3^{2-} \rightleftharpoons CaCO_3^0$$
$$Mg^{2+} + CO_3^{2-} \rightleftharpoons MgCO_3^0$$
$$Na^+ + CO_3^{2-} \rightleftharpoons NaCO_3^-$$
$$Ca^{2+} + HCO_3^- \rightleftharpoons CaHCO_3^+$$
$$Mg^{2+} + HCO_3^- \rightleftharpoons MgHCO_3^+$$
$$Na^+ + HCO_3^- \rightleftharpoons NaHCO_3^0 \,.$$

Thus, in addition to the diminution of the activity of the CO_3^{2-} ion due to electrostatic interaction with all the other ions in solution, the carbonate ion in seawater is not 'free' - it forms pairs with oppositely charged ions. With respect to our illustration of the traffic analogy this might correspond to a vehicle in New York City traffic hauling a trailer.

The concentrations of HCO_3^- and CO_3^{2-} used in the definitions of the stoichiometric constants (Eqs. 1.1.5 and 1.1.6) refer to total concentrations, i.e. free ions plus ion pairs; they are sometimes denoted by $[HCO_{3T}^-]$ and $[CO_{3T}^{2-}]$, respectively (Skirrow, 1975).

Considering the effects of ion pairing on activities, it is useful to talk about 'free' and 'total' activity coefficients. In case of a dilute solution, there are no differences between free and total activity coefficients because all ions are thought to be 'free' (i.e. no ion pairing or complex formation). In seawater, however, the total activity coefficient can be dramatically lower than the free activity coefficient due to e.g. formation of ion pairs.

The free activity coefficient of an ion, or single-ion activity, of simple electrolytes, γ_f, might be calculated using the Debye-Hückel limiting law (see, e.g. Stumm and Morgan, 1996):

$$\log(\gamma_f) = -Az^2\sqrt{I} \qquad\qquad I < 10^{-2.3} \qquad\qquad (1.1.18)$$

or the Davies equation:

$$\log(\gamma_f) = -Az^2\left(\frac{\sqrt{I}}{1+\sqrt{I}} - 0.2I\right) \qquad I < 0.5 \qquad (1.1.19)$$

with $A = 1.82 \times 10^6 (\epsilon T)^{-3/2}$, where $\epsilon \approx 79$ is the dielectric constant of water, and T is the temperature in Kelvin. At 25°C, A is about 0.5 for water. The charge of the ion is denoted as z and I is the ionic strength of the medium. Using the approximations given in Eqs. (1.1.18) and (1.1.19), activity coefficients for simple dilute solutions can then be calculated (Table 1.1.2). The question to be addressed is: can those approximations also be used to calculate total activity coefficients of ions in seawater?

Table 1.1.2: Free activity coefficient γ_f of some ions in dilute solutions at $T_c = 25°C$.

Ion	$\gamma_f{}^a$	$\gamma_f{}^b$	$\gamma_f{}^b$	$\gamma_f{}^b$
Ionic strength:	$I = 10^{-3}$	10^{-2}	0.1	0.5
$Cl^-, Na^+, HCO_3^-, H^+,$				
$OH^-, B(OH)_4^-$	0.96	0.90	0.77	0.69
$Mg^{2+}, SO_4^{2-}, Ca^{2+}, CO_3^{2-}$	0.86	0.66	0.36	0.23
PO_4^{3-}	0.72	0.40	0.10	0.04

[a] Calculated using the Debye-Hückel limiting law, Eq. (1.1.18).
[b] Calculated using the Davies equation, Eq. (1.1.19).

The ionic strength of seawater is approximately 0.7 which is only slightly higher than the ionic strength limit up to which the Davies equation should hold. Thus, one might be tempted to use the Davies equation to calculate activities of ions in seawater. However, as already said, for seawater the Debye-Hückel limiting law or the Davies equation no longer apply since they only hold for dilute solutions and simple electrolytes (as opposed to concentrated solutions and electrolyte mixtures of unlike charge type). In seawater, ion pairing and complex formation occur which makes it necessary to consider total activity coefficients. The models that are widely used to describe total activity coefficients of ionic solutes in natural waters including seawater are the 'ion pairing model' (e.g. Bjerrum, 1926; Garrels and Thompson, 1962; Sillén, 1961; Millero and Schreiber, 1982) and the 'Pitzer chemical equilibrium model' (e.g. Pitzer, 1973; Millero and Pierrot 1998). The description of these rather elaborated models is beyond the scope of this book. A model for natural waters based on the Pitzer model has recently been published by Millero and Pierrot (1998), yielding good results for activity coefficients in seawater when compared to measured values (Table 1.1.3; compare also Millero, 2001).

Table 1.1.3: Total activity coefficient γ_t of some ions in seawater at $T_c = 25°C$ and $S = 35$.

Ion	γ_t (measured)[a]	γ_t (calculated)[b]	γ_f (Davies eq.)[c]
Cl^-	0.666	0.666^d	0.69
Na^+	0.668	0.664	0.69
H^+	0.590	0.581	0.69
OH^-	0.255	0.263	0.69
HCO_3^-	0.570	0.574	0.69
$B(OH)_4^-$	0.390	0.384	0.69
Mg^{2+}	0.240	0.219	0.23
SO_4^{2-}	0.104	0.102	0.23
Ca^{2+}	0.203	0.214	0.23
CO_3^{2-}	0.039	0.040	0.23
$H_2PO_4^-$	0.453	0.514	0.69
HPO_4^{2-}	0.043	0.054	0.23
PO_4^{3-}	0.00002	0.00002	0.04

[a] For references, see Millero and Pierrot (1998).
[b] Millero and Pierrot (1998).
[c] Free activity coefficient, Davies equation, Eq. (1.1.19).
[d] Assigned value.

Exercise 1.3 (*)
What is the fundamental difference between NaCl solutions and seawater at the same ionic strength?

Exercise 1.4 (**)
Calculate the ionic strength of seawater using Eq. (1.1.17) and Table A.12.3 (see Appendix A). How many percent of the total ionic strength is due to the sum of Na^+, Cl^-, Mg^{2+}, and SO_4^{2-} ions?

1.1.3 Thermodynamic, hybrid, and stoichiometric equilibrium constants

Several concepts are in use in order to describe proton transfer reactions in aqueous media. Here we shall only briefly summarize the different equilibrium constants, also called acidity or dissociation constants, associated with these concepts (for review see Dickson, 1984). The second dissociation constant of carbonic acid, K_2, may serve as an example.

(1) The standard acidity, or thermodynamic equilibrium constant K_2 for the proton-transfer reaction:

$$HCO_3^- \rightleftharpoons CO_3^{2-} + H^+$$

is expressed entirely in terms of activities:

$$K_2 = \frac{\{H^+\}\{CO_3^{2-}\}}{\{HCO_3^-\}} . \qquad (1.1.20)$$

From a theoretical point of view, activity is the appropriate quantity in equilibrium thermodynamics.[4] A severe disadvantage in using activity is, however, the fact that activities are not as easy (if at all) measurable as concentrations, and the conversion from concentrations to activities at typical seawater salinities requires rather laborious and uncertain calculations of activity coefficients. Fortunately, equilibrium relations can also be formulated in terms of concentrations.

(2) A useful concept employed in oceanography is based on so-called hybrid (mixed) constant where concentrations and activities occur at the same time:

$$K_2' = \frac{a_H[CO_3^{2-}]}{[HCO_3^-]} . \qquad (1.1.21)$$

where a_H is operationally defined under the infinite dilution convention (for a detailed discussion of this approach, cf. Dickson, 1984; Skirrow, 1975).

(3) The constant entirely expressed in terms of concentrations is the classical mass action product for the acid dissociation reaction, also called stoichiometric constant as introduced in Section 1.1:

$$K_2^* = \frac{[H^+][CO_3^{2-}]}{[HCO_3^-]} \qquad (1.1.22)$$

where $[H^+]$ is operationally defined under the constant ionic medium convention. Stoichiometric constants are conventionally denoted by K^*. Please note that some authors omit the star for the sake of simplicity (cf. DOE (1994) and Millero (1995)).

An apparent constant was defined with apparent activities of the proton based on measurements using NBS buffers. These constants are no longer

---

[4]Note that concentration (not activity) is the appropriate quantity in chemical kinetics (see Chapter 2) because the reaction rate depends on the number of colliding molecules per volume and thus on concentrations (Lasaga, 1981).

used in chemical oceanography since all the constants are determined using the total proton scale or seawater scale (compare Section 1.3). The following quotes by Skirrow (1975) that refer to apparent constants apply to stoichiometric and hybrid constants as well. The occurrence of ion pair formation "is the principal reason for the dependence of apparent constants not only on the ionic strength of a solution, but also on the ionic composition" (Skirrow, 1975, p. 76).

1.1.4 Effect of ionic strength on pK^* values

Having introduced the concept of the ionic strength, activity, and the different acidity constants, we can now quantify the effect of ionic strength (and therefore salinity) on the pK^* values as described earlier.

Combining Eqs. (1.1.20) and (1.1.22), K_2^* can be expressed as (note that $\{CO_3^{2-}\} = \gamma_{CO_3^{2-}}[CO_3^{2-}]$):

$$K_2^* = K_2 \frac{\gamma_{HCO_3^-}}{\gamma_{H^+}\,\gamma_{CO_3^{2-}}} \,. \tag{1.1.23}$$

As is obvious from Eq. (1.1.23), relative changes in the *total* activity coefficients with higher ionic strength lead to changes in the dissociation constant K_2^*. Using values for γ_t given in Table 1.1.3 and the fresh water value of 10.33 for K_2 at 25°C, the value of pK_2^* in seawater at $S = 35$ can be estimated (evaluate the negative common logarithm of Eq. (1.1.23)):

$$
\begin{aligned}
pK_{2,\text{seawater}}^* &= 10.33 - \log\left(\frac{0.570}{0.590 \times 0.039}\right) \\
&= 8.94
\end{aligned}
$$

which is close to the experimentally determined value of 8.92 (DOE, 1994; Table 1.1.1).

In summary, considering a solution that changes from fresh water to seawater, the number of ions in solution increases, the relative proportion of ions changes, and the ion activity decreases. Since the activities of different ions are affected differently, the ratio of their activity coefficients changes (cf. Eq. (1.1.23)). Considering the example of pK_2^* as discussed above, the activity coefficient of CO_3^{2-} is decreasing more strongly (from 1 at infinite dilution to 0.039 in seawater) than, e.g. the activity coefficient of HCO_3^- (from 1 to 0.570). Consequently, pK_2^* in seawater is smaller (K_2^* is greater) than in fresh water. As a result, seawater has a higher CO_3^{2-} concentration (relative to the concentrations of CO_2 and HCO_3^-) than fresh water at the same pH (cf. Figure 1.1.4).

1.1.5 Effect of chemical composition on pK^* values

In general, the stoichiometric equilibrium constants of an electrolyte (including seawater) depend on the composition of the solution. Fortunately, the composition of seawater is fairly constant. Thus, the pK^*'s are functions of P, T, and S (or I) only. In certain regions of the oceans (for example, the Baltic Sea), in pore waters or in many laboratory experiments the salt composition differs from standard mean ocean values. Ben-Yaakov and Goldhaber (1973) estimated the variation in pK_1^* and pK_2^* with changing salt composition. Their approach made use of a seawater model and calculation procedure similar to that described by Garrels and Thompson (1962) and Berner (1971). Ben-Yaakov and Goldhaber (1973) provide sensitivity parameters:

$$s_{K^*} = \frac{\Delta K^* / K^*}{\Delta c_i / c_i}$$

where ΔK^* is the change in K^* due to relative change in concentration, $\Delta c_i / c_i$, of component i. As expected from its known tendency to form ion pairs in seawater, magnesium has the largest sensitivity parameters: $s_{K_1^*, Mg^{2+}} = 0.155$, $s_{K_2^*, Mg^{2+}} = 0.442$ (at 19‰ chlorinity, which corresponds to a salinity of 34.3, and $T_c = 25°C$). A doubling of the Mg^{2+} concentration, for example, will reduce pK_1^* from 5.86 to 5.80 and pK_2^* from 8.93 to 8.77. The consequences of changes in the salt composition on the equilibrium partial pressure will be investigated in Exercise 1.5. Nowadays it is recommended to calculate pK^*'s directly from a state-of-the-art seawater model such as the one developed by Millero and Roy (1997).

Exercise 1.5 (**)
Consider seawater at $T_c = 25°$, $S = 34.3$, $P = 1$ bar with DIC = 2 mmol kg^{-1} and TA = 2.35 mmol kg^{-1}. Calculate the equilibrium partial pressure of CO_2 (hint: use formulas given in the Appendix). Estimate the change in pCO$_2$ when the natural concentration of Mg^{2+} ions is doubled.

1.1.6 The choice of equilibrium constants

At this stage we have to discuss an area of the carbonate chemistry which, we feel, is somewhat unsatisfactory at its present state. The equilibrium constants for the dissociation of carbonic acid, boric acid, water and so forth, have been measured by different authors in different media, i.e. natural seawater and artificial seawater, and on different pH scales (cf. Section 1.3 and

Appendix A). Using dissociation constants for the calculation of carbonate system parameters given by different authors, one obtains different results. Particularly huge differences may arise when the conversion between pH scales is not taken into account. Although there have been attempts to agree on a single consistent set of constants and a single pH scale (cf. e.g. Dickson (1984), UNESCO (1987)), this approach has hitherto not become standard.

An example of the potential problems associated with the calculations of the concentrations of the carbonate species is listed in Table 1.1.4. Let us assume that the pH and the DIC of a sample have been measured but that the pH scale on which the measurements were made, has not been reported. In other words, it is not clear whether the seawater scale, the total scale, or the free pH scale has been used - for definitions, see Section 1.3. With pH = 8.08 and DIC = 2 mmol kg^{-1}, the carbonate system parameters can be calculated for all three scales (Table 1.1.4). Whereas the differences between the seawater and the total scale are rather small (~ 10 μatm in terms of pCO_2), the differences between these two scales and the free scale are huge (> 100 μatm in terms of pCO_2). It is emphasized that the calculated differences result from the fact that the scale on which the pH was measured has not been defined. If the pH scale used in the measurements was known and the pH for the calculations would then be converted to the correct pH scale (corresponding to the scale of the dissociation constants), one would obtain exactly the same results in each case. This scenario requires, however, that a single set of constants is used, see below. In practice, one could use any pH scale for the calculations - the crucial point is that the pH scale and the dissociation constants used, have to agree.

Table 1.1.4: Calculated carbonate system parameters when the scale on which pH was measured is not known (pH = 8.08, DIC = 2 mmol kg^{-1}, $S = 35$, $T_c = 25°C$).[a]

pH scale	pH	pCO_2 μatm	$CO_2(aq.)$ μmol kg^{-1}	HCO_3^- μmol kg^{-1}	CO_3^{2-} μmol kg^{-1}
Seawater	8.08	354	10.0	1735	255
Total	8.08	363	10.3	1739	250
Free	8.08	478	13.6	1786	201

[a] The constants of Roy et al. (1993a) as converted to different pH scales (see Section 1.3) have been used in all calculations.

Even when the calculations are carried out on the same pH scale, differences in the values may occur because dissociation constants are used which have been determined by different authors. The differences are, however, usually smaller than in the case mentioned above. Table 1.1.5 shows values for the concentrations of the carbonate species calculated from DIC and TA using constants (all converted to the seawater scale) referred to as 'Roy', 'Hansson', and 'Mehrbach' (for definition, see footnote to Table 1.1.5). The calculated values for pCO_2 differ by up to ~ 30 μatm in this particular example.

Very recently, it has been demonstrated that using Mehrbach constants for the dissociation of carbonic acid yields the best results when determining pCO_2 from DIC and TA (Wanninkhof et al., 1999; Johnson et al., 1999; Lee et al., 2000; Lueker et al., 2000). One could therefore agree to use Mehrbach constants in this particular case, which we actually recommend. However, which set of constants should we use when, e.g. determining pCO_2 from pH and DIC or $[CO_3^{2-}]$ from pCO_2 and TA? The fact that Mehrbach constants do a good job in the important case mentioned above does not imply that they do so in every case. As Wanninkhof et al. (1999) put it: " ... the good agreement between $pCO_2(\text{SST})$ and $pCO_2(\text{TA, DIC})$ using the constants of Mehrbach does not necessarily imply that these constants yield the best agreement with other carbon system parameters". Here is the problem: when calculating various parameters of the carbonate system using different input variables (as is done in this book, for example) one cannot use different sets of constants since this would lead to inconsistent results.

Table 1.1.5: Calculated carbonate system parameters using dissociation constants given by different authors (DIC = 2 mmol kg^{-1}, TA = 2.35 mmol kg^{-1}, $S = 35$, $T_c = 25°$C).

Author	pH_{SWS}	pCO_2 μatm	CO_2(aq.) μmol kg^{-1}	HCO_3^- μmol kg^{-1}	CO_3^{2-} μmol kg^{-1}
Roy[a]	8.08	354	10.0	1735	255
Hansson[b]	8.10	343	9.7	1739	251
Mehrbach[c]	8.11	327	9.3	1742	249

[a] Roy et al. (1993a) converted to seawater scale (Millero, 1995).
[b] Mehrbach et al. (1973) as refit by Dickson and Millero (1987).
[c] Hansson (1973b) as refit by Dickson and Millero (1987).

Another issue that should be kept in mind when choosing a particular set of dissociation constants is to make sure that all constants have been determined in the same medium, i.e. natural or artificial seawater. The constants referred to as 'Mehrbach' have been determined in natural seawater, whereas 'Roy' and 'Hansson' have been determined in artificial seawater. It is likely that this is the reason for the fact that Mehrbach constants do a good job when measurements in the ocean, i.e. in natural seawater are considered. Precise calculations of the carbonate system parameters include (besides the dissociation of carbonic acid) the dissociation of water, boric acid, hydrogen sulfate, hydrogen fluoride, phosphoric acid, silicic acid and more. In order to achieve internal consistency of the calculations, all the constants used should be determined in a single medium, i.e. in natural seawater *or* in artificial seawater.

Regarding the calculations presented throughout this book, we have decided to use the set of constants summarized in DOE (1994), including the dissociation constants of carbonic acid referred to as 'Roy', see Appendix for values. These constants, which have been determined in artificial seawater, are all expressed in terms of the total hydrogen ion concentration (i.e., total pH scale, see Section 1.3) and in units of moles per kilogram of solution. For the reasons given above, it appears that this set is the most appropriate one for our current purposes. We note, however, that e.g. Mehrbach constants may be more appropriate under particular circumstances.

It is extremely desirable that the problems described here will be solved in the future. We believe that major steps towards this goal are (a) the agreement on a single pH scale by the community and (b) the high-precision determination of all relevant dissociation constants in natural seawater.

1.1.7   CaCO$_3$ solubility and the saturation state of seawater

Formation and dissolution of calcium carbonate (CaCO$_3$) in the ocean are important players in the global carbon cycle and are intimately related to the control of atmospheric CO$_2$ on various time scales. For instance, anthropogenic CO$_2$, which is currently accumulating in the atmosphere, is thought to be mostly absorbed by the oceans and ultimately neutralized by the reaction with CaCO$_3$ in marine sediments (the so-called fossil fuel neutralization, e.g. Broecker and Takahashi, 1977; Sundquist, 1986; Archer et al., 1998). A determining factor in the context of formation and dissolution of calcium carbonate is the CaCO$_3$ saturation state of seawater, which is a function of the carbonate ion concentration. In other words, the equilibrium between the solid state and the solution is controlled by the corrosiveness of seawater. The $CaCO_3$ saturation state of seawater is therefore an important aspect of the seawater carbonate chemistry.

In this section, $CaCO_3$ solubility in seawater, the seawater saturation state, and the distribution of $CaCO_3$ in marine sediments are discussed. The effect of $CaCO_3$ production on the carbonate system is examined in Section 1.6.1 while potential effects of increasing atmospheric CO_2 on calcification in the future are mentioned in Section 1.6.5. For further reading, see e.g. Berger et al. (1976), Broecker and Peng (1982), Mucci (1983), Morse and Mackenzie (1990), Holligan and Robertson (1993), Millero (1996).

The vast majority of marine calcium carbonate is produced by organisms which secrete calcitic or aragonitic shells and skeletons. The major calcite producers in the open ocean are coccolithophorids and foraminifera, while the most abundant pelagic aragonite organisms are pteropods. (For a review on $CaCO_3$ production including coral reefs and other environments, see Milliman, 1993; Milliman and Droxler, 1996.) Calcite and aragonite minerals both consist of $CaCO_3$, but differ in their mineralogy. The crystal structure of calcite is rhombohedral whereas the structure of aragonite is orthorhombic (for review cf. e.g. Hurlbut (1971) or Reeder (1983)). The different structures of the two minerals lead to different physical and chemical properties (Table 3.2.4, Chapter 3) of which the solubility is of major importance here.

The stoichiometric solubility product is defined as:

$$K_{sp}^* = [Ca^{2+}]_{sat} \times [CO_3^{2-}]_{sat} , \qquad (1.1.24)$$

where e.g. $[CO_3^{2-}]_{sat}$ refers to the equilibrium total (free + complexed) carbonate ion concentration in a seawater solution saturated with $CaCO_3$. Aragonite is more soluble than calcite at a given temperature, salinity and pressure: $K_{sp}^* = 10^{-6.19}$ and $10^{-6.37}$ mol^2 kg^{-2} for aragonite and calcite, respectively ($T = 25°C$, $S = 35$, $P = 1$ atm). Formulas for the calculation of K_{sp}^* for aragonite and calcite as determined by Mucci (1983) are given in Appendix A.10. At equilibrium, the product of the Ca^{2+} and CO_3^{2-} concentration in solution is given by the left-hand side of Eq. (1.1.24). If $[Ca^{2+}] \times [CO_3^{2-}]$ is larger than this, the solution is supersaturated with respect to $CaCO_3$ - otherwise, the solution is undersaturated. The $CaCO_3$ saturation state of seawater, Ω, is expressed as:

$$\Omega = \frac{[Ca^{2+}]_{sw} \times [CO_3^{2-}]_{sw}}{K_{sp}^*} , \qquad (1.1.25)$$

where $[Ca^{2+}]_{sw}$ and $[CO_3^{2-}]_{sw}$ are the concentrations of Ca^{2+} and CO_3^{2-} in seawater, respectively, and K_{sp}^* is the solubility product at the *in situ* conditions of temperature, salinity, and pressure. $\Omega > 1$ corresponds to supersaturation, whereas $\Omega < 1$ corresponds to undersaturation.

In the open ocean, $[Ca^{2+}]$ variations are rather small and closely related to variations in salinity. The $CaCO_3$ saturation state is therefore mainly determined by the carbonate ion concentration. Table 1.1.6 summarizes values for the carbonate ion concentration at saturation, $[CO_3^{2-}]_{sat}$, for calcite and aragonite as a function of temperature and pressure (Mucci, 1983).

Table 1.1.6: Saturation carbonate ion concentration at $S = 35$ and $[Ca^{2+}] = 10.28$ mmol kg^{-1} (cf. Broecker and Peng, 1982).[a]

Temperature (°C)	Pressure (atm)	$[CO_3^{2-}]_{sat}$ (μmol kg^{-1})	
		Calcite	Aragonite
25	1	41.6	63
2	1	41.9	67
2	250	69.4	107
2	500	111.7	167

[a] Values for K_{sp}^* after Mucci (1983). Pressure correction after Millero (1995). See Appendix A.10 for formulas.

Calcium carbonate is an unusual salt: the solubility increases at lower temperature! The effect of temperature on the solubility is, however, rather small (see Table 1.1.6). More important is the fact that the solubility increases with pressure. This is of great significance for the distribution of $CaCO_3$ in marine sediments. Assuming a constant temperature in the deep ocean, $[CO_3^{2-}]_{sat}$ increases with pressure and hence with depth in the ocean (Figure 1.1.6).

The crossover between *in situ* and saturation carbonate ion concentration is called the saturation horizon or saturation depth/level. Calcium carbonate falling from the surface to the deep ocean is mainly preserved in supersaturated waters above the saturation horizon, and starts to dissolve below the saturation horizon in undersaturated waters.[5] We might therefore think of the ocean floor as a landscape with snow-covered mountains (Sillén, 1967; Broecker and Peng, 1982). The upper parts of the ocean floor, such as ridge crests, are covered with $CaCO_3$ of a light color, whereas the

---

[5] This description is highly simplified, see Broecker and Takahashi (1978), Broecker and Peng (1982) or Morse and Mackenzie (1990) for details.

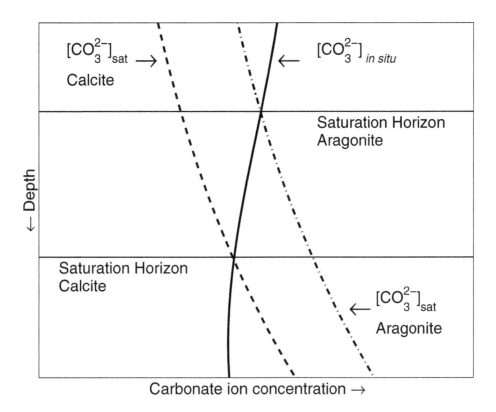

Figure 1.1.6: Illustration of the calcite and aragonite saturation horizon (SH) in the ocean (after Broecker and Peng, 1987). As pressure increases with depth, the solubility of calcite and aragonite increases as well ($[CO_3^{2-}]_{sat}$). The crossover between the *in situ* carbonate ion concentration (solid curve) and the saturation concentration for calcite (dashed curve) and aragonite (dot-dashed curve) determines the saturation horizon of the different mineral phases.

valleys are mostly free of $CaCO_3$ and covered with clay minerals of a darker color.

Figure 1.1.7 shows the calcite and aragonite saturation state of seawater in the North Pacific and North Atlantic Ocean. Surface seawater is roughly 6 and 4 times supersaturated with respect to calcite and aragonite, respectively. The supersaturation decreases with depth until it crosses the line $\Omega = 1$. The crossover occurs at shallower depth for aragonite than for calcite because aragonite is more soluble than calcite. In addition, the saturation state of the North Pacific Ocean is smaller than that of the North Atlantic Ocean. These regional differences are discussed below.

The depth of rapid increase in the rate of dissolution as observed

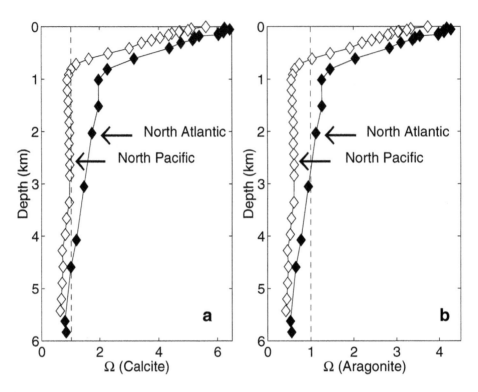

Figure 1.1.7: Saturation state of seawater, Ω, with respect to calcite (a) and aragonite (b) as a function of depth. Ω was calculated using Eq. (1.1.25) and DIC-TA profiles from the North Pacific and North Atlantic Ocean. The dashed vertical line separates areas of supersaturation ($\Omega > 1$) from undersaturation ($\Omega < 1$). Data are from WOCE Section P14N, Stn. 70 (F. Millero and Ch. Winn) and WOCE Section A05, Stn. 84, (F. Millero and S. Fiol), http://cdiac.esd.ornl.gov/oceans/CDIACmap.html.

in sediments is called the lysocline. The term lysocline stems from the Greek words for solution and inclination - in analogy to thermocline, see Berger (1968). The lysocline therefore separates poorly preserved and well-preserved $CaCO_3$ assemblages. Because of the higher solubility of aragonite in comparison to calcite, the aragonite lysocline is always found at lower depth than the calcite lysocline. The aragonite lysocline can be as shallow as 500 m in the Pacific Ocean and ~ 3 km in the Atlantic Ocean. The calcite lysocline is found at ~ 4 km in the Pacific and Indian Ocean and between 4 and 5 km in the Atlantic Ocean. The reason for the differences in the lysocline depth between Pacific and Atlantic Ocean is that deep water of the Pacific Ocean has a lower carbonate ion concentration (and a higher CO_2 content) than deep water of the Atlantic Ocean. In other words, the deep water of the Pacific Ocean is more corrosive (Figure 1.1.7). This is mainly a result of the interaction of ocean circulation and biological activity: Deep water of the Pacific Ocean is 'older' than that of the Atlantic Ocean and has therefore taken up more CO_2 from remineralization of organic matter which lowers its carbonate ion content. The age of a water mass here refers to the time elapsed since its last contact with the atmosphere.

The depth at which the sediments are virtually free of calcium carbonate is called the calcium carbonate compensation depth (CCD). Usually, a transition zone occurs between the lysocline and the CCD in which the calcium carbonate content of sediments gradually decreases with depth. The transition zone can be as large as several hundred meters which has to do with the supply rate of $CaCO_3$, i.e. the production in the surface ocean, and the kinetics of $CaCO_3$ dissolution in the deep ocean (Broecker and Peng, 1982; Morse and Mackenzie, 1990; Millero, 1996).

We will return to the production and dissolution of calcium carbonate in the sections to follow. Stable isotope fractionation of carbon, oxygen, and boron in marine calcium carbonate is examined in Chapter 3.

1.2 Alkalinity

"I found at least 20 different definitions of alkalinity!"
An ocean carbon cycle modeler from Hamburg.

"... alkalinity, one of the most central but perhaps not the best understood concept in aquatic chemistry."
Morel and Hering (1993, p.157)

Alkalinity is an important and very useful concept in the context of the carbonate system in seawater. It is also called total alkalinity or titration alkalinity and is denoted by TA. From the knowledge of TA and DIC together with T and S, all other quantities of the carbonate system, i.e. $[CO_2]$, $[HCO_3^-]$, $[CO_3^{2-}]$, and pH can be calculated. The definition of DIC as the sum of the concentrations of CO_2, HCO_3^-, and CO_3^{2-} is so simple that the meaning is understood in a second and one hesitates to call it a concept. In contrast, understanding the alkalinity concept takes much more time as we know from our own experience and from discussions with many colleagues: the meaning is not obvious from its expression in terms of concentrations of certain compounds. In what follows, the alkalinity concept will be introduced step by step. Approximations of increasing complexity will be presented that finally lead to the most exact definition given by Dickson (1981). Aqueous solutions of simple chemical compositions - compared to seawater - are discussed to clarify the alkalinity concept.

The alkalinity concept in natural waters is a non-trivial one and it therefore takes a little time to work through it. For the fast reader, who is not interested in all the details, a brief summary is given in the following subsection.

1.2.1   A shortcut to alkalinity

Dickson gives the following expression for alkalinity in seawater (DOE, 1994; compare also Dickson, 1981)

$$
\begin{aligned}
\text{TA} \;=\; & [\text{HCO}_3^-] + 2[\text{CO}_3^{2-}] + [\text{B(OH)}_4^-] + [\text{OH}^-] \\
& + [\text{HPO}_4^{2-}] + 2[\text{PO}_4^{3-}] + [\text{H}_3\text{SiO}_4^-] \\
& + [\text{NH}_3] + [\text{HS}^-] - [\text{H}^+]_\text{F} - [\text{HSO}_4^-] - [\text{HF}] - [\text{H}_3\text{PO}_4]
\end{aligned}
\tag{1.2.26}
$$

where $[\text{H}^+]_\text{F}$ is the free concentration of hydrogen ion (compare Section 1.3).

TA is derived from titration with a strong acid,[6] which explains the term 'titration alkalinity'. The expression (1.2.26) contains several species whose contribution to TA is very small at typical surface seawater pH values around 8.2. They have to be included, however, to derive TA from inverse calculation using titration data over a large pH range as their contribution to TA becomes increasingly important at low pH.

Total alkalinity is a conservative quantity such as mass, salt, and dissolved inorganic carbon (see box on page 29 and Section 1.2.6). For example, if total alkalinity of a sample is expressed in units of mol kg^{-1}, then the total alkalinity will stay constant during changes of temperature and pressure. Moreover, when two water samples are mixed, then the resulting alkalinity is simply given by the weighted mean

$$
M_m \cdot \text{TA}_m = M_1 \cdot \text{TA}_1 + M_2 \cdot \text{TA}_2
\tag{1.2.27}
$$

---

[6]In aqueous solutions, strong acids may be defined as follows. Consider an acid HA that can provide protons according to the equilibrium $\text{HA} \rightleftharpoons \text{A}^- + \text{H}^+$ with acidity constant $K_A^* = [\text{H}^+][\text{A}^-]/[\text{HA}]$. A strong acid is a strong proton donor, and its acidity constant is very large in water, i.e. the acid is virtually completely dissociated (examples: HCl, HBr). A strong base is a strong proton acceptor, and it is virtually fully protonated in water (an example is the O^{2-} ion, which does not exist as such in water because it is fully protonated: $\text{O}^{2-} + \text{H}_2\text{O} \longrightarrow 2\text{OH}^-$; Atkins, 1990).

where M_1, M_2, and M_m and TA_1, TA_2, and TA_m are the masses and alkalinities of the two samples and the mixture, respectively.

At pH values above 8, the alkalinity of natural seawater is given to a very good approximation (for almost all practical purposes) by

$$TA \simeq [HCO_3^-] + 2[CO_3^{2-}] + [B(OH)_4^-] + [OH^-] - [H^+] = PA, \quad (1.2.28)$$

i.e. by the sum of carbonate alkalinity ($[HCO_3^-] + 2\,[CO_3^{2-}]$), borate alkalinity ($[B(OH)_4^-]$), and water alkalinity ($[OH^-] - [H^+]$).[7] This approximate expression will be denoted as practical alkalinity, PA. It is the sum of the charges of the major weak acids in seawater plus the charge of OH^- minus the charge of H^+. If not stated otherwise this approximation is used in calculations involving alkalinity.

Conservative quantities.
The term 'conservative ions' can be defined according to Drever (1982, p.52):

"Ions such as Na^+, K^+, Ca^{2+}, Mg^{2+}, Cl^-, SO_4^{2-}, and NO_3^- can be regarded as 'conservative' in the sense that their concentrations are unaffected by changes in pH, pressure, or temperature (within the ranges normally encountered near the earth's surface and assuming no precipitation or dissolution of solid phases, or biological transformations)."

When a larger pH range is considered, say down to a value of 4 which may be reached during titration, the species SO_4^{2-} has to be removed from Drever's list (see below). Other quantities such as total phosphate or total ammonia whose the concentrations are unaffected by changes in pH, pressure, and temperature will also be called 'conservative'. Later on, it will be shown that alkalinity is a conservative quantity that can be expressed in terms of non-conservative quantities (the traditional expression) or in terms of conservative quantities only (explicitly conservative expression).

In the following, the concept of alkalinity is introduced, starting with the carbonate alkalinity. Adding water and borate alkalinity, this will then lead us to a formal definition of total alkalinity. The first two sections of this chapter (Sections 1.2.2 to 1.2.3) consider alkalinity from a chemical point

---

[7]The interpretation of the symbol $[H^+]$ depends on the pH scale used (compare Section 1.3)

of view. This part is of central importance to the use of alkalinity in chemical oceanography, because it provides a definition of total alkalinity and a procedure to measure total alkalinity. In the context of CO_2 measurements in the ocean this is a crucial issue. However, we believe that more is required to understand the concept of total alkalinity. This is elaborated in subsequent sections in which the relationship between charge balance and alkalinity is derived and variations of total alkalinity in the ocean are discussed. One may say that the latter part of this chapter considers alkalinity from a physical or geochemical point of view.

1.2.2   Carbonate alkalinity

In order to illustrate the concept of alkalinity, the carbonate alkalinity is discussed first. We will study the carbonate chemistry of a seawater sample of simple chemistry during titration with strong acid. In contrast to natural seawater this 'simplified seawater' considered here does not contain any boron, phosphate, silicate, and ammonia[8]. In our example, it will be shown that the carbonate alkalinity measures the charge concentration of the anions of carbonic acid present in solution. Approaching the concept of alkalinity from this side, we might say that:

'Alkalinity keeps track of the charges of the ions of weak acids'

Consider one kilogram of 'simplified seawater' with pH of 8.2 and $\Sigma CO_2 = 2$ mmol kg^{-1}. We can use the equations derived in Section 1.1 to calculate the concentrations of the dissolved carbonate species. At $T = 25°C$, and $S = 35$ we have: $[CO_2] \simeq 8$ μmol kg^{-1}, $[HCO_3^-] \simeq 1.7$ mmol kg^{-1}, and $[CO_3^{2-}] \simeq 0.3$ mmol kg^{-1} (see left vertical axis in Figure 1.2.8).

Now let us add strong acid to the sample, i.e. we titrate the 'simplified seawater' by, say HCl. Hydrochloric acid is a strong acid and dissociates completely into H^+ and Cl^-. Upon addition of HCl to the sample, the protons will mainly combine with carbonate ion to form bicarbonate. As a result, $[CO_3^{2-}]$ decreases and $[HCO_3^-]$ initially increases (Figure 1.2.8). (The small change of volume during titration is neglected because it is not relevant to the current discussion.) After addition of roughly 0.5 mmol kg^{-1} HCl, $[HCO_3^-]$ is strictly decreasing while CO_2 is increasing because bicarbonate is converted to carbon dioxide. After addition of ~ 2.3 mmol kg^{-1} HCl,

---

[8]The species HSO_4^-, HF, and HS^- are ignored, as well.

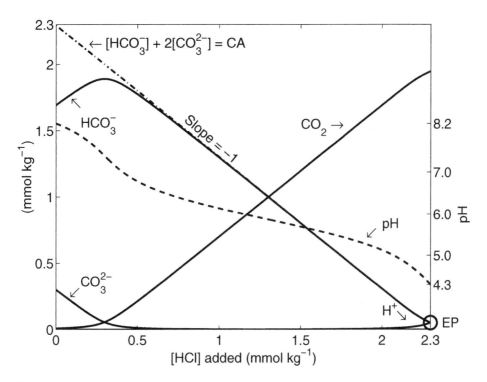

Figure 1.2.8: Titration of 'simplified seawater' with strong acid. Left vertical axis: concentrations of HCO_3^-, CO_3^{2-}, and CO_2. Right vertical axis: pH. Bicarbonate and carbonate ion have almost completely been converted to CO_2 when ~ 2.3 mmol kg^{-1} HCl have been added. This number is just equal to the initial carbonate alkalinity, CA $=$ $[HCO_3^-] + 2[CO_3^{2-}]$, at the intercept with the left vertical axis. The second equivalence point is indicated by the circle (EP).

virtually all carbonate ion and bicarbonate have been converted to CO_2. In other words, the charge of the anions of carbonic acid have been neutralized by H^+.

If the quantity $[HCO_3^-] + 2[CO_3^{2-}]$ is plotted as a function of acid added, a linear relationship is found over a wide range with a slope of -1 (Figure 1.2.8, dot-dashed line). This quantity is called the carbonate alkalinity:

$$CA = [HCO_3^-] + 2[CO_3^{2-}] . \qquad (1.2.29)$$

For the 'simplified seawater' considered here, the number of moles of HCl that need to be added in order to neutralize the anions of the weak acid is approximately equal to the carbonate alkalinity. Graphically, this corresponds to the point where the graph of CA approaches the horizontal axis, that is, where the carbonate alkalinity goes to zero. More precisely, this point is approached at the so-called second equivalence point where

$$[H^+] = [HCO_3^-] + 2[CO_3^{2-}] + [OH^-] \,, \tag{1.2.30}$$

(Dyrssen and Sillén, 1967) which is also referred to as a proton condition (see circle in Figure 1.2.8).[9] The proton condition is very important because it can be determined from titration data, see below. The number of moles of HCl added to reach this point is just equal to the initial carbonate alkalinity of our sample (~ 2.3 mmol kg^{-1}).

We see that the carbonate alkalinity in our example measures the charge concentration of the anions of the weak acid present in solution. If, for a given ΣCO_2, the sample initially had a higher pH and thus a higher carbonate ion content, more acid would have been required to neutralize the charge of the weak acid - and vice versa. The carbonate alkalinity here is the number of equivalents of strong acid required to neutralize 1 kg of seawater until the second equivalence point is reached.

As mentioned above, the proton condition is very useful because it can be determined from titration data. As a result, the definition of alkalinity is based on an analytical procedure that measures alkalinity in terms of acid added and is therefore an operational definition. Historically, this concept was widely used to determine alkalinity, for example, by the Gran method (cf. e.g. Gran, 1952; Dyrssen and Sillén, 1967). We will see that the modern definition of total alkalinity is explicitly based on a chemical model of the acid-base processes occurring in seawater (Section 1.2.3). The result is that total alkalinity can be defined unambiguously and a procedure is provided to measure it - the methods frequently used are the modified Gran method or curve-fitting procedures (for summary, see Barron et al., 1983; Dickson,

---

[9]The second equivalence point is located at a pH value of about 4.3. At this pH the concentrations of CO_3^{2-} and OH^- are much smaller than those of HCO_3^- and H^+ (compare Fig. 1.1.2). Thus the second equivalence point condition can be stated as $[H^+] \simeq [HCO_3^-]$.

1992; DOE, 1994; Anderson et al., 1999).

Alkalinity as a master variable.

Given the alkalinity and ΣCO_2, carbonate system parameters can be determined. Assume that the total alkalinity and ΣCO_2 of a seawater sample have been determined to be 2.3 and 2.0 mmol kg^{-1}, respectively. To a first approximation we may take the total alkalinity equal to the carbonate alkalinity:

$$\text{TA} \simeq [\text{HCO}_3^-] + 2[\text{CO}_3^{2-}] \,.$$

Furthermore, because $[CO_2]$ is small, we may set

$$\Sigma CO_2 \simeq [\text{HCO}_3^-] + [\text{CO}_3^{2-}] \,.$$

Thus, the carbonate ion concentration in seawater is roughly given by the difference between TA and ΣCO_2:

$$[\text{CO}_3^{2-}] \simeq \text{TA} - \Sigma CO_2 \,,$$

which yields $[\text{CO}_3^{2-}] = 300$ μmol kg^{-1} and thus $[\text{HCO}_3^-] = 1700$ μmol kg^{-1}. Including CO_2, and using $\text{TA} = [\text{HCO}_3^-] + 2[\text{CO}_3^{2-}] + [\text{B(OH)}_4^-] + [\text{OH}^-] - [\text{H}^+]$ (see below), one obtains an equation of fifth order in $[\text{H}^+]$ which can be solved numerically (see Appendix B). The result is $[\text{CO}_3^{2-}] = 220$ μmol kg^{-1}, $[\text{HCO}_3^-] = 1768$ μmol kg^{-1}, and $[CO_2] = 12$ μmol kg^{-1} at $T = 25°C$ and $S = 35$. The difference between the two calculations is mainly due to the fact that the contribution of boron to the total alkalinity has been ignored in the first case.

Including borate and water alkalinity

Considering the titration outlined in the previous section, it is clear that in natural seawater not only the carbonic acid system but every similar acid-base system present in solution will contribute to the alkalinity as determined by titration. Upon addition of strong acid, all proton acceptors will take up protons, regardless whether the anions of, for example, carbonic acid, boric acid, or phosphoric acid are considered. If we want to use alkalinity as a master variable of the carbonate system in natural seawater, we therefore have to be careful because the total alkalinity is not equal to the carbonate alkalinity. In order to calculate carbonate system parameters correctly, our definition of alkalinity has to include additional terms arising from the presence of acid-base systems other than carbonic acid.

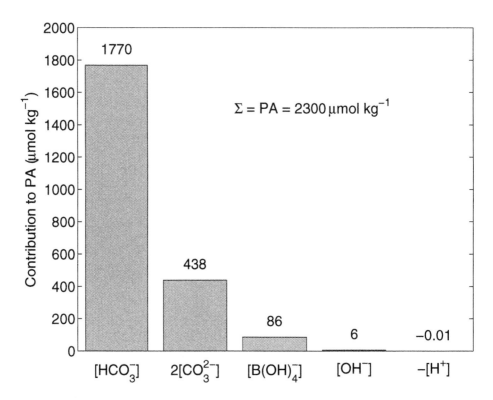

Figure 1.2.9: Contribution of the various species to PA (alkalinity for most practical purposes) at PA = 2300 μmol kg^{-1}, DIC = 2000 μmol kg^{-1}, $T_c = 25°C$, and $S = 35$. It is often sufficient to consider this subset of total alkalinity.

In many ocean waters, the most important acid-base systems that contribute to total alkalinity in addition to carbonic acid are boric acid and water itself. The borate alkalinity and water alkalinity are:

$$\text{borate alkalinity} = [B(OH)_4^-]$$
$$\text{water alkalinity} = [OH^-] - [H^+] .$$

Adding these contributions to the carbonate alkalinity, we have:

$$PA = [HCO_3^-] + 2[CO_3^{2-}] + [B(OH)_4^-] + [OH^-] - [H^+] . \qquad (1.2.31)$$

In many cases this is a good approximation to the total alkalinity at seawater pH and we denote it by PA (alkalinity for most practical purposes). This approximation will be used in the vast majority of calculations given in the current book. Typical contributions of the various compounds to PA are

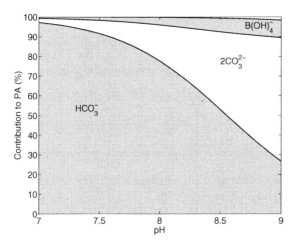

Figure 1.2.10: Relative contribution of various compounds to PA as a function of pH (DIC = 2000 μmol kg^{-1}, T_c = 25°C, $S = 35$). Areas show the percentage of each compound HCO$_3^-$ (lower shaded area and left vertical axis), 2CO$_3^{2-}$ (large white area), B(OH)$_4^-$ (upper shaded area), and OH$^-$ (upper small white area).

shown in Figure 1.2.9. In our example, the carbonate alkalinity contributes 96% to PA, demonstrating that the carbonate alkalinity is by far the most important part of PA. This statement also holds true for a wide range of pH (Figure 1.2.10).

Proton acceptors and donors

The necessity for the inclusion of further acid-base systems such as the dissociation of boric acid and water into the alkalinity shows that a more general concept of alkalinity is required. It turns out that considering a balance between proton acceptors and proton donors is a very useful approach which will finally be included in the definition of the total alkalinity (Section 1.2.3). For example, carbonate ion and bicarbonate are proton acceptors that act as bases. On the other hand, H$^+$ is a proton donor that acts as an acid (recall that the symbol H$^+$ actually represents species such as H$_3$O$^+$). We can thus state that the alkalinity defined by Eq. (1.2.31) is the excess of bases (proton acceptors) over acids (proton donors) in seawater, where the proton acceptors are [HCO$_3^-$], [CO$_3^{2-}$], [B(OH)$_4^-$], and [OH$^-$] and the proton donor is H$^+$ (cf. Rakestraw, 1949). The point at which PA equals zero defines the following proton condition:

$$\underbrace{[\text{H}^+]}_{\text{proton donor}} = \underbrace{[\text{HCO}_3^-] + 2[\text{CO}_3^{2-}] + [\text{B(OH)}_4^-] + [\text{OH}^-]}_{\text{proton acceptors}} . \qquad (1.2.32)$$

This expression is analogous to the proton condition defined in Eq. (1.2.30) for the 'simplified seawater'.

Having introduced the basic concept of alkalinity, proton acceptors and donors, and the proton condition, we now move on to the definition of total alkalinity. The task of defining total alkalinity essentially is how to correctly expand Eqs. (1.2.31) and (1.2.32) in order to include further acid-base systems (Dickson, 1981).

1.2.3   Dickson's definition of alkalinity

The currently most precise definition of titration or total alkalinity was given by Dickson (DOE, 1994; compare also Dickson, 1981): "The total alkalinity of a natural water is thus defined as the number of moles of hydrogen ion equivalent to the excess of proton acceptors (bases formed from weak acids with a dissociation constant $K \leq 10^{-4.5}$, at 25°C and zero ionic strength) over proton donors (acids with $K > 10^{-4.5}$) in one kilogram of sample."[10] For the compounds found in seawater, the expression for total alkalinity following from Dickson's definition reads:

$$
\begin{aligned}
\text{TA} \;=\; & [\text{HCO}_3^-] + 2[\text{CO}_3^{2-}] + [\text{B(OH)}_4^-] + [\text{OH}^-] \\
& + [\text{HPO}_4^{2-}] + 2[\text{PO}_4^{3-}] + [\text{H}_3\text{SiO}_4^-] \\
& + [\text{NH}_3] + [\text{HS}^-] - [\text{H}^+]_\text{F} - [\text{HSO}_4^-] - [\text{HF}] - [\text{H}_3\text{PO}_4]
\end{aligned}
\tag{1.2.33}
$$

where $[\text{H}^+]_\text{F}$ is the free concentration of hydrogen ion (compare Section 1.3). The possible contribution of unknown protolytes to the total alkalinity is discussed in Bradshaw and Brewer (1988a).

The appropriate proton condition that defines the equivalence point is given by:

$$
\begin{aligned}
& [\text{H}^+]_\text{F} + [\text{HSO}_4^-] + [\text{HF}] + [\text{H}_3\text{PO}_4] \\
=\; & [\text{HCO}_3^-] + 2[\text{CO}_3^{2-}] + [\text{B(OH)}_4^-] + [\text{OH}^-] \\
& + [\text{HPO}_4^{2-}] + 2[\text{PO}_4^{3-}] + [\text{H}_3\text{SiO}_4^-] \\
& + [\text{HS}^-] + [\text{NH}_3] \,,
\end{aligned}
\tag{1.2.34}
$$

where the proton donors appear on the left-hand side and the proton acceptors appear on the right-hand side.

---

[10]It is noted that the thermodynamic constant K is used in this case, and not the stoichiometric constant K^* which is usually employed for seawater. The reason for this is that K is independent of salinity. Thus, the constraint on the dissociation constants of the weak acids is unambiguous.

The total alkalinity of a sample according to Eq. (1.2.33) is determined as follows. The sample is titrated with strong acid, usually HCl, and the pH is recorded as a function of acid added. Then TA is calculated by non-linear curve fitting of the theoretical titration curve based on Eq. (1.2.33) to the actual data using all the titration points. This procedure is also called inverse calculation (e.g. Dickson, 1981; Johansson and Wedborg, 1982; Anderson et al., 1999). It is important that this determination of TA uses an explicit model of the acid-base reactions occurring in seawater. This modern definition of alkalinity is conceptually different from the operational definition introduced in Section 1.2.2. That is, the right-hand side of Eq. (1.2.33) explicitly states which acid-base reactions contribute to TA, whereas an operational definition simply provides a numerical value for TA based on the analytical procedure.

Dickson's definition proposes that bases formed from weak acids with $pK \geq 4.5$ (e.g. HCO_3^-) are to be considered proton acceptors, while acids with $pK < 4.5$ (e.g. H_3PO_4) are to be considered proton donors. This definition unambiguously separates proton acceptors from proton donors. Figure 1.2.11 illustrates this separation by the vertical dashed line at $pH = 4.5$ for the acid-base systems included in Eq. (1.2.33). With this definition at hand, the inclusion of other acid-base systems into the definition of TA is straightforward. Note that the choice of the value 4.5 which defines proton acceptors and donors is arbitrary, though thoughtfully chosen (Dickson, 1981).

Contribution of uncharged species

It may appear surprising that Eq. (1.2.33) contains also uncharged species such as NH_3, HF, and H_3PO_4. It is noted that Eq. (1.2.33) therefore cannot be derived from a charge balance because this would involve charged species only. The relationship between charge balance and alkalinity is examined in detail in Section 1.2.4. We shall here briefly explain why uncharged species occur in the definition of TA, using phosphoric acid as an example.

The proton condition (1.2.34) is a mass balance for hydrogen ion which defines the so-called zero level of protons. (Of course, this does not mean there are zero protons in solution!) With respect to a particular acid-base system, the proton condition defines which species are to be considered proton donors and acceptors. Figure 1.2.12 illustrates the situation for phosphoric acid. In this case, $H_2PO_4^-$ defines the zero level of protons

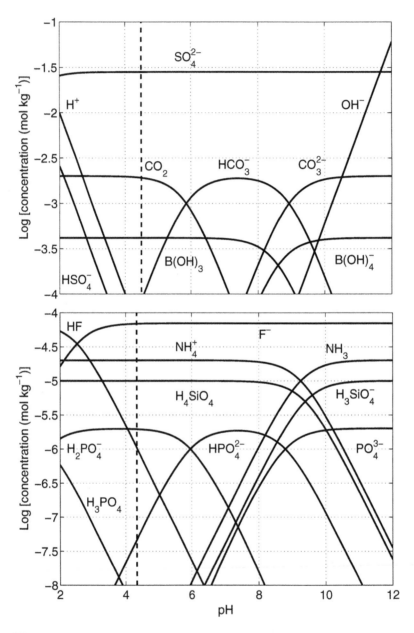

Figure 1.2.11: Extended Bjerrum plot showing the acid-base systems relevant to TA in seawater. The vertical dashed line at $pH = 4.5$ indicates which species are to be considered proton donors and acceptors (cf. Figure 1.2.12).

because it is the dominant species at $pH = 4.5$.[11] As a result, H_3PO_4 is

<hr>

[11]A significant confusion arises in the literature in some papers where the zero level

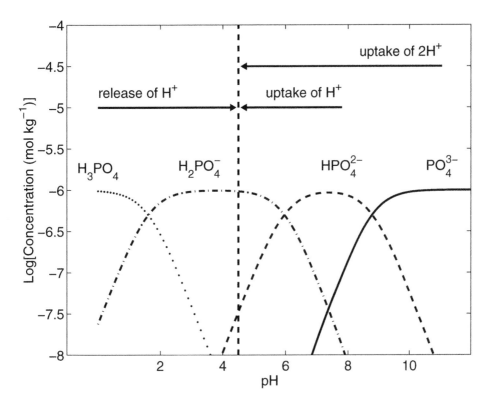

Figure 1.2.12: Application of the definition of TA to the phosphoric acid system. The species $H_2PO_4^-$ defines the zero level of protons. As a result, PO_4^{3-} is a proton acceptor with two protons less than $H_2PO_4^-$. H_3PO_4 is a proton donor with one proton more than $H_2PO_4^-$.

a proton donor that provides a proton when converted to $H_2PO_4^-$. This is the reason why also uncharged species appear in the definition of total alkalinity.

With respect to carbonic acid, HCO_3^- and CO_3^{2-} are to be considered proton acceptors because both species take up protons to form H_2CO_3. The dominant chemical species at $pH = 4.5$ which defines the zero level of protons is carbonic acid (H_2CO_3 and $CO_2 + H_2O$) and we can state that HCO_3^-, for instance, contains less protons than the defined zero level, H_2CO_3 (Figure 1.2.13). Generally, the right-hand side of Eq. (1.2.34) consists of those species which contain less protons than the defined zero level, while

for phosphoric acid has been chosen to be H_3PO_4. This is not 'wrong' just confusing as it implies that adding H_3PO_4 will not change the total alkalinity - though you can acidify the solution enough to reduce the carbonate alkalinity to zero (Dickson, personal communication).

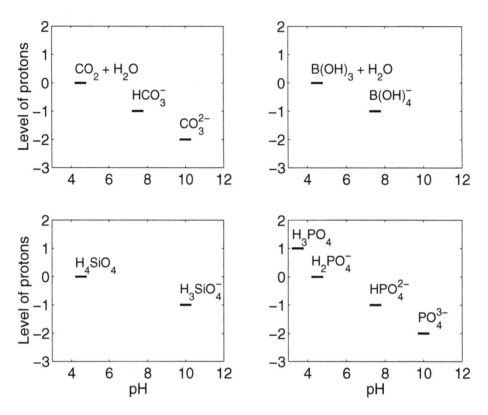

Figure 1.2.13: Chemical species of several acid-base systems and their proton levels; the dominant species over a given pH range is indicated. The zero level of protons of a particular acid-base system is defined by the dominant species at pH = 4.5. For carbonic, boric, silicic, and phosphoric acid, these species are H_2CO_3, $B(OH)_3$, H_4SiO_4, and $H_2PO_4^-$. H_2O (not shown) is the dominant species over the whole pH range from below 4 to above 10 for the system encompassing H_2O, H^+, and OH^-, and thus defines the zero level of protons for the dissociation of water.

the left-hand side of Eq. (1.2.34) consists of those species with more protons than the defined zero level (Dickson, 1981).

Summary. The concept of alkalinity was introduced, starting with the carbonate alkalinity (CA). Summarizing this preliminary discussion of alkalinity, we may say that 'alkalinity keeps track of the charges of the ions of weak acids and can be determined by titration'. Further acid-base reactions occurring in natural seawater were included, i.e. the borate and water alkalinity were added to yield PA (alkalinity for most practical purposes). Finally, a formal definition of total alkalinity (TA) was given by Eq. (1.2.33) which is based on a balance of proton acceptors and donors. In addition, the procedure to measure TA was outlined (for details, see Dickson, 1981;

Johansson and Wedborg, 1982; Anderson et al., 1999). Equations to calculate carbonate system parameters using measured TA as a master variable are given in Appendix B. The subjects discussed so far are of central importance for CO_2 measurements in the ocean. This concludes our treatment of alkalinity from a chemical point of view.

The remainder of this chapter will add nothing new to the definition of total alkalinity. The formal definition may often suffice for the application of the total alkalinity concept but usually a deeper understanding is required in order to answer questions such as: Does alkalinity change due to nitrogen assimilation by algae and - if yes - how does it change? The answer is yes, it does - although NO_3^- and NH_4^+ do not show up in the formal definition of TA, see below.

1.2.4 Total alkalinity and charge balance

While the preceding sections considered total alkalinity from a chemical point of view, we will now approach the concept of alkalinity from a physical or geochemical point of view. It will be shown that alkalinity is closely related to the charge balance in seawater which makes it an appealing concept because properties such as the conservation of total alkalinity can be derived.

Sillén (1961) argued that considering the origin of the ocean, we might say that the ocean is the result of a gigantic acid-base titration in which acids such as HCl, H_2SO_4, and CO_2 that have leaked out from the interior of the earth are titrated with bases derived from the weathering of primary rock.[12] As we have seen in the previous chapters, the buffer capacity and alkaline properties of seawater can be mainly attributed to the dissolved carbonate species. One may therefore state that the ocean owes its buffer properties to the presence of carbonic acid, which is correct in some respect.

Sillén (1967) pointed out, however, that the buffer capacity of the carbonates in solution is pitifully small compared with the amounts of acids and bases that have passed through the ocean system in the course of time. With respect to the definition of total alkalinity in terms of an acid-base balance, this view suggests a more fundamental reason for the alkaline property of seawater. We will see that the origin of alkalinity in the ocean has to do with the charge balance of the major conservative ions in seawater.

---

[12] Sillén continues: "In this acid-base titration, volcanoes against weathering, it would seem that we are about 0.5 per cent from the equivalence point. This, by the way, is better than most students of chemistry do in their first titrations."

The sum of the charges of the major cations Na^+, K^+, Mg^{2+}, and Ca^{2+} are not exactly balanced by the major anions Cl^- and SO_4^{2-}. This small charge imbalance is responsible for the total alkalinity in the ocean and is mainly compensated for by the anions of carbonic acid. We may say that:

> *'Total alkalinity is equal to the charge difference between conservative cations and anions'*

Total alkalinity in simple systems

The close relation between total alkalinity and charge balance will firstly be demonstrated by considering simple aqueous solutions which are less complicated than natural seawater. Seawater is a very complicated system because it contains many different cations and anions and a number of weak acids and bases. McClendon et al. (1917) stated that:

> *'The sea is of too complex a composition to admit of any simple mathematical relations'*

(cf. Dickson, 1992). In order to get rid of some of the complications associated with natural seawater, simple systems that contain only Cl^-, Na^+, and one weak acid are discussed first.

System I: Na^+, Cl^-, and one weak acid

Let us consider a solution that contains only Na^+, Cl^-, and a monoprotic weak acid, represented by HA:

$$HA \rightleftharpoons A^- + H^+; \qquad K_A$$

with $pK_A = 6$. The following initial conditions are prescribed: $pH_i = 8.2$, $[Na^+]_i = 0.6$ mol kg^{-1}, and the total concentration of A, $A_T = [A^-]_i + [HA]_i = 2300 \, \mu$mol kg^{-1}. The pH value is therefore similar to typical surface ocean conditions, the concentration of sodium is similar to the total charge of all cations in seawater at $S = 35$, and the total concentration of A is numerically similar to typical alkalinity values in surface seawater.

From the prescribed values, the initial concentrations of OH^- and A^- can be calculated:

$$[OH^-]_i = K_W/[H^+]_i = 9.6 \, \mu\text{mol kg}^{-1}$$
$$[A^-]_i = A_T/(1 + [H^+]_i/K_A), = 2286 \, \mu\text{mol kg}^{-1}$$

It is important to note that the concentration of Cl^- was not prescribed. It is set, however, by requiring electroneutrality:

$$[Cl^-] = [Na^+] + [H^+] - [A^-] - [OH^-] = 0.5977 \text{ mol kg}^{-1} .$$

Thus, the concentration of Cl^- is slightly smaller than that of Na^+. The difference of the concentration of the cations and anions of strong bases and acids, that is, the charge imbalance of the conservative ions, is:

$$[Na^+] - [Cl^-] = TA = [A^-] + [OH^-] - [H^+] . \tag{1.2.35}$$

The right-hand side of Eq. (1.2.35) is analogous to the definition of TA in terms of acid-base reactions in Eq. (1.2.33). However, we have now expressed the alkalinity entirely in terms of the conservative ions, the left-hand side of Eq. (1.2.35). It is this property that makes TA a conservative quantity. The expression on the left-hand side of Eq. (1.2.35) will be denoted as 'explicitly conservative alkalinity' or $TA^{(ec)}$ because each term and not only the whole expression is conservative; the expression on the right-hand side of Eq. (1.2.35) will be represented by the symbol TA. We will see that considering the $TA^{(ec)}$ is extremely useful when determining changes of TA due to biogeochemical process (Section 1.2.7).

The relationship between the charge balance of the conservative ions and TA for the simple system can be illustrated by considering the titration of the system with HCl (Figure 1.2.14). $TA^{(ec)}$ is plotted in Figure 1.2.14a, while TA, i.e. proton acceptors and donors, is included in Figure 1.2.14b. Initially, there is a small deficit of $[Cl^-]$ compared to $[Na^+]$ of about 2.3 mmol kg^{-1} which is exactly equal to the initial total alkalinity (Figure 1.2.14a). At the endpoint of the titration, the initial deficit has been compensated by addition of HCl. At this point, the pH is 4.32, the concentrations of chloride and sodium ions are equal, and the total alkalinity is zero.

The point at which $[Cl^-] = [Na^+]$ corresponds to the zero level of protons, i.e. the equivalence point of the acid-base system:

$$[H^+] = [A^-] + [OH^-] \tag{1.2.36}$$

which is indicated by the circle in Figure 1.2.14b. At this point, A^- has almost completely been neutralized by H^+ and converted to HA. The total alkalinity of the simple system considered here, can be equally well defined by the excess of proton acceptors over proton donors (TA) or by the charge difference of the conservative ions ($TA^{(ec)}$). A glance at Figure 1.2.14 shows that it may be much more simple to consider the conservative ions instead of dealing with all the details of the acid-base system.

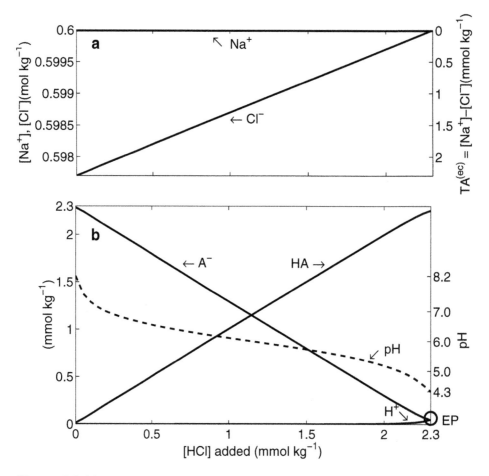

Figure 1.2.14: Titration of System I by hydrochloric acid. (a) Concentrations of the conservative ions. (b) Concentrations of the compounds of the acid-base system and pH. The difference $[Na^+] - [Cl^-]$ is exactly equal to the total alkalinity, $TA = [A^-] + [OH^-] - [H^+]$.

System II: Table salt and a weak acid dissolved in water

It was mentioned above that the total alkalinity of System I is similar to typical values observed in surface seawater. It was then shown that the alkalinity is due to the small but important charge imbalance of the conservative ions. In order to emphasize the significance of this small imbalance we will now consider a system in which the imbalance is missing. System II consists of an aqueous solution containing 2300 $\mu mol\ kg^{-1}$ of the weak acid discussed in System I and of 0.6 mol kg^{-1} Na^+ and Cl^-. In other words, System II is made of pure water to which the weak acid and table salt (NaCl) have been added. System II is very similar to System I. However, the important difference is that the concentrations of Na^+ and Cl^- are identical in System II. In contrast to System I, the pH of System II is not an independent initial parameter but can be calculated from electroneutrality:

$$
\begin{aligned}
0 &= TA = [Na^+] - [Cl^-] \\
&= [A^-] + [OH^-] - [H^+] \\
&= \frac{A_T}{1 + \dfrac{[H^+]}{K_A}} + \frac{K_W}{[H^+]} - [H^+].
\end{aligned}
$$

The numerical solution of the resulting cubic equation for $[H^+]$ yields $pH \simeq 4.32$. Thus the addition of the weak acid to an otherwise neutral table salt solution at $pH = 7$ leads to a low pH value. This value is reached in System I only after addition of an appreciable amount of strong acid. Actually, the titration of System I to the pH of the proton condition transforms System I into System II. Therefore it is not surprising that the pH of System II is equal to the pH of System I after titration to the equivalence point where the total alkalinity equals zero (see Figure 1.2.14). In System II, most of the weak acid is in the uncharged form, i.e. $pH < pK$, and therefore the system is unable to buffer any addition of acids.

From the comparison of System I and II it is concluded that *the imbalance of the charge of the conservative ions is responsible for alkalinity.* As a result, the system is buffered. In seawater, TA varies with salinity, see Section 1.2.7. This variation is due to the total charge imbalance of the conservative ions which varies with salinity. (Note that the composition of sea salt is virtually constant). Seawater therefore owes its buffer capacity to the charge imbalance of the conservative ions. Freshwater systems usually contain only small amounts of conservative ions and are therefore only weakly buffered. Small changes in acid concentrations due to, for example, uptake of CO_2 by plants or input of acid rain thus result in large shifts of pH.

1.2.5 The charge balance in natural seawater

Natural seawater contains many more ions than the simple systems discussed above. Nevertheless, the same principle of electroneutrality applies to seawater, as well. The sum of the charges of all ions present in seawater must equal zero. This is mathematically expressed by:

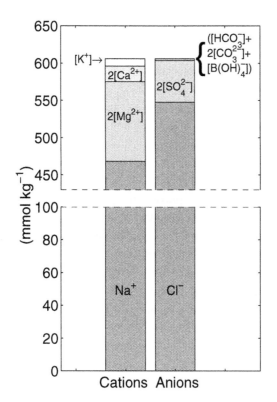

Figure 1.2.15: Charge balance of the major ions in seawater (cf. Broecker and Peng,
1998). The small excess charge
of the conservative cations over
anions is mainly balanced by
$[HCO_3^-]+2[CO_3^{2-}]+[B(OH)_4^-]$.

$$[Na^+] + 2[Mg^{2+}] + 2[Ca^{2+}] + [K^+] + ... + [H^+]_F$$
$$-[Cl^-] - 2[SO_4^{2-}] - [NO_3^-]$$
$$-[HCO_3^-] - 2[CO_3^{2-}] - [B(OH)_4^-] - [OH^-] - ... = 0 \qquad (1.2.37)$$

or

$$\sum_j z_j[c_j] = 0, \qquad (1.2.38)$$

where $[c_j]$ and z_j are the concentration and charge of compound j, respectively.

The charge balance of the major ions in seawater is illustrated in
Figure 1.2.15, where all minor species have been neglected. First of
all, it shows that the charge imbalance between the conservative cations
and anions is very small compared to their total concentration (~ 2 vs.
~ 600 mmol kg^{-1}). We also see that the small excess of positive charge

Table 1.2.7: Concentrations, $[c_i]$ (mmol kg^{-1}), and charge concentrations, $[q_i] = z_i \cdot [c_i]$ (mmol kg^{-1}), of conservative ions in seawater at $S = 35$ (after Bakker, 1998).

Cations	$[c_i]$	$[q_i]$	Anions	$[c_i]$	$[q_i]$
Na$^+$	467.8	467.8	Cl$^-$	545.5	545.5
Mg^{2+}	53.3	106.5	SO$_4^{2-}$	28.2	56.4
Ca^{2+}	10.3	20.6	Br$^-$	0.8	0.8
K$^+$	9.9	9.9	F$^-$	0.1	0.1
Sr^{2+}	0.1	0.2		.	.
Total	.	605.0	Total	.	602.8

$([Na^+] + 2[Mg^{2+}] + 2[Ca^{2+}] + [K^+])$ over negative charge $([Cl^-] + 2[SO_4^{2-}])$ is compensated by $[HCO_3^-] + 2[CO_3^{2-}] + [B(OH)_4^-]$, where the latter sum represents the most important contribution to TA. Table 1.2.7 lists the concentrations and charge concentrations of the major conservative cations and anions in seawater at $S = 35$. The total charge concentration of the conservative cations yields 605.0 mmol kg^{-1}, while the total charge concentration of the conservative anions yields 602.8 mmol kg^{-1}. The small charge imbalance of 2.2 mmol kg^{-1} is equal to the total alkalinity in seawater (strictly, this is only correct when phosphate, ammonia and bisulfate ion are neglected, cf. Appendix C.1).

The fact that the alkalinity can be expressed in terms of the charge imbalance of the conservative ions allows us to derive a very useful equation. Equation (1.2.37) may be rewritten such that all conservative ions appear on the left-hand side and all non-conservative ions appear on the right-hand side:

$$\sum \text{conservative cations} - \sum \text{conservative anions}$$
$$= [HCO_3^-] + 2[CO_3^{2-}] + [B(OH)_4^-] + [OH^-] - [H^+] \qquad (1.2.39)$$
$$\pm \text{ minor compounds}$$

This expression is very similar to the one given for PA in Eq. (1.2.31) (alkalinity for most practical purposes). Neglecting the term 'minor compounds', we can write:

$$[Na^+] + 2[Mg^{2+}] + 2[Ca^{2+}] + [K^+] + \dots$$
$$-[Cl^-] - 2[SO_4^{2-}] - [NO_3^-] - \dots$$
$$= PA =$$
$$[HCO_3^-] + 2[CO_3^{2-}] + [B(OH)_4^-] + [OH^-] - [H^+] \tag{1.2.40}$$

or:

$$TA^{(ec)} = PA$$

The right-hand side of Eq. (1.2.40) simply expresses PA in terms of proton acceptors and donors as discussed in the first part of this chapter. However, the left-hand side of Eq. (1.2.40) expresses PA entirely in terms of the conservative ions (explicitly conservative alkalinity, $TA^{(ec)}$). This is very useful as will be elaborated in the following.

1.2.6 Conservation of total alkalinity

The fact that total alkalinity can be expressed by the charge balance of the conservative ions in seawater allows us to make important inferences without using the formal definition of total alkalinity (Eq. 1.2.33).[13]

Firstly, if total alkalinity of a seawater sample is expressed in units of mol kg^{-1}, then the total alkalinity will remain exactly conserved during changes of temperature and pressure. This follows directly from $TA^{(ec)}$ because the concentration of the conservative ions in 1 kg seawater is unaffected by changes of temperature and pressure. This is not obvious from the alkalinity expression in terms of weak acids (right-hand side of Eq. (1.2.40)) because the individual concentrations of the chemical species in this expression do change as a function of temperature and pressure. The reason for this is that the chemical equilibria (pK's) depend on temperature and pressure.

Secondly, the charge of the conservative ions does not change when, for instance, CO_2 is exchanged between water and air or when CO_2 is taken up or respired by algae. During uptake and release of CO_2, the individual concentrations on the right-hand side of Eq. (1.2.40) may change dramatically because of variations in ΣCO_2 and pH. However, $TA^{(ec)}$ is constant

---

[13]Note that the relationship between charge balance and alkalinity in Eq. (1.2.40) was strictly derived here only for PA and not TA. The derivation for TA can be found in Appendix C.1.

because the conservative ions are not affected and therefore alkalinity is constant. On the other hand, if a strong base such as NaOH is added to the sample, the alkalinity increases because $[Na^+]$ and therefore also $TA^{(ec)}$ increases. If strong acid such as HCl is added to the sample, the alkalinity decreases because $[Cl^-]$ (with a negative sign in the $TA^{(ec)}$ expression) increases. Note that TA changes by one unit for every unit of strong base or acid added because these substances dissociate completely. Analogously, changes of TA due to biogeochemical processes in the ocean can be understood (Section 1.2.7).

Mixing and conservation of mass, salt, dissolved inorganic carbon, and total alkalinity.
When two bodies ($i = 1, 2$) of seawater with different properties are mixed, the amounts of mass (M_i), salt ($M_i \cdot S_i$), dissolved inorganic carbon ($M_i \cdot DIC_i$), and total alkalinity ($M_i \cdot TA_i$), respectively, add up to yield the exact property values of the mixture (index m):

$$M_m = M_1 + M_2 \tag{1.2.41}$$
$$M_m \cdot S_m = M_1 \cdot S_1 + M_2 \cdot S_2 \tag{1.2.42}$$
$$M_m \cdot DIC_m = M_1 \cdot DIC_1 + M_2 \cdot DIC_2 \tag{1.2.43}$$
$$M_m \cdot TA_m = M_1 \cdot TA_1 + M_2 \cdot TA_2 \tag{1.2.44}$$

whereby S_i, DIC_i, and TA_i are in gravimetric (per mass) units. In contrast, neither volume nor single components of DIC or TA such as $[CO_2]$, $[HCO_3^-]$, or $[CO_3^{2-}]$ are conserved when two water bodies are mixed (see Exercises 1.9 and 1.10). The conservation property of DIC and TA is the reason for using these two quantities in oceanic models of the carbon cycle as prognostic variables.

Thirdly, total alkalinity is conserved during mixing. When two water masses are mixed with different properties (salinity S_i, temperature T_i, mass M_i, $i = 1, 2$), the charges of conservative ions are additive, i.e. the following relation is obeyed (cf. box on mixing):

$$M_m \cdot \sum_s [q_{m,s}] = M_1 \cdot \sum_s [q_{1,s}] + M_2 \cdot \sum_s [q_{2,s}], \tag{1.2.45}$$

where the summation index s runs over all conservative ions; the index m denotes quantities after mixing, and $M_m = M_1 + M_2$. As a consequence, total alkalinity is also additive and the the relation

$$M_m \cdot TA_m = M_1 \cdot TA_1 + M_2 \cdot TA_2 \tag{1.2.46}$$

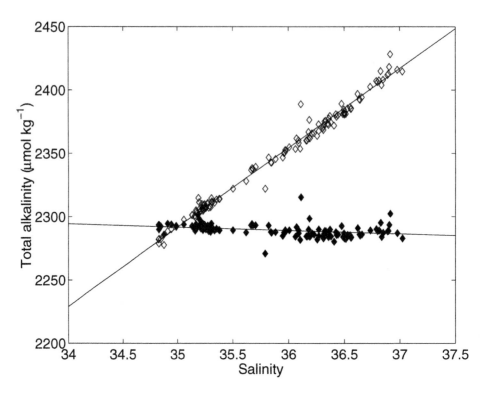

Figure 1.2.16: Total alkalinity (open diamonds) and total alkalinity normalized to $S = 35$ (closed diamonds) in surface waters of the Indian Ocean (WOCE Section I07N, F. Millero and K. Johnson, http://cdiac.esd.ornl.gov/oceans/CDIACmap.html). In this case, total alkalinity is linearly related to salinity.

holds. In other words, total alkalinity is a conserved quantity just like mass, salt, or dissolved inorganic carbon.

1.2.7   Total alkalinity in the ocean

There are several processes which lead to changes of total alkalinity.[14] First of all, the charge difference between conservative cations and anions varies with salinity. Therefore TA in the ocean is closely related to salinity (Figure 1.2.16). Salinity changes are due to precipitation, evaporation, fresh water input, and formation or melting of sea ice and cause corresponding changes in total alkalinity.

The second major change in total alkalinity in the ocean is due to biogenic precipitation of calcium carbonate ($CaCO_3$) by marine organisms such

---

[14] A summary of this subject can also be found in Stoll (1994).

as coccolithophorids, foraminifera, pteropods, and corals, and dissolution of calcareous shells or skeletons. Precipitation leads to a decrease in Ca^{2+} concentration. Thus the charge difference between conservative cations and anions decreases and so does the total alkalinity. Precipitation of 1 mol $CaCO_3$ reduces DIC by 1 mol and TA by 2 mol. Please note that these changes are independent of the carbon source (HCO_3^-, CO_3^{2-}, or even CO_2) used by the organisms for calcification. As a result, the carbon source cannot be inferred from observations of changes in DIC and TA in seawater.

Minor changes of total alkalinity in the ocean are related to nitrogen assimilation by plants and release of dissolved inorganic nitrogen from organic compounds during remineralization. The change of alkalinity due to assimilation of nitrate, ammonia, or urea by plants has been investigated by Brewer and Goldman (1976) and Goldman and Brewer (1980). They found that alkalinity changes were consistent with a simple stoichiometric model in which NO_3^- uptake is balanced by OH^- production (that is, alkalinity increases) and NH_4^+ uptake is balanced by H^+ production (that is, alkalinity decreases); no change of alkalinity occurs when urea is assimilated.

The TA$^{(ec)}$ expression can be used to summarize changes of alkalinity due to certain biogeochemical processes. This approach is very convenient because each single component of TA$^{(ec)}$ (conservative ions) can be varied without affecting any other component in that expression. On the other hand, changes of a single component of the expression in terms of weak acids for alkalinity (proton acceptors and donors) affects all other components via changes in, for instance, ΣCO_2 and pH. Alkalinity changes due to several biogeochemical processes based on the TA$^{(ec)}$ expression, Eq. (1.2.40), are as follows:

1. Precipitation of one mole $CaCO_3$: alkalinity decreases by two moles (term: $2[Ca^{2+}]$).

2. Dissolution of one mole $CaCO_3$: alkalinity increases by two moles (term: $2[Ca^{2+}]$).

3. Uptake of DIC by algae under the assumption that electroneutrality of algae is ensured by parallel uptake of H^+ or release of OH^-: no change in alkalinity.

4. Uptake of one mole of nitrate (NO_3^-) by algae under the assumption that electroneutrality of algae is ensured by parallel uptake of H^+ or release of OH^-: alkalinity increases by one mole (term: $-[NO_3^-]$). This is consistent with laboratory experiments with algae grown on

different nitrogen sources (Brewer and Goldman (1976); Goldman and Brewer (1980)).

5. Remineralization of algal material has the reverse effects on alkalinity.

Determining changes of alkalinity due to algal uptake of ammonia, phosphate, etc. requires a definition of the left-hand side of TA corresponding to the full right-hand side of TA as given by Dickson (Eq. (1.2.33)). This can be found in Appendix C.1.

Summary. Total alkalinity is defined as the number of moles of hydrogen equivalent to the excess of proton acceptors over proton donors in 1 kg sample. For the most practical purposes total alkalinity in seawater can be defined as PA = $[HCO_3^-]+2[CO_3^{2-}]+[B(OH)_4^-]+[OH^-]-[H^+]$. From the fact that TA is equal to the charge difference between conservative cations and anions, it follows that TA is a conservative quantity. Total alkalinity does not change when CO_2 is exchanged with the atmosphere or is taken up or released by plants. However, total alkalinity does change with salinity (precipitation, evaporation), precipitation and dissolution of calcium carbonate, and with the assimilation and release of dissolved inorganic nitrogen.

Further reading: Dickson (1992), Bradshaw and Brewer (1988b), Anderson et al. (1999).

Exercise 1.6 (**)

In laboratory experiments biologists often add nutrients at concentrations much higher than observed in the ocean. Does this lead to changes in alkalinity? If yes, how much does alkalinity change when you add nutrients to a natural seawater sample in order to match the concentrations of the f/2-medium? The f-medium contains 882 μmol kg^{-1} nitrate, 36 μmol kg^{-1} phosphate and 30 μmol kg^{-1} silicate (Guillard and Ryther, 1962). The f/2-medium contains half as much. The natural seawater sample shall already contain 15 μmol kg^{-1} nitrate, 1 μmol kg^{-1} phosphate, and 15 μmol kg^{-1} silicate. The nutrients are added in the form of NaNO$_3$, Na$_2$HPO$_4$ and Na$_2$SiO$_3$.

Exercise 1.7 (**)

Consider a culture of noncalcifying algae in a closed bottle at $T_c = 25°C$ and $S = 35$. The initial conditions are DIC = 2 mmol kg^{-1}, TA = 2.3 mmol kg^{-1} and NO$_3^-$ = 20 μmol kg^{-1}. After some time the algae have taken up carbon and have assimilated all the nitrate. Calculate the change in [CO$_2$] assuming a Redfield ratio C:N = 106:15. How much of this variation is due to a change in alkalinity?

Exercise 1.8 (**)

As carbon dioxide is taken up or lost from sea water the total alkalinity is conserved. What happens to the carbonate alkalinity?

Exercise 1.9 (**)

Consider two bodies of seawater with equal masses (1000 kg) and temperatures (25°C) but different salinities ($S_1 = 30$, $S_2 = 40$). Calculate the corresponding volumes of the water bodies before and after mixing at surface pressure. Is volume conserved? Hint: the density of seawater as a function of salinity, temperature, and pressure ($\rho(S, T, P)$) is given in Appendix A.13 (cf. Millero and Poisson, 1981; Gill, 1982). A MATLAB routine including $\rho(S, T, P)$ can be found on our web-page; 'http://www.awi-bremerhaven.de/Carbon/co2book.html'.

Exercise 1.10 (**)

Consider two bodies of seawater with equal masses (1 kg) and temperatures (25°C) but different salinities ($S_1 = 30$, $S_2 = 40$), dissolved inorganic carbon ($DIC_1 = 1800$ μmol kg^{-1}, $DIC_2 = 2000$ μmol kg^{-1}), and total alkalinity ($TA_1 = 2100$ μmol kg^{-1}, $TA_2 = 2300$ μmol kg^{-1}). Calculate the corresponding concentrations of CO_2, HCO_3^-, and CO_3^{2-} of the water bodies before and after mixing at surface pressure. Are the total amounts of CO_2, HCO_3^-, and CO_3^{2-} conserved?

1.3 *p*H scales

> **"The field of *p*H scales ... in sea water is one of the more confused areas of marine chemistry."**
> Dickson (1984) p. 2299.

> **"The choice of a suitable *p*H scale for the measurement of oceanic *p*H remains for many people a confused and mysterious topic ..."**
> Dickson (1993a) p. 109.

The goal of this section is to introduce the different *p*H scales used in chemical oceanography and briefly discuss the origin of these scales. This requires a little work but it will show that there is nothing mysterious about seawater *p*H scales. Based on a simple inter-conversion numerical differences between the scales will be estimated. By means of an example, we shall finally illustrate possible errors in the calculation of *p*CO$_2$ if the differences between the *p*H scales are ignored.

In the following, we address two main questions: Why and when should we care about the different *p*H scales (definitions)?

Why? The equilibrium constants of proton-transfer reactions such as the first and second acidity constants of H_2CO_3 must be defined consistently with pH scales (see Dickson (1984; 1993a) for details). If this point is disregarded in calculations of the carbonate system, serious errors can occur (cf. Section 1.1.6).

When? The values of different pH scales in seawater differ by up to 0.12 units (see Eqs. (1.3.57), (1.3.58)) corresponding to a comparable difference in pK_1^* and pK_2^* (see Eq. (1.3.60)). Because these differences are much larger than the desired accuracy of pH measurements, we should care about the different pH scales whenever pH is a master variable. Please note that the differences between acidity constants from different publications may be mainly due to the use of different pH scales.

Various definitions

In high school we learned that pH (lat.: *potentia hydrogenii* or also *pondus hydrogenii*) is the negative common logarithm of the concentration of hydrogen ions:

$$pH_{\text{high school}} := -\log[H^+]. \tag{1.3.47}$$

This definition dates from Sørensen (1909). Things are a bit more complicated for several reasons:

- "It is safe to say that free protons" (hydrogen ions) "do not exist in any significant amount in aqueous solutions. Rather the proton is bonded to a water molecule thus forming an H_3O^+ ion; this in turn is hydrogen bonded to three other water molecules to form an $H_9O_4^+$ ion." (Dickson, 1984, p. 2299); compare also Marx et al. (1999) and Hynes (1999). Thus, the symbol 'H^+' represents hydrate complexes rather than the concentration of free hydrogen ions. As noted in Section 1.1, it is however convenient to refer to $[H^+]$ as the hydrogen ion concentration.

- In a refined theory one should use activity (an 'effective' concentration; denoted by a) instead of concentration

$$pH_a = -\log a_{H^+} . \tag{1.3.48}$$

Unfortunately, it is not possible to measure pH according to Eq. (1.3.48). The reason for this is that individual ion activities cannot be determined experimentally since the concentration of a single ion cannot be varied independently because electroneutrality is required (e.g. Klotz and Rosenberg, 2000).

An operational definition of the NBS[15] *p*H scale was given by the International Union of Pure and Applied Chemistry (IUPAC). The NBS *p*H scale is defined by a series of standard buffer solutions with assigned *p*H values close to the best estimates of $-\log a_{\rm H+}$, i.e. $p{\rm H}_{\rm NBS}$ is close to but not identical to $p{\rm H}_{\rm a}$

$$p{\rm H}_{\rm NBS} \approx p{\rm H}_{\rm a}. \tag{1.3.49}$$

The reference state for $p{\rm H}_{\rm a}$ and $p{\rm H}_{\rm NBS}$ is the infinite dilute solution, i.e. the activity coefficient of H^+, $\gamma_{\rm H+}$, approaches unity when $[H^+]$ approaches zero in pure water.

NBS standard buffer solutions have very low ionic strength, ~ 0.1. In contrast, seawater has high ionic strength, ~ 0.7. The use of NBS buffers in *p*H measurements using electrodes in seawater is therefore not recommended because the large differences in ionic strength between the buffer and the seawater causes significant changes in the liquid junction potential[16] between calibration and sample measurement. To make it worse, these changes depend on the electrode system used. The error due to this change in liquid junction potential is larger than the desired accuracy of 0.01 - 0.001 *p*H units (Wedborg et al., 1999).

Hansson (1973a) greatly improved the situation by the adoption of seawater as the standard state and the introduction of a new *p*H scale based on artificial seawater. On this *p*H scale, called the 'total' scale, the activity coefficient $\gamma_{\rm H_T^+}$ approaches unity when $[H^+]_{\rm T}$ approaches zero in the pure synthetic seawater on which the scale is based[17] (for definition of $[H^+]_{\rm T}$, see below). Hansson (1973a) introduced a new set of standard buffers based on artificial seawater and assigned *p*H values to these buffers according to

---

[15]NBS: National Bureau of Standards, now NIST: National Institute of Standards and Technology.

[16]The liquid junction potential is the electric potential difference between the solution in the electrode and the measurement solution. Ideally it depends only on the composition of the electrode solution and measurement solution - in reality, however, it also depends on the practical design of the liquid junction (Wedborg et al., 1999).

[17]The chemical potential μ_i of species i is related to the concentration c_i and the activity a_i by $\mu_i - \mu_i^0 = RT \ln a_i = RT \ln (c_i \gamma_i)$ where μ_i^0 is the chemical potential of the reference state and γ_i is the activity coefficient. The relation $\gamma_i \to 1$ in seawater corresponds to a certain choice of the values of the μ_i^0.

his new pH scale. One of the great advantages of this approach is that the changes of liquid junction potential between the buffer and the sample are greatly reduced because of the very similar composition of the two solutions.

In addition to the total scale, the free scale and the seawater scale have been proposed for the use in seawater which will be discussed below. We denote the free scale by pH$_F$, the total (or Hansson) scale by pH$_T$, and the seawater scale by pH$_{SWS}$. They are defined by:

$$pH_F = -\log[H^+]_F \tag{1.3.50}$$

$$pH_T = -\log\left([H^+]_F + [HSO_4^-]\right) = -\log[H^+]_T \tag{1.3.51}$$

$$pH_{SWS} = -\log\left([H^+]_F + [HSO_4^-] + [HF]\right) = -\log[H^+]_{SWS}. \tag{1.3.52}$$

where $[H^+]_F$ is the 'free' hydrogen ion concentration, including hydrated forms, see above.

The free scale

Certainly, the free scale is conceptually the clearest. However, it will become apparent very soon that this concept suffers from the fact that the stability constant of HSO_4^-, K_S^*, has to be determined accurately in seawater which is not a simple task (Dickson, 1984; 1993b). In seawater, protonation of ions such as SO_4^{2-} occurs:

$$HSO_4^- \rightleftharpoons H^+ + SO_4^{2-}; \qquad K_S^* .$$

Thus, in a seawater medium containing sulphate ions, the total hydrogen ion concentration is:

$$[H^+]_T = [H^+]_F + [HSO_4^-] .$$

Analytically, only $[H^+]_T$ can be determined (Dickson, 1993b; Wedborg et al., 1999). As a result, $[H^+]_F$ has to be calculated according to

$$[H^+]_F = [H^+]_T - [HSO_4^-]$$
$$= [H^+]_T \left(1 + [SO_4^{2-}]/K_S^*\right)^{-1}$$

which involves K_S^*. The use of the free scale therefore requires an accurate value of K_S^* in seawater which is difficult to obtain.

The total scale

A medium containing sulphate ions was used by Hansson (1973a) who therefore defined the total scale as given in Eq. (1.3.51). This scale includes the effect of sulfate ion in its definition and therefore circumvents the problem of determining K_S^*. On this scale, the activity coefficient approaches unity when ($[H^+]_F + [HSO_4^-]$) approaches zero in the ionic medium.

The seawater scale

If the medium additionally contains fluoride ions, we also have to take the protonation of F^- into account:

$$HF \quad \rightleftharpoons \quad H^+ + F^-; \qquad K_F^* .$$

Such a medium was used, for instance, by Goyet and Poisson (1989). The total hydrogen ion concentration then is:

$$[H^+]_F + [HSO_4^-] + [HF]$$

which leads to the definition of the seawater scale, Eq. (1.3.52). The difference between the total and the seawater scale therefore simply arises from the fact whether the medium on which the scale is based contains fluoride ions or not. This difference is, however, small because the concentration of HSO_4^- in seawater is much larger than the concentration of HF.

The bottom line is as follows. Three pH scales have been proposed for the use in seawater. The use of the free scale requires an accurate determination of the stability constant of HSO_4^- in seawater which is not a simple matter. This problem is circumvented by the definition of the total and the seawater scale. The difference between the total and seawater scale is explained by differences in the laboratory protocols of the different researchers.

1.3.1 Conversion between pH scales

The various pH scales are inter-related by the following equation (after conversion to the same concentration unit, say mol kg^{-1}):

$$
\begin{aligned}
[H^+]_F &= \frac{[H^+]_T}{1 + S_T/K_S^*} = \frac{[H^+]_{SWS}}{1 + S_T/K_S^* + F_T/K_F^*} \\
&\quad \text{or} \\
pH_F &= pH_T + \log\left(1 + S_T/K_S^*\right) \\
&= pH_{SWS} + \log\left(1 + S_T/K_S^* + F_T/K_F^*\right)
\end{aligned}
\tag{1.3.53}
$$

where K_S^* and K_F^* are the stability constants of hydrogen sulfate and hydrogen fluoride, respectively (see Appendix A). Furthermore,

$$S_T = [\mathrm{SO_4^{2-}}] + [\mathrm{HSO_4^-}] \simeq [\mathrm{SO_4^{2-}}] \qquad (1.3.54)$$

is the total sulfate concentration and

$$F_T = [\mathrm{F^-}] + [\mathrm{HF}] \simeq [\mathrm{F^-}] \qquad (1.3.55)$$

is the total fluoride concentration. Equation (1.3.53) shows that scale conversion is basically simple. It also shows that in order to convert accurately between the different scales, the stability constants K_S^* and K_F^* in seawater have to be known precisely. As said above, this appears to be difficult (Dickson, 1993a). Thus, uncertainties are introduced into calculations when values are compared between different scales. However, such uncertainties are small compared to those arising when scale conversion is simply ignored.

Let us estimate the numerical differences between the pH scales in seawater using the data available for K_S^* and K_F^*. The numerical differences are given by the terms $\log(1 + S_T/K_S^*)$ and $\log(1 + S_T/K_S^* + F_T/K_F^*)$. Values of S_T and F_T for seawater at $S = 35$ may be found in DOE (1994):

$$
\begin{aligned}
S_T &= 28.24\,\mathrm{mmol\,(kg\text{-}soln)^{-1}} \\
F_T &= 70\,\mu\mathrm{mol\,(kg\text{-}soln)^{-1}}\ .
\end{aligned}
\qquad (1.3.56)
$$

At $S = 35$ and $T_c = 25°\mathrm{C}$, $K_S^* \simeq 0.10\ \mathrm{mol\ kg^{-1}}$ and $K_F^* \simeq 0.003\ \mathrm{mol\ kg^{-1}}$, and one obtains:

$$
\begin{aligned}
\log(1 + S_T/K_S^*) &\simeq 0.11 & (1.3.57) \\
\log(1 + S_T/K_S^* + F_T/K_F^*) &\simeq 0.12 & (1.3.58)
\end{aligned}
$$

In other words, under standard conditions, the pH value of a sample reported on the free scale is about 0.11 and 0.12 units higher than on the total and seawater scale, respectively. For example, if $p\mathrm{H_F} = 8.22$, then $p\mathrm{H_T} = 8.11$ and $p\mathrm{H_{SWS}} = 8.10$. This gives us a feeling of the potential error occuring in calculations when scale conversion is ignored. Such differences are huge when compared to the accuracy and precision of pH measurements. Millero et al. (1993b) give an accuracy in measuring pH of ± 0.002 (for recent advances in measurement techniques see, for instance, Dickson (1993b), Bellerby et al. (1995), and Wedborg et al. (1999)). The currently used pH scales and the estimated differences between them are summarized in Table 1.3.8.

Table 1.3.8: The *p*H scales and the differences between them.

*p*H scale	applicable in	reference state	difference to free scale[a] $pH_F - pH_i$
pH_{NBS}	freshwater	pure water	
pH_F	seawater	artificial seawater	
pH_T	seawater	artificial seawater	~ 0.11
pH_{SWS}	seawater	artificial seawater	~ 0.12

[a] At $S = 35$, $T_c = 25°C$.

The most important lesson to be learned here is: check which *p*H scale is used! If this is ignored, large uncertainties can occur. Compared to these uncertainties, errors arising from the actual procedure of scale conversion, Eq. (1.3.53), which are due to uncertainties in the values of K_S^* and K_F^*, are small. Potential problems arising from the conversion between the different *p*H scales could be eliminated by agreement of the community on a single *p*H scale. One promising candidate for this is certainly the total scale.

1.3.2   Conversion of acidity constants

The conversion of acidity constants between different *p*H scales is described in Dickson and Millero (1987). A typical equilibrium relation reads:

$$[CO_3^{2-}]/[HCO_3^-] = K_{2T}^*/[H^+]_T \qquad (1.3.59)$$

where the index 'T' indicates that equilibrium constant and *p*H refer to the total scale. Now convert from total to free scale. Taking the negative logarithm results in:

$$
\begin{aligned}
- \log\ &([CO_3^{2-}]/[HCO_3^-]) \\
= \ &pK_{2T}^* - pH_T \\
= \ &\underbrace{pK_{2T}^* + \log(1 + S_T/K_S^*)}_{=:\,pK_{2F}^*} - pH_F \\
= \ &pK_{2F}^* - pH_F\ . \qquad (1.3.60)
\end{aligned}
$$

When we change the *p*H-scale, the left-hand side of Eq. (1.3.60) does not change because the concentrations of carbonate ion and bicarbonate are physical quantities that cannot depend on our choice of the *p*H scale. Thus the right-hand side of Eq. (1.3.60) does not change either. Consequently, if we switch from the total to the free *p*H scale, and the *p*H is higher on

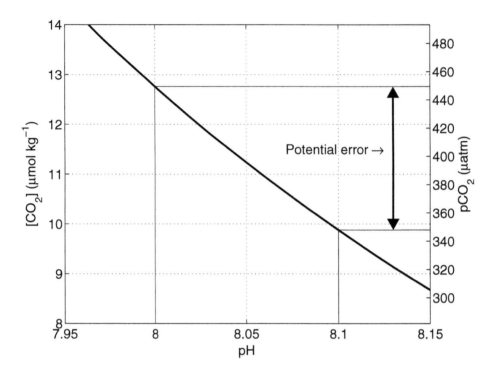

Figure 1.3.17: $[CO_2]$ and pCO_2 as a function of pH ($T_c = 25°C$, $S = 35$, DIC = 2 mmol kg^{-1}). If conversion between pH scales is ignored, the error in pH may be as large as 0.1 units which causes a difference in the calculated pCO_2 of about 100 μatm.

this scale by say 0.1 units, then pK_{2F}^* must be higher by 0.1 units as well. In order to illustrate the potential error that may occur if such differences are ignored, consider the CO_2 concentration as a function of pH at constant DIC (Figure 1.3.17). An increase of pH by 0.1 results in a decrease of $[CO_2]$ by ca. 3 μmol kg^{-1} at $pH = 8.0$, $S = 35$, $T_c = 25°C$ and DIC = 2 mmol kg^{-1}. The corresponding difference in the calculated pCO_2 would be about 100 μatm! This figure was already given in Section 1.1.6, where the choice of equilibrium constants was discussed.

In the handbook of Dickson and Goyet (DOE, 1994) it is stated that all acidity constants, with the exception of that for bisulfate ion, are expressed in terms of 'total' (Hansson) hydrogen ion concentration (DOE, 1994, Chapter 5, p. 12). This is very useful because confusion may arise when the pH scale used it is not explicitly stated.

In summary, the equilibrium constants are defined consistently with certain pH-scales. Currently, marine chemists use the scales pH_T or pH_{SWS}

which differ by about 0.01; both scales differ from the free scale (pH_F) by approximately 0.1. The NBS scale is not recommended for measurements in seawater. Following DOE (1994) and Wedborg et al. (1999) the total pH scale will be used in the current book (unless stated otherwise).

Further reading: Dickson (1984, 1993a), Millero et al. (1993a), Wedborg et al. (1999).

1.4 Partial pressure and fugacity

The four carbonate system parameters that can be determined analytically are ΣCO_2, TA, pH, and pCO_2. The knowledge of any two of them allows us to calculate the carbonate chemistry of a seawater sample. The three variables ΣCO_2, TA, and pH have been discussed in previous sections. This section deals with the partial pressure of CO_2. It is to point out that the CO_2 partial pressure assigned to a seawater sample refers to the partial pressure of CO_2 in the gas phase that is in equilibrium with that seawater. Once the pCO_2, or more precisely the fugacity fCO_2, has been determined, one may use Henry's law to calculate the concentration of dissolved CO_2 in solution and use it as a master variable (Appendix B). One may also calculate differences in pCO_2 between the ocean and atmosphere and use the difference to estimate the net air-sea gas CO_2 flux (e.g. Takahashi et al., 1997).

Marine chemists report the amount of CO_2 in the surface ocean as fugacity which is related to the more widely known partial pressure. The quantity that is measured, however, is the mole fraction of CO_2. In the following we will explain the differences between these quantities.

1.4.1 Mole fraction and partial pressure

The mole fraction, x, of a gas A is the number of moles of A divided by the total number of moles of all components in the sample: $x_A = n_A/\Sigma_i n_i$. The mole fraction is expressed in units of mol mol^{-1}; xCO_2, for example, is usually expressed in μmol mol^{-1}. For perfect gases, the mole fraction is equal to the volume mixing ratio (volume per volume), expressed in ppmv (parts per million by volume).

The partial pressure of a gas A is proportional to its mole fraction: $p_A = x_A \cdot p$, where p is the total pressure of the mixture. This serves as the definition of the partial pressure even when the gas is not behaving perfectly (Atkins, 1998). The unit of partial pressure is atmosphere (atm). For a mixture of gases we may write:

$$p_A + p_B + p_C + ... = (x_A + x_B + x_C + ...)\, p = p$$

which shows that the mole fraction of a gas is numerically only equal to its partial pressure when the total pressure is 1 atm. The various quantities used to describe CO_2 in the gas phase are summarized in Table 1.4.9.

Table 1.4.9: The different quantities used to describe CO_2 in the gas phase.

Quantity	Symbol used in literature	Unit	Value[a]	Remark
mole fraction	xCO_2	μmol mol^{-1}	360.0	in dry air
mixing ratio/ concentration[b]	xCO_2	ppmv	360.0	in dry air
partial pressure	pCO_2	μatm	349.0[c]	at 100% humidity
fugacity	fCO_2	μatm	347.9[d]	at 100% humidity

[a] Values refer to, or are calculated from $xCO_2 = 360\ \mu$mol mol^{-1} measured in dry air.
[b] Also used for mole fraction, assuming that all gases in the mixture behave perfectly.
[c] At $T = 25°C$ and $S = 35$.
[d] At $T = 25°C$.

Atmospheric CO_2

When measuring the CO_2 content of a gas sample of the atmosphere, it is usually reported as the mole fraction, xCO_2, or simply as the CO_2 concentration. If values were reported in terms of the partial pressure of CO_2, the values would rapidly decrease with height because the partial pressure depends on the total pressure which decreases as a function of height. In contrast, the mole fraction of CO_2 may be fairly constant at different altitudes. Obviously, the mole fraction and the partial pressure are quite different quantities and even at the surface, where the total pressure is approximately 1 atm, they are not the same.[18]

At a given mole fraction, the partial pressure of CO_2 at the surface depends, for example, on the local atmospheric pressure that varies from

---

[18] Note that in many publications the terms partial pressure and mole fraction of CO_2 and their units are used interchangeably which may cause confusion.

place to place. The partial pressure of CO_2 is lower in areas of low pressure and higher in areas of high pressure because the total pressure changes accordingly. In addition, if the mole fraction is determined in dry air, the calculated partial pressure of two samples with the same xCO_2 depends on the local *in situ* humidity because the partial pressure of H_2O contributes to the total pressure. In order to avoid such complications, atmospheric CO_2 is reported in terms of mole fraction. Note, however, that from a marine perspective, the thermodynamically important quantity is the partial pressure. Only in dry air and at standard pressure, the partial pressure is numerically equal to the mole fraction.

Surface seawater pCO_2

With respect to the CO_2 of the surface ocean, the quantity that matters is the partial pressure, or more precisely the fugacity, see next section. Consider, for example, the exchange of CO_2 between ocean and atmosphere. The quantity that drives this physical process is the partial pressure and not the mole fraction.

The CO_2 partial pressure of a seawater sample is usually determined by equilibrating a large volume of seawater with a small volume of gas. Then the mole fraction of CO_2 in the gas phase is determined from which the partial pressure is calculated. Most of the measurements are made at a temperature that is higher or lower than the *in situ* temperature. Because pCO_2 varies strongly with temperature, large corrections of up to 150 μatm may have to be applied in order to calculate pCO_2 at *in situ* conditions (Figure 1.4.18). This problem can be minimized if measurements are made close to the *in situ* temperature (Wanninkhof and Thoning, 1993; Körtzinger, 1999). In addition, a correction for water vapor pressure has to be applied if measurements are made in dry air (Appendix C.2).

In summary, the calculation of pCO_2 from xCO_2 involves several steps which include corrections due to (1) differences between *in situ* sea surface temperature and equilibration temperature, (2) the water vapor pressure at the equilibration temperature, and (3) the barometric pressure (e.g. Broecker and Takahashi, 1966; Copin-Montegut, 1988; DOE, 1994). The uncertainties in pCO_2 associated with the analytical procedures and the corrections can be on the order of 10 μatm for discrete systems but may be reduced to about 2 μatm for continuous systems (Wanninkhof and Thoning, 1993; Körtzinger, 1999). These uncertainties are likely to be much larger than those arising from the difference between partial pressure and fugacity

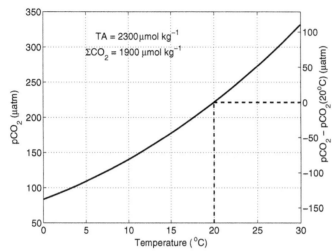

Figure 1.4.18: Seawater pCO_2 as a function of temperature (T). If measurements are made at $20°C$ (so-called discrete systems), large corrections may have to be applied to calculate pCO_2 at *in situ* T (cf. Goyet et al., 1993). The problem is less severe in so-called continuous systems where equilibration and *in situ* T are similar.

(~ 1 μatm, see next section).

When the atmospheric pCO_2 is calculated from the measured mole fraction of CO_2 in dry air, similar corrections have to be applied. The corrections involve the barometric pressure and the water vapor pressure at *in situ* temperature.

The bottom line of this section is as follows. In order to determine the amount of CO_2 in a gas sample, the mole fraction is measured. The partial pressure, which is thermodynamically important, is calculated from the mole fraction using several corrections. The mole fraction of a gas is numerically only equal to its partial pressure, when the total pressure is 1 atm.

1.4.2　Fugacity

"**What is the difference between partial pressure and fugacity?**"
(question posed at a meeting)
"**It's the same number!**"
(answer by a well-known marine chemist; this is almost true)

"**In most natural applications which do not require accuracies greater than $\sim 0.7\%$, the fugacity ... may be taken as equal to the partial pressure.**" (Weiss, 1974).

Partial pressure is a concept appropriate for ideal gases. According to Dalton's law, the total pressure of an ideal gas mixture is given by the sum of the partial pressures of the gases, where the partial pressure of a perfect gas is the pressure it would exert if it occupied the container alone. The chemical potential of gas species i, μ_i, reads:

$$\mu_i = \mu_i^0 + RT \ln p_i \quad \text{(ideal gas)} \tag{1.4.61}$$

where μ_i^0 is the standard potential, R is the gas constant, T is the absolute temperature, and p_i is the partial pressure. For real gases Dalton's law is an approximation and the chemical potential is strictly given by:

$$\mu_i = \mu_i^0 + RT \ln f_i \quad \text{(real gas)} \tag{1.4.62}$$

where f_i is the fugacity of gas species i. Eqs. (1.4.61) and (1.4.62) are of the same form where fugacity has taken the role of partial pressure. Fugacity approaches the partial pressure in the limit of infinitely dilute mixtures:

$$\frac{f_i}{p_i} \to 1 \quad \text{as} \quad p \to 0. \tag{1.4.63}$$

The relationship between fugacity and partial pressure is analogous to the relationship between activity and concentration of ions in aqueous solutions (see Section 1.1.2). The activity, a, approaches the concentration, $[c]$, in the limit of infinitely dilute solutions:

$$\frac{a}{[c]} \to 1 \quad \text{as} \quad [c] \to 0. \tag{1.4.64}$$

For very accurate calculations, the fugacity of CO_2, fCO_2, may be used instead of the partial pressure. The fugacity can be calculated from its partial pressure (e.g. Körtzinger, 1999):

$$fCO_2 = pCO_2 \cdot \exp\left(p \frac{B + 2\delta}{RT}\right) \tag{1.4.65}$$

where fCO_2 and pCO_2 are in μatm, the total pressure, p, is in Pa (1 atm = 101325 Pa), the first virial coefficient of CO_2, B, and parameter δ are in $m^3\,mol^{-1}$, $R = 8.314$ J K^{-1} mol^{-1} is the gas constant and the absolute temperature, T, is in K. B has been determined by Weiss (1974):

$$B(m^3\,mol^{-1}) = \left(-1636.75 + 12.0408\,T - 3.27957 \cdot 10^{-2}\,T^2\right.$$
$$\left. +3.16528 \cdot 10^{-5}\,T^3\right) 10^{-6}. \tag{1.4.66}$$

The parameter δ is the cross virial coefficient:

$$\delta(m^3\,mol^{-1}) = (57.7 - 0.118\,T)\,10^{-6} \tag{1.4.67}$$

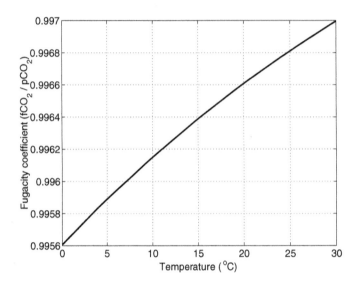

Figure 1.4.19: The fugacity coefficient, fCO_2 / pCO_2, as a function of temperature. The fugacity is about 3 to 4‰ smaller than the partial pressure. At pCO_2 = 360 μatm, the difference is about 1 μatm.

The fugacity coefficient, i.e. the ratio of fugacity and partial pressure of CO_2, varies between ~ 0.996 and ~ 0.997 over the range $0° \leq T_c \leq 30°C$ at 1 atm (Figure 1.4.19). At $pCO_2 = 360\ \mu$atm, the difference is therefore about 1 μatm.

The fugacity of a pure gas can be calculated from its equation of state (Appendix C.3). The calculation of fugacity for gas mixtures is beyond the scope of this book (compare, for example, Guggenheim (1967) or Weiss (1974) and references therein). The relationship between chemical potential, fugacity, and partial pressure at the air-sea interface is discussed in Appendix C.4.

In summary, CO_2 in the gas phase can be characterized as mole fraction, xCO_2 (μmol mol^{-1}), which is also often denoted as concentration or mixing ratio (ppmv). The mole fraction or the partial pressure, pCO_2 (μatm), may be used depending on the application. The mole fraction of CO_2 in the gas phase is determined analytically and pCO_2 is calculated from it using several corrections. The values of the mole fraction in dry air (in μmol mol^{-1}) and the partial pressure (in μatm) are the same numbers, provided that the partial pressure also refers to dry air at a pressure of 1 atm. The fugacity, fCO_2 (μatm), and the partial pressure, pCO_2 (μatm), are almost the same numbers (a few per mil difference).

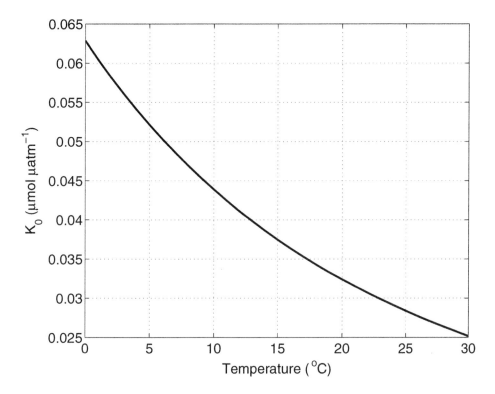

Figure 1.5.20: The solubility of CO_2, K_0, as a function of temperature at $S = 35$.

1.5 The Revelle factor

In order to demonstrate some properties of the carbonate system, we will answer the following question: *How does DIC change in response to increasing atmospheric CO_2 concentrations?*

The CO_2 concentration in surface waters, $[CO_2]$, currently increases due to invasion of CO_2 from the atmosphere. The increase of atmospheric CO_2 is a result of anthropogenic CO_2 emissions as, for instance, burning of fossil fuel releases CO_2. It is very important to note in this context that invasion of CO_2 does not change the total alkalinity in seawater (cf. Section 1.2). In equilibrium, the net exchange of CO_2 between air and sea is zero. This is the case when the partial pressure of CO_2 in the atmosphere, pCO_2, equals the partial pressure of CO_2 in the surface ocean, PCO_2, which is related to the concentration of CO_2 by Henry's law:

$$[CO_2] = K_0(T, S) \cdot PCO_2 \tag{1.5.68}$$

where $K_0(T, S)$ is the mainly temperature-dependent solubility (Figure 1.5.20). Note that for very accurate studies the fugacity may be used (see Section 1.4 and Appendix A.3). The CO_2 concentration in seawater increases proportionally to the increase of CO_2 in the atmosphere. But what happens to the other components of the carbonate system? Before addressing this question we will remind the reader of pH-buffering.

1.5.1   Titration of a weak acid by a strong base: pH-buffering

When a strong base is added to pure water ($pH = 7$) the pH of the solution will strongly increase (dashed line in Figure 1.5.21). However, when the initial solution contains a weak acid, the pH-response to the addition of a strong base is quite different: the pH increase is much less. Addition of OH^- leads to a more complete dissociation of the weak acid. The resulting H^+ ions combine with OH^- ions and form water. The system is buffered. This stabilization of a dynamic equilibrium against an outside perturbation can be regarded as another example of Le Chatelier's principle (Atkins, 1990): A system at equilibrium, when subjected to a disturbance, responds in a way that tends to minimize the effect of the disturbance.

Consider a volume V_A of a weak acid ($HA \rightleftharpoons H^+ + A^-$) with concentration $[A_0]$ that will be titrated with a strong base ($BOH \rightleftharpoons B^+ + OH^-$) of concentration $[B_0]$. We will calculate the pH after adding the volume V_B of the strong base. Conservation of electric charge leads to

$$[B^+] + [H^+] = [A^-] + [OH^-] \tag{1.5.69}$$

The total number of A groups (in HA and A^-) is constant. The concentration decreases, however, due to the addition of the strong base (V_B):

$$[HA] + [A^-] = \frac{[A_0]V_A}{V_A + V_B}. \tag{1.5.70}$$

The concentration of the base group B^+ in the titration volume reads (for a strong base: $[BOH] \approx 0$):

$$[B^+] = \frac{[B_0]V_B}{V_A + V_B}. \tag{1.5.71}$$

In addition we may exploit the equilibrium relations

$$K_A^* = \frac{[H^+][A^-]}{[HA]} \tag{1.5.72}$$

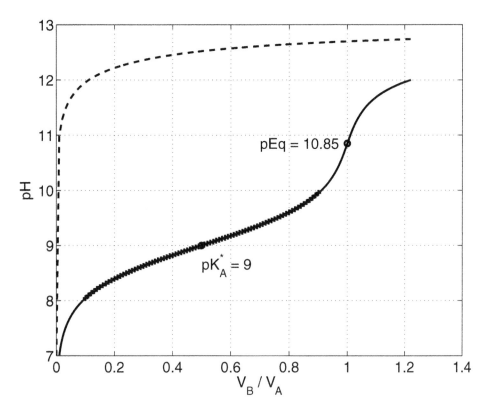

Figure 1.5.21: Titration of a weak acid ($pK_A^* = 9$, $[A_0] = 0.1$ mol kg^{-1}) with a strong base ($[B_0] = 0.1$ mol kg^{-1}). The exact relation between pH and V_B/V_A, Eq. (1.5.76), can be approximated (in the range $0.1 < V_B/V_A < 0.9$; thick line) by the Henderson-Hasselbalch equation (1.5.77) which gives an explicit expression for pH as a function of V_B/V_A. The dashed line shows the titration curve for an unbuffered solution (pure water). pEq is the equivalence point defined as the pH at which $V_B = V_A$.

and

$$K_w^* = [H^+][OH^-]. \tag{1.5.73}$$

Eliminating [HA] between Eq. (1.5.70) and Eq. (1.5.72) and solving for [A$^-$] yields

$$[A^-] = \frac{[A_0]V_A}{V_A + V_B} \frac{K_A^*}{K_A^* + [H^+]}. \tag{1.5.74}$$

Replacing the various terms in Eq. (1.5.69) by the expressions given in Eqs. (1.5.71), (1.5.73), and (1.5.74) leads to an equation for [H$^+$]:

$$\frac{[B_0]V_B}{V_A + V_B} + [H^+] = \frac{[A_0]V_A}{V_A + V_B} \frac{K_A^*}{K_A^* + [H^+]} + \frac{K_w^*}{[H^+]}. \tag{1.5.75}$$

Unfortunately, Eq. (1.5.75) is a cubic equation in $[H^+]$ which is very awkward to solve. However, it can be solved readily for V_B as a function of $[H^+]$:

$$\frac{V_B}{V_A} = \frac{(K_A^* + [H^+])\,(K_w^* - [H^+]^2) + K_A^*[A_0][H^+]}{(K_A^* + [H^+])\,([B_0][H^+] + [H^+]^2 - K_w^*)}. \qquad (1.5.76)$$

Thus one can find the volume of base needed to achieve any pH; this is sufficient to produce the solid line in Figure 1.5.21. The exact relation between pH and V_B/V_A, Eq. (1.5.76), can be approximated (in the range $0.1 < V_B/V_A < 0.9$) by the Henderson-Hasselbalch equation

$$pH = pK_A^* - \log \frac{[A_0]V_A - [B_0]V_B}{[B_0]V_B}. \qquad (1.5.77)$$

Exercise 1.11 (*)
Derive the equation for the unbuffered titration curve (Figure 1.5.21).

Exercise 1.12 (**)
Which approximations have to be made in order to obtain the Henderson-Hasselbalch equation (1.5.77)? Is the Henderson-Hasselbalch equation applicable at $V_B/V_A = 1$? Hint: start with Eq. (1.5.72).

1.5.2 CO_2-buffering

When CO_2 dissolves in seawater, the CO_2 concentration in solution changes only slightly because the system is buffered by CO_3^{2-} ions. In case of pH-buffering discussed in the previous section, the OH^- ions are neutralized by the H^+ ions provided by the weak acid. In case of CO_2-buffering, the CO_2 is scavenged by the CO_3^{2-} ions according to the reaction:

$$CO_2 + CO_3^{2-} + H_2O \rightarrow 2HCO_3^- . \qquad (1.5.78)$$

However, a small part of the resulting HCO_3^- will dissociate into CO_3^{2-} and H^+ and therefore lower the pH (CO_2 is a weak acid!). In the following, we will derive a quantitative expression for this CO_2-buffering, called the Revelle factor.

To analyze the problem we first reduce the system of equations (1.1.7), (1.1.9) − (1.1.11), and (1.2.31) in that we express DIC and TA as functions of

$h := [\mathrm{H}^+]$ and $s := [\mathrm{CO_2}]$ (note that the abbreviations are used here only to simplify the expressions):

$$\mathrm{DIC} = s \left(1 + \frac{K_1^*}{h} + \frac{K_1^* K_2^*}{h^2} \right) \tag{1.5.79}$$

$$\mathrm{TA} = s \left(\frac{K_1^*}{h} + 2\frac{K_1^* K_2^*}{h^2} \right) + \frac{B_T K_B^*}{K_B^* + h} + \frac{K_w^*}{h} - h. \tag{1.5.80}$$

The change of DIC and TA is given by the total differentials:

$$d\mathrm{DIC} = D_s ds + D_h dh \tag{1.5.81}$$

$$d\mathrm{TA} = A_s ds + A_h dh \tag{1.5.82}$$

where[19]

$$D_s := \left(\frac{\partial \mathrm{DIC}}{\partial s} \right)_h = 1 + \frac{K_1^*}{h} + \frac{K_1^* K_2^*}{h^2} \tag{1.5.83}$$

$$D_h := \left(\frac{\partial \mathrm{DIC}}{\partial h} \right)_s = -s \left(\frac{K_1^*}{h^2} + 2\frac{K_1^* K_2^*}{h^3} \right) \tag{1.5.84}$$

$$A_s := \left(\frac{\partial \mathrm{TA}}{\partial s} \right)_h = \frac{K_1^*}{h} + 2\frac{K_1^* K_2^*}{h^2} \tag{1.5.85}$$

$$A_h := \left(\frac{\partial \mathrm{TA}}{\partial h} \right)_s = -s \left(\frac{K_1^*}{h^2} + 4\frac{K_1^* K_2^*}{h^3} \right) - \frac{K_B^* B_T}{(K_B^* + h)^2} - \frac{K_w^*}{h^2} - 1. \tag{1.5.86}$$

We know the change of $\mathrm{CO_2}$ from Henry's law. When $\mathrm{CO_2}$ invades the surface ocean, the *pH* decreases. To calculate the change in hydrogen ion concentration we make use of the differential of alkalinity. The uptake of $\mathrm{CO_2}$ from the atmosphere does not change alkalinity ($d\mathrm{TA} = 0$) and therefore

$$\begin{aligned} \frac{dh}{ds} &= -A_s A_h^{-1} \\ &= \frac{\left(\dfrac{K_1^*}{h} + 2\dfrac{K_1^* K_2^*}{h^2} \right)}{1 + \dfrac{K_w^*}{h^2} + \dfrac{B_T K_B^*}{(K_B^* + h)^2} + s \left(\dfrac{K_1^*}{h^2} + 4\dfrac{K_1^* K_2^*}{h^3} \right)} \end{aligned} \tag{1.5.87}$$

---

[19] D_s is the partial derivative of DIC with respect to the carbon dioxide concentration, s, while the H^+ concentration, h, is kept constant (indicated by the index h at the foot of the right bracket).

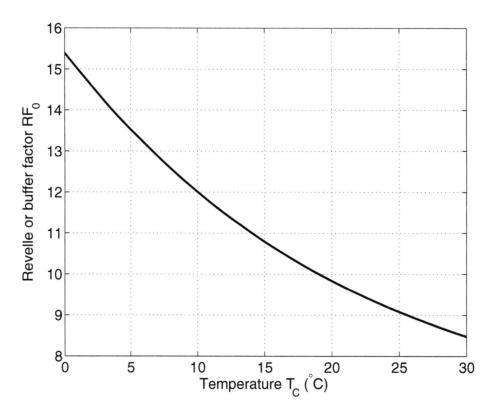

Figure 1.5.22: The Revelle factor RF_0 as a function of temperature ($S = 35$, $pCO_2 = 360$ μatm, TA = 2300 μmol kg^{-1}).

and

$$\frac{d\mathrm{DIC}}{ds} = \left[D_s - D_h A_s A_h^{-1}\right]$$

$$= \frac{4\dfrac{K_1^* K_2^*}{h^2} + \dfrac{K_1^*}{h} + \dfrac{(K_1^*)^2 K_2^*}{h^3} + \dfrac{h}{s}}{4\dfrac{K_1^* K_2^*}{h^2} + \dfrac{K_1^*}{h} + \dfrac{h}{s}\left[1 + \dfrac{K_w^*}{h^2} + \dfrac{B_T K_B^*}{(K_B^* + h)^2}\right]}$$

$$+ \frac{\left(1 + \dfrac{K_1^* K_2^*}{h^2} + \dfrac{K_1^*}{h}\right)\left(1 + \dfrac{K_w^*}{h^2} + \dfrac{B_T K_B^*}{(K_B^* + h)^2}\right)}{4\dfrac{K_1^* K_2^*}{h^2} + \dfrac{K_1^*}{h} + \dfrac{h}{s}\left[1 + \dfrac{K_w^*}{h^2} + \dfrac{B_T K_B^*}{(K_B^* + h)^2}\right]} \ .$$

$$(1.5.88)$$

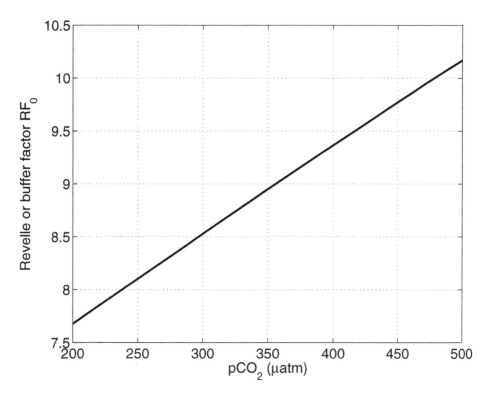

Figure 1.5.23: The Revelle factor RF_0 as a function of pCO_2 ($T_c = 25°C$, $S = 35$, TA $= 2300$ μmol kg^{-1}). With increasing partial pressure of CO_2 the Revelle factor increases and thus the buffering capacity of the ocean decreases.

The right-hand side of Eq. (1.5.87) is always positive, i.e. with increasing CO_2 concentration the hydrogen ion concentration increases and the pH decreases: the water becomes more acidic. As a consequence the concentration ratios of CO_2, HCO_3^-, and CO_3^{2-} change: while the portions of CO_2 and HCO_3^- increase, the portion of CO_3^{2-} decreases (compare Figure 1.1.2). The DIC concentration increases with increasing CO_2 concentration (all terms of the right-hand side of Eq. (1.5.88) are positive). However, the increment is not proportional to the increment of $[CO_2]$. This effect is quantified by the differential Revelle[20] or buffer factor RF_0:

$$RF_0 := \left(\frac{d[CO_2]}{[CO_2]} \Big/ \frac{dDIC}{DIC} \right)_{TA\ =\ \text{const.}} \tag{1.5.89}$$

[20] Roger Revelle (1909-1991)

where the index 0 indicates that the alkalinity is constant. The Revelle factor for more general conditions is discussed in Section 1.6.2.

In other words, the Revelle factor is given by the ratio of the relative change of CO_2 to the relative change of DIC. Typical values of RF_0 in the ocean are between 8 and 15 (see Figures 1.5.22 and 1.5.23), depending on the atmospheric CO_2 concentration and seawater temperature (Broecker et al., 1979). Thus, the relative change of CO_2 is larger than the relative change of DIC by about one order of magnitude. As a consequence, a doubling of atmospheric CO_2 leads to a change of DIC by only 10% ($\simeq 200\ \mu$mol kg^{-1}), provided that all other parameters including temperature are kept constant.

In summary, the carbonate system in seawater comprises only a few components (CO_2, HCO_3^-, CO_3^{2-}, H^+, OH^-) which are relevant for buffering. In addition, boron compounds ($B(OH)_3$, $B(OH)_4^-$) also serve as a pH buffer and have to be taken into account in quantitative calculations. The response of the system due to uptake of CO_2 is not easily predictable and differentials of DIC and TA have been used to derive an expression for the Revelle factor. Using the Revelle factor, one calculates that when the ocean takes up CO_2, the relative increase in DIC is approximately only one tenth of the relative increase in dissolved CO_2.

Exercise 1.13 (**)
Derive Eqs. (1.5.79) and (1.5.80).

1.6 Worked out problems

In this section, some problems are discussed which shall illustrate the properties of the carbonate system. The discussion includes important topics such as the formation of calcium carbonate, the Revelle factor as a function of biological processes, air-sea gas exchange, and changes of atmospheric CO_2 concentrations on glacial-interglacial time scales and in the future. Another worked out problem that deals with the manipulation of seawater chemistry for the purpose of culture experiments with phytoplankton can be found in Appendix C.5.

1.6.1 Formation of $CaCO_3$ leads to higher CO_2 levels

As already mentioned in the introduction, formation of $CaCO_3$ increases the concentration of CO_2. In this section three different ways of reasoning will be given to explain this counterintuitive behavior.

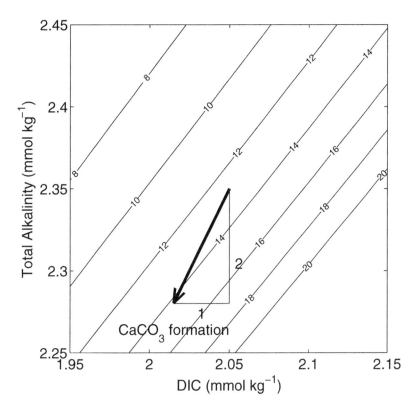

Figure 1.6.24: Isocontours of $[CO_2]$ in μmol kg^{-1} as a function of DIC and TA ($T_c = 25°C$, $S = 35$). The arrow shows the path during calcification: for the change of DIC by one unit, the alkalinity changes by two units. As a result, the concentration of CO_2 increases.

The first is not much more than a donkey-bridge. Consider the chemical reaction

$$Ca^{2+} + 2\,HCO_3^- \rightarrow CaCO_3 + CO_2 + H_2O \qquad (1.6.90)$$

in which two bicarbonate ions are consumed and one CO_2 molecule is produced for each $CaCO_3$ molecule precipitated. There are several possible interpretations of this reaction equation:

- $CaCO_3$ formation results in production of CO_2. This is correct. Consider the backward reaction of (1.6.90). Carbonic acid and other acids can be neutralized by $CaCO_3$. For example, tons of lime have been added to acidified lakes in order to restore their pH by neutralizing sulfuric acid (Mackenzie and Mackenzie, 1995). This procedure is analogous to taking sodium bicarbonate tablets to treat heartburn. The inverse reaction must liberate CO_2.

- For each mole of $CaCO_3$ formed the amount of CO_2 in the water increases by one mole. This is wrong because of buffering: most of the newly formed CO_2 will be converted to bicarbonate. Note that per mole $CaCO_3$ formed, indeed 1 mole CO_2 is produced. However, as this CO_2 is mainly converted to HCO_3^-, the concentration of CO_2 in the water does not increase by 1 mole. Reaction scheme (1.6.90) is therefore incomplete.

- TA and DIC are reduced by 2 and 1 units, respectively, for each unit of $CaCO_3$ formed. This is correct.

- Organisms use bicarbonate for $CaCO_3$ formation. This might be correct; for sure carbonate alkalinity is consumed.

For the second way of reasoning consider the isocontours of $[CO_2]$ as a function of DIC and alkalinity (Figure 1.6.24); the values have been calculated with formulas derived in Appendix B. The arrow describes the shift in the carbonate system in response to calcification. For each unit of $CaCO_3$ produced, DIC is reduced by one unit and alkalinity by two units. According to this ratio the arrow crosses the isocontours from lower to higher CO_2 concentrations.

In the next subsection the change of concentration will be calculated explicitly (third way of reasoning).

1.6.2   The Revelle factor as a function of rain ratio

The Revelle or buffer factor, RF, has been introduced already in Section 1.5.2 in the context of changing atmospheric CO_2 concentrations. The buffer factor RF is defined as the ratio of the relative change in $[CO_2]$ to the relative change in DIC:

$$RF := \frac{d[CO_2]}{[CO_2]} \bigg/ \frac{d\text{DIC}}{\text{DIC}}. \qquad (1.6.91)$$

In Section 1.5.2, RF has been calculated for the case when the alkalinity is kept constant and was denoted by RF_0. Now the Revelle factor will be discussed for more general conditions including $CaCO_3$ precipitation.

Dissolved inorganic carbon is removed or exported from the upper mixed layer of the ocean either as $CaCO_3$ or as particulate organic carbon (POC).

The ratio of these two forms is called the rain rate ratio, or for short the rain ratio, r:

$$r := \frac{\text{rate of CaCO}_3 \text{ export}}{\text{rate of POC export}} \qquad (1.6.92)$$

(see, for example, Heinze et al., 1991, p. 402).[21] This ratio may vary between zero and infinity, whereas the rain ratio parameter γ, defined by

$$\gamma := \frac{\text{rate of CaCO}_3 \text{ export}}{\text{rate of carbon export (POC plus CaCO}_3)} \qquad (1.6.93)$$

is in the range between zero and one. For $\gamma = 0$, there is only POC export and no CaCO$_3$ export, for $\gamma = 1$ there is only CaCO$_3$ export and no POC export. The quantities r and γ are related by

$$r = \frac{1}{1-\gamma} \quad \text{and} \quad \gamma = \frac{r}{1+r}. \qquad (1.6.94)$$

Consider the change in DIC and TA due to the production of CaCO$_3$ and POC. Let U be the export of carbon per time interval. The change of DIC reads:

$$d\text{DIC} \;=\; D_s ds + D_h dh = -U\,dt \qquad (1.6.95)$$

where $D_s ds + D_h dh$ is the total differential of DIC; D_s and D_h are the partial derivatives of DIC with respect to $[\text{CO}_2]$ and $[\text{H}^+]$, respectively (Eqs. 1.5.83 and 1.5.84), and ds and dh are the infinitesimal changes in $[\text{CO}_2]$ and $[\text{H}^+]$, respectively. Only the part γU which is in the form of CaCO$_3$ reduces TA:

$$d\text{TA} \;=\; A_s ds + A_h dh = -2\gamma U\,dt \qquad (1.6.96)$$

The factor two stems from the fact that TA changes by two units for each unit of DIC change. The partial derivatives of DIC and TA are given by Eqs. (1.5.83) - (1.5.86).

From Eqs. (1.6.95) and (1.6.96) the Revelle factor can be calculated. Multiplying (1.6.95) by 2γ ($\gamma \neq 0$) leads to

$$2\gamma(D_s ds + D_h dh) = A_s ds + A_h dh \qquad (1.6.97)$$

which may be solved for dh

$$dh = \frac{A_s - 2\gamma D_s}{2\gamma D_h - A_h} ds. \qquad (1.6.98)$$

---

[21]Some authors use the inverse of r as the rain rate ratio.

Eliminating dh in Eq. (1.6.95) leads to

$$dDIC/ds = D_s + D_h \frac{A_s - 2\gamma D_s}{2\gamma D_h - A_h}. \tag{1.6.99}$$

Note that Eq. (1.6.99) also contains the special case where only CO_2 is taken up (set $\gamma = 0$ and compare with Eq. (1.5.88)).

The Revelle factor as a function of the rain ratio parameter γ reads:

$$
\begin{aligned}
RF \quad &:= \quad \frac{ds}{s} \bigg/ \frac{dDIC}{DIC} \\[2mm]
&= \quad \frac{DIC \, (2\gamma D_h - A_h)/s}{D_s \, (2\gamma D_h - A_h) + D_h \, (A_s - 2\gamma D_s)} \\[2mm]
&= \quad \frac{DIC}{s} \frac{2\gamma D_h - A_h}{D_h A_s - D_s A_h} \tag{1.6.100}
\end{aligned}
$$

Some remarks are in order.

- The Revelle factor is a linear function of the rain ratio parameter γ:

$$
\begin{aligned}
RF(\gamma) \quad &= \quad c_0 + c_1 \gamma \\[2mm]
c_0 \quad &= \quad -\frac{DIC}{s} \frac{A_h}{D_h A_s - D_s A_h} \\[2mm]
c_1 \quad &= \quad \frac{DIC}{s} \frac{2D_h}{D_h A_s - D_s A_h}
\end{aligned}
$$

- For $\gamma = 0$, the Revelle factor for constant alkalinity, RF_0, is recovered from Eq. (1.6.100) which has been derived in Section 1.5.

- The denominator $D_h A_s - D_s A_h$ is always positive:

$$
\begin{aligned}
D_h A_s - D_s A_h \quad = \quad &\frac{s K_1^*}{h^2} \left(1 + 4\frac{K_2^*}{h} + \frac{K_1^* K_2^*}{h^2} \right) \\[2mm]
&+ \left(1 + \frac{K_1^*}{h} + \frac{K_1^* K_2^*}{h^2} \right) \left(\frac{K_B^* B_T}{(K_B^* + h)^2} + \frac{K_W^*}{h^2} + 1 \right)
\end{aligned}
$$

- The Revelle factor is

  - positive for $\gamma < \dfrac{A_h}{2 D_h}$

  - zero for $\gamma = \gamma^{(0)} := \dfrac{A_h}{2 D_h}$

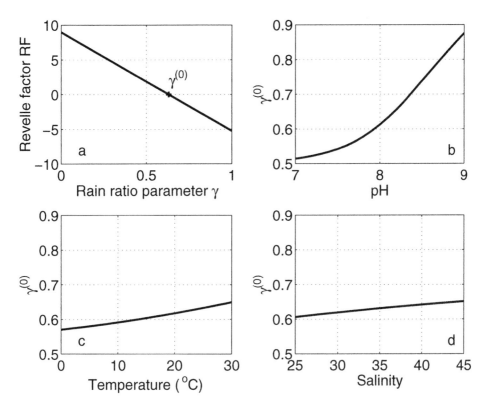

Figure 1.6.25: Revelle factor at $T_c = 25°$, $S = 35$, DIC = 2000 μmol kg^{-1}, pCO$_2$ = 360 μatm. (a) The Revelle factor, RF, as a function of the rain ratio parameter γ ($\gamma = 0$: no CaCO$_3$, only organic carbon; $\gamma = 1$: only CaCO$_3$, no organic carbon; see text). RF vanishes at the rain ratio parameter $\gamma^{(0)} \approx 0.63$. (b) $\gamma^{(0)}$ as a function of pH, (c) temperature, and (d) salinity.

$$- \text{ negative for } \gamma > \frac{A_h}{2D_h}$$

A typical value for $A_h/2D_h$ is 0.63 at $T_C = 25°$C, $S = 35$, DIC = 2000 μmol kg^{-1}, pCO$_2$ = 360 μatm (Figure 1.6.25). Export of carbon with a rain ratio parameter $\gamma = 0.63$ does not change [CO$_2$]; at lower rain rate ratios ($\gamma < 0.63$) [CO$_2$] decreases (dominance of POC production), whereas at higher ratios ($\gamma > 0.63$) [CO$_2$] increases (dominance of alkalinity decrease due to CaCO$_3$ production). The variations of $\gamma^{(0)}$ with pH, temperature, and salinity are shown in Figure 1.6.25.

Exercise 1.14 (**)

In the calculations given above the (small) change in alkalinity due to assimilation of nitrate has been neglected. How can this effect be taken into account?

1.6.3   Equilibration time for air-sea gas exchange

At the air-sea interface, gases are exchanged between atmosphere and ocean. On seasonal time scales, the troposphere is well mixed whereas only the oceanic mixed layer (and not the deep ocean) is involved in the exchange process, except for regions of strong vertical convection or upwelling. The equilibration of partial pressure of O_2 between the atmosphere is governed by the equation:

$$\frac{d[O_2]^{oce}}{dt} = \frac{k_{ge}}{d_{ML}\,\alpha_{O_2}} \left(pO_2{}^{atm} - pO_2{}^{oce}\right)$$

$$= \frac{k_{ge}}{d_{ML}} \left([O_2]^{atm} - [O_2]^{oce}\right) \tag{1.6.101}$$

where $k_{ge} = 4.2$ m d^{-1} is the gas exchange coefficient (derived from radiocarbon measurements, Siegenthaler (1986)), $d_{ML} = 50$ m is the typical depth of the mixed layer, and α_{O_2} is the solubility of O_2 in seawater. The partial pressure and concentration of O_2 are denoted by pO_2 and $[O_2]$, respectively; superscripts 'atm' and 'oce' refer to the values in the atmosphere and ocean. Note that the relations $[O_2]^{atm} = pO_2{}^{atm}/\alpha_{O_2}$ and $[O_2]^{oce} = pO_2{}^{oce}/\alpha_{O_2}$ have been used.

The characteristic time constant, or equilibration time, can be defined as:

$$\tau_{O_2} := \left([O_2]^{atm} - [O_2]^{oce}\right) \left(\frac{d[O_2]^{oce}}{dt}\right)^{-1} = \frac{d_{ML}}{k_{ge}} \approx 12 \text{ d.}$$

After the time span τ_{O_2} the perturbation $([O_2]^{atm} - [O_2]^{oce})_{t=0}$ has decreased to 37% ($\simeq 1/e$) of its initial value.

The equilibration time for CO_2 is much longer because CO_2 is a small part (roughly 0.5%) of the dissolved inorganic carbon (DIC) which buffers changes in CO_2 concentration and thereby slows down equilibration. The change in DIC is driven by partial pressure differences of CO_2 between air and water:

$$\frac{d\mathrm{DIC}}{dt} = \left(\frac{d\mathrm{DIC}}{d[CO_2]^{oce}}\right)\left(\frac{d[CO_2]^{oce}}{dt}\right)$$

$$= \frac{k_{ge}}{d_{ML}} \left([CO_2]^{atm} - [CO_2]^{oce}\right)$$

The equilibration time for CO_2 is defined analogously to that for oxygen:

$$
\begin{aligned}
\tau_{CO_2} \; &:= \; \left([CO_2]^{atm} - [CO_2]^{oce}\right) \left(\frac{d[CO_2]^{oce}}{dt}\right)^{-1} \\
&= \; \frac{d_{ML}}{k_{ge}} \left(\frac{dDIC}{d[CO_2]^{oce}}\right) \\
&= \; \underbrace{\frac{d_{ML}}{k_{ge}}}_{\approx 12\,d} \underbrace{\frac{DIC}{[CO_2]^{oce}}}_{\approx 200} \underbrace{\frac{1}{RF_0}}_{\approx 0.1} \approx 240\,d \qquad (1.6.102)
\end{aligned}
$$

where RF_0 is the Revelle factor at constant TA (Eq. (1.5.89)). The factor (d_{ML}/k_{ge}) is similar for all gases. The quotient $DIC/[CO_2]^{oce}$, which is unique for CO_2, expresses the fact that CO_2 dissociates in seawater and hence builds up a large reservoir of dissolved inorganic carbon. In summary, the extraordinary long equilibration time of CO_2 as opposed to O_2 (about 20 times longer) is due to the large concentration of DIC which can only be exchanged via a bottleneck, namely the low concentration of CO_2.

Exercise 1.15 (**))

How does the equilibration time τ_{CO_2} vary with pCO_2 while TA is kept constant?

Exercise 1.16 (**))

Find an interpretation of the factor $1/RF_0$ in Eq. (1.6.102).

1.6.4 Glacial to interglacial changes in CO_2

The atmospheric CO_2 concentration at the last glacial maximum (LGM; $\sim 18,000$ years before present) has been about 100 ppmv lower compared to the year 1800 (preindustrial value $\simeq 280$ ppmv). Can this change be explained simply by a change in the ocean surface temperature?

Based on the results of box models (Knox and McElroy, 1984; Sarmiento and Toggweiler, 1984; Siegenthaler and Wenk; 1984), Broecker and Peng (1998) argue that the atmospheric CO_2 concentration depends on the state of the surface ocean in subpolar regions where most of the deep water is formed. The sea surface temperatures in these regions were most likely about 1 K lower during the last glacial. The cooling by 1 K leads to a decrease of CO_2 by about 20 ppmv. During the LGM the mean salinity was $\sim 3\%$ higher because large amounts of fresh water were deposited in ice sheets in North America, Scandinavia and northern Russia (the sea level was more than 100 m below the current level). This increase of salinity compensates almost half of the CO_2 decrease due to cooling. If we assume

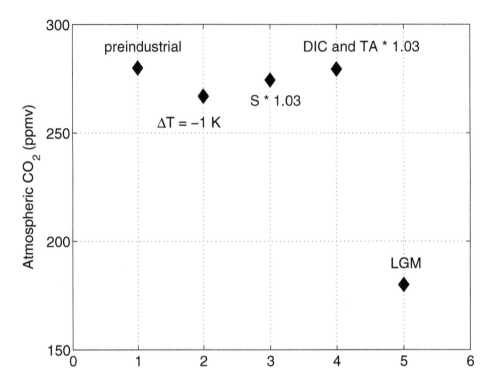

Figure 1.6.26: Changes in atmospheric CO_2 due to changes in temperature, salinity, DIC, and TA (preindustrial values: $CO_2 = 280$ ppmv, DIC $= 2000$ μmol kg^{-1} at $T_c = 10°$C and $S = 35$). The glacial-interglacial change in atmospheric CO_2 (from 180 ppmv at the LGM to 280 ppmv in preindustrial time) cannot be explained simply by changes in temperature and sea level.

that DIC and TA were also higher by 3%, which is an oversimplification of the glacial situation, the change due to cooling is almost compensated (Figure 1.6.26).

Thus the glacial-interglacial change in atmospheric CO_2 cannot be explained simply by differences in temperature and sea level. During the past two decades several scenarios have been proposed for the state of the glacial ocean (see, for instance, Broecker (1982), Berger and Keir (1984), Boyle (1988), Broecker and Peng (1989), Heinze et al. (1991), Broecker and Henderson (1998), Sigman and Boyle (2000)) but none of these hypotheses appears to be without contradictions to accepted facts. While the composition of the paleoatmosphere can be determined quite well from air contained in gas bubbles in glacial ice, there is no such tool for the ocean. Furthermore, whereas the atmosphere is fairly homogeneous due to short mixing time scales on the order of years, the ocean is rather inhomogeneous and

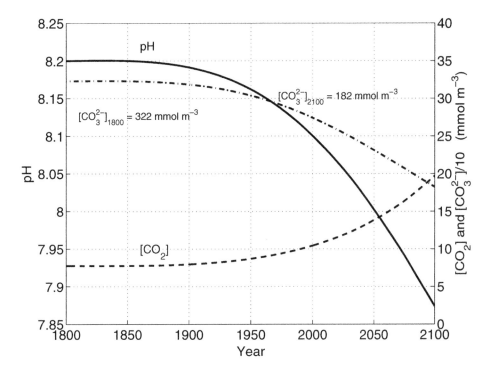

Figure 1.6.27: Changes of CO_2, CO_3^{2-}, and pH in the surface ocean calculated according to the business as usual scenario IS92a ($T_c = 25°C$, $S = 35$).

has mixing time scales of the order of thousand years. The reconstruction of the state of the paleocean is therefore much more intricate. One has to rely on so-called paleoproxies which are used to approximate the inhomogeneous distributions of temperature, salinity, nutrients, carbonate system parameters and other variables of the past ocean. Some of those proxies will be discussed in Chapter 3.

1.6.5 Future CO_2 emissions and change in pH

The atmospheric CO_2 concentration has increased from the preindustrial value of 280 ppmv to a value of 364 ppmv in 1997 (Indermühle et al., 1999). According to a business as usual scenario, the CO_2 concentration will reach 700 ppmv in the year 2100 (Scenario IS92a; IPCC, 1995). What are the consequences of this rapid increase for the ocean?

The carbonate system in the surface ocean will follow the forcing by the atmosphere with a time lag of less than one year (compare Section 1.6.3

on air-sea equilibration time). The mean surface ocean pH today is already about 0.1 units lower than the preindustrial value. In the future, the pH and the concentration of carbonate ions will further decrease due to the invasion of CO_2 from the atmosphere (Figure 1.6.27).

These changes may have important consequences for growth, calcification rates, and isotopic composition of marine plankton and corals. For instance, regarding $CaCO_3$ production in the surface ocean, it is to be expected that marine calcification in corals (e.g. Gattuso et al., 1999; Kleypas et al., 1999; Langdon et al., 2000), foraminifera (Wolf-Gladrow et al., 1999b), and coccolithophorids (Riebesell et al., 2000; Zondervan et al., 2001) will decrease in the future. As atmospheric CO_2 concentrations increase, CO_2 concentrations in the surface ocean increase as well, leading to a reduction of the $CaCO_3$ saturation state in surface seawater. Calcification rates in marine organisms have been found to be sensitive to changes of the saturation state and, as a result, showed reduced production of $CaCO_3$ at higher CO_2 concentrations. On a global scale, this is a potential negative feedback effect on atmospheric pCO_2 which has been estimated to lead to an additional storage of ~ 6 to ~ 32 Gt C in the surface ocean until the year 2100 (Riebesell et al., 2000; Zondervan et al., 2001).

Chapter 2

Kinetics

In Chapter 1, the properties of the carbonate system in thermodynamic equilibrium were studied. It was demonstrated how those properties can be used to understand changes of the seawater chemistry resulting from processes such as the invasion of anthropogenic CO_2 into the ocean or from the formation of $CaCO_3$. Equilibrium properties of the carbonate system are applicable to the description of those processes because the characteristic time and length scales involved are on the order of months to years and meters to kilometers. On small length and time scales, however, disequilibrium of the carbonate system has to be taken into account. It will be shown in Section 2.4 that the time required to establish chemical equilibrium is on the order of a minute. Consequently, processes faster than this cannot be adequately described by equilibrium concepts. One example is the transport or the supply of chemical substances on length scales where chemical conversion and diffusion are the dominant mechanisms ($\lesssim 10^{-3}$ m). This is the case, e.g. within the diffusive boundary layer at the ocean-atmosphere interface (air-sea gas-exchange) or within the microenvironment of larger marine plankton.

In this chapter some basic features of chemical kinetics and their mathematical description are introduced (Section 2.1). Then the values and the temperature dependence of the rate constants of the carbonate system are summarized (Section 2.3). The kinetics of the carbonate system are described in detail (Section 2.4), including an analysis of the time scales involved in the relaxation of the system towards equilibrium. In this context, the stable carbon isotopes ^{12}C, ^{13}C, and ^{14}C are also considered (Section 2.5). Finally, diffusion-reaction equations in plane and in spherical geometry are discussed (Section 2.6).

It is not possible to discuss all aspects of the kinetics of the carbonate chemistry and their application in this book. For further reading on subjects such as CO_2 exchange between ocean and atmosphere, the kinetics of $CaCO_3$ precipitation/dissolution, and biochemical and physiological aspects of the CO_2 kinetics in living organisms, see for example, Emerson (1995), Wanninkhof (1996), Mucci et al. (1989), Wollast (1990), Morse and Mackenzie (1990), Forster et al. (1969).

2.1   Basic concepts of kinetics

Chemical kinetics deal with the change of chemical properties through time. Unlike the chemical equilibrium in which, by definition, the concentrations of the reactants do not change through time, chemical kinetics describe the process of changing concentrations in the course of the reaction. Chemical kinetics also examine the details of the transitions from one chemical species into the other on a molecular level. It is interesting to note that even at chemical equilibrium, reactions take place all the time - the unique property being that the rate of the forward reaction and the backward reaction are equal. In this sense, thermodynamic equilibrium may be considered as a special case of chemical kinetics in which the system has approached the steady-state. However, it is important here that concentration is the appropriate quantity in chemical kinetics, whereas activity is the fundamental quantity in thermodynamics (cf. Chapter 1). Concentrations are used in kinetics because the reaction rate depends on the number of colliding molecules per volume and thus on concentrations (Lasaga, 1981).

In a chemical reaction, a chemical species A might react with species B to form C

$$A + B \longrightarrow C \tag{2.1.1}$$

As is the case, for instance, for the combination of a proton and a hydroxyl ion to form water:

$$H^+ + OH^- \longrightarrow H_2O$$

or for the hydroxylation of CO_2:

$$CO_2 + OH^- \longrightarrow HCO_3^- \ .$$

Central to the description of the temporal development of a reaction is the rate of reaction, which we shall denote as r. For reaction (2.1.1), the rate of consumption of A is given by the derivative of A with respect to time:

$$r = -\frac{d[A]}{dt} \ . \tag{2.1.2}$$

Note that the minus sign ensures that the rate of reaction is positive. The unit of r is concentration per time, e.g. mol kg^{-1} s^{-1}. Usually, the reaction rate is a function of the concentrations of the reactants. In case of simple or elementary reactions, the reaction rate is proportional to the product of algebraic powers of individual concentrations, e.g.:

$$r = -\frac{d[A]}{dt} \propto [A]^1 [B]^1 = k [A][B] . \tag{2.1.3}$$

The proportionality constant k is called the rate coefficient or rate constant. Equation (2.1.3) mathematically expresses the observation that the rate of consumption of A in the reaction A + B $\rightarrow$ C is proportional to the concentrations of A and B. This might be illustrated by considering a volume in which the reaction takes place. If the concentrations of A and B are high, the probability for molecules of species A and B to meet, and ultimately to react, is high, whereas the opposite is true when the concentrations are small.

The unit of the rate constant k depends on the order of the reaction (see below). In the considered example (Eq. (2.1.3)) the unit of k is kg mol^{-1} s^{-1}. This is because the unit of the right-hand side has to be equal to the unit of the left-hand side (mol kg^{-1} s^{-1}). Since the unit of $[A] \times [B]$ is mol^2 kg^{-2} - the unit of k has to be kg mol^{-1} s^{-1}, yielding an overall unit of mol kg^{-1} s^{-1}. On the other hand, the unit of k according to the following rate law

$$r = -\frac{d[C]}{dt} = k [C]$$

is s^{-1}.

In order to illustrate the rate of a chemical reaction, let us consider Eq. (2.1.3) with A = CO_2 and B = OH^-. Using typical concentrations of $[CO_2] = [OH^-] = 1 \times 10^{-5}$ mol kg^{-1} at $pH = 8.2$ and 25°C in surface seawater and a rate constant of ~ 4000 kg mol^{-1} s^{-1}, the reaction rate is 4×10^{-7} mol kg^{-1} s^{-1}. Thus, within 1 second, 4% of the CO_2 and OH^- undergoes chemical reaction - or, in other words: the turnover time of CO_2 and OH^- is about 25 s. This is a very slow reaction rate compared to e.g. the recombination of H^+ and OH^-, ($H^+ + OH^- \rightarrow H_2O$), which has a rate constant k on the order of 10^{11} kg mol^{-1} s^{-1}. At $pH = 8.2$ and 25°C the reaction rate is

$$\begin{aligned} r &= k \times [H^+] \times [OH^-] \\ &= 10^{11} \times 6.3 \times 10^{-9} \times 1 \times 10^{-5} \\ &= 6.3 \times 10^{-3} \text{ mol kg}^{-1} \text{ s}^{-1} . \end{aligned}$$

In other words, in 1 kg of seawater, 6.3×10^{-3} moles OH^- undergo chemical reaction within 1 second. This amount is equal to 630 times the total concentration of OH^- - the turnover time of OH^- for this reaction is $1/630$ s.

When reversible reactions are considered, rate constants for the forward and backward reaction are necessary to describe the kinetics:

$$A \; \underset{k_-}{\overset{k_+}{\rightleftarrows}} \; B \;.$$

If the rate law is given by:

$$\frac{d[A]}{dt} = -k_+[A] + k_-[B]$$

then the equilibrium constant K^* is given by:

$$K^* = \frac{[B]}{[A]} = \frac{k_+}{k_-} \tag{2.1.4}$$

since $d[A]/dt = 0$ at equilibrium. Equation (2.1.4) is a fundamental relationship between the equilibrium constant of a reaction (usually denoted by an uppercase K^*) and the rate constants (denoted by lowercase k's).

Reaction (2.1.1) is referred to as an overall second-order reaction between two unlike species because the reaction rate is proportional to the product of $[A]^1$ and $[B]^1$ (1+1=2). However, reaction (2.1.1) is first order in each of the reactants A and B. In general, the overall order of a reaction according to the rate law

$$r \propto [A]^m[B]^n$$

is $m + n$. For example, the reaction

$$2A + B \; \overset{k}{\longrightarrow} \; C$$

is second order in reactant A and first order in reactant B, the overall order being 3. The rate law for this reaction may be written as:

$$r = -\frac{1}{2}\frac{d[A]}{dt} = -\frac{d[B]}{dt} = \frac{d[C]}{dt} = k\,[A]^2[B] \;. \tag{2.1.5}$$

In this example it was assumed that the order of the reaction can be directly inferred from the stoichiometry of the reaction. Unfortunately, this applies only to simple or elementary reactions. No reaction, even monomolecular or bimolecular can be assumed to be *a priori* elementary. For instance, the rate expression for the reaction

$$Br_2 + H_2 \; \rightleftharpoons \; 2HBr$$

is

$$r = \frac{k\,[H_2][Br_2]^{1/2}}{1 + k'\,[HBr]/[Br_2]}$$

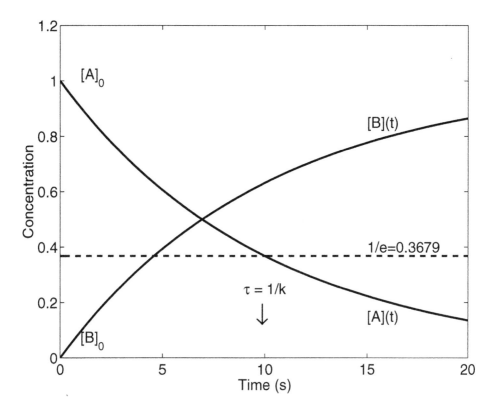

Figure 2.1.1: Change of the concentrations of A and B for the reaction $A \to B$. The inverse of the rate constant $k = 0.1 \text{ s}^{-1}$ is the characteristic time scale or decay time τ of the reaction (see text).

(Moore and Pearson, 1981) for a series of complicated steps are involved in this reaction. In general, the reaction order is an experimentally determined quantity that can also take on noninteger values. However, for a number of reactions it follows from:

$$n_1 A_1 + n_2 A_2 + ... + n_N A_N \xrightarrow{k} m_1 B_1 + m_2 B_2 + ... + m_M B_M$$

that the rate law is given by:

$$r = k\,[A_1]^{n_1}[A_2]^{n_2}...[A_N]^{n_N} \ .$$

When studying chemical kinetics - theoretically and experimentally - the change of the concentrations of the chemical species through time is of great interest. The mathematical description of those changes is based on the solution of differential equations such as Eqs. (2.1.3) and (2.1.5). As an example consider the simple reaction:

$$A \xrightarrow{k} B$$

for which the rate law reads:

$$r = -\frac{d[A]}{dt} = \frac{d[B]}{dt} = k\,[A]\,.$$

The differential equation to be solved for A is:

$$\frac{d[A]}{dt} = -k\,[A] \qquad\qquad\qquad (2.1.6)$$

which is a well known mathematical equation, occurring in many problems such as radioactive decay, light absorption and more. It says that the change of a quantity is proportional to the value of the quantity itself. Specifying initial conditions for the concentrations of A and B, say $[A]_0$ and $[B]_0$, the solution is found to be (see Exercise):

$$[A](t) = [A]_0\,\exp(-kt)$$
$$[B](t) = [B]_0 + [A]_0\,\{1 - \exp(-kt)\}\,.$$

The temporal development of [A] and [B] for initial conditions $[A]_0 = 1$ and $[B]_0 = 0$ (arbitrary units) is shown graphically in Figure 2.1.1; the rate constant is $k = 0.1$ s^{-1}. Whereas the concentration of A is exponentially decreasing, [B] is increasing and approaches the asymptotic value of 1.0 as $t \to \infty$. The concentration of A reaches about 37% ($\simeq 1/e$) of its initial concentration after 10 s, corresponding to the inverse of the rate constant ($\tau = 1/k$) which has the unit of time.[1] This so-called decay time is a very useful quantity, characterizing the relaxation time of a system approaching equilibrium.

We will return to the solution of differential equations in reaction kinetics in Section 2.4 where the solution of the dynamical system describing the kinetics of the carbonate chemistry in seawater will be discussed. Various examples of other reactional systems and recipes for their mathematical solution can be found in textbooks such as Steinfeld et al. (1999) and Atkins (1998).

Exercise 2.1 (*)
Explain the factor 1/2 occurring in Equation (2.1.5).

Exercise 2.2 (**)
The solution to Eq. (2.1.6) can be found by separating the variables:

$$\frac{d[A]}{[A]} = -k\,dt\,.$$

---

[1]Provided that the decrease is exponential one can also calculate the time to achieve 99% of the equilibrium value, which is about 4.6 ($\approx -\ln(0.01)$) times larger than the relaxation time.

Integration from $t' = 0$ to $t' = t$ yields:

$$\int_{[A]_0}^{[A]} \frac{d[A]'}{[A]'} = -k \int_0^t dt'$$

$$\ln\left(\frac{[A]}{[A]_0}\right) = -kt$$

and finally:

$$[A](t) = [A]_0 \exp(-kt) .$$

Derive the solution for $[B](t)$.

2.2 Temperature dependence of rate constants

In many chemical reactions the reaction rate increases strongly when the temperature of the system is raised. This behavior is mainly a consequence of a higher mean energy of the molecules leading to higher numbers of reactions per time interval. A similar behavior is observed in the metabolic rates of biological systems since temperature also affects enzyme-catalyzed reactions. However, most biological systems have optimal temperatures above which their enzymes will not perform ($\gtrsim 40°C$). This is because heat inactivates enzymes at higher temperatures and eventually denatures all proteins above 120°C (e.g. Beck et al., 1991). The relation between temperature and the rate of an enzymatic pathway is expressed in terms of the so-called Q_{10} factor (see below).

In a non-enzymatic chemical reaction, the reaction rate is a function of the concentration of the reactants and of the rate constant. Since concentrations are independent of temperature, the dependence on temperature is expressed in the rate constant. The relationship between temperature and rate constant can often be fitted by an Arrhenius equation:

$$k = A \exp(-E_a/RT) \qquad (2.2.7)$$

where A is called the pre-exponential factor or Arrhenius factor, $R = 8.3145$ J mol^{-1} K^{-1} is the gas constant, T is the absolute temperature in Kelvin, and E_a is the activation energy (see below).

The increase of the rate constant with temperature is a function of E_a. This may be demonstrated by comparing values of the rate constant at two different temperatures, e.g. at 15°C and 25°C. The ratio of the rate constants at T_1 and T_2 is given by:

$$\frac{k(T_1)}{k(T_2)} = \exp\left[-\frac{E_a}{R}\left(\frac{1}{T_1} - \frac{1}{T_2}\right)\right]$$

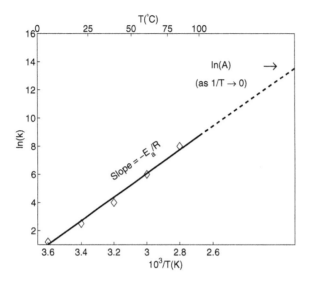

Figure 2.2.2: Schematic illustration of an Arrhenius plot, i.e. $\ln(k)$ vs. $1/T$. The Arrhenius parameters A and E_a in this representation can be found by linear interpolation (line) to the data (diamonds). $\ln(A)$ is given by the intercept with the vertical axis at $1/T \to 0$, whereas E_a is given by the slope ($= -E_a/R$).

Using a value of 70 kJ mol^{-1} for the activation energy E_a, the ratio is about 3. In other words, the rate constant increases by a factor of 3 between 15°C and 25°C. As mentioned above, the corresponding increase of the rate of an enzymatic reaction in biological systems is expressed in terms of the Q$_{10}$ factor, which expresses the rate of increase for every 10°C rise in temperature. Hence Q$_{10}$ would be approximately 3 for this example.

The logarithmic form of Eq. (2.2.7):

$$\ln(k) = \ln(A) - E_a/RT$$

suggests a linear relationship between $\ln(k)$ and $1/T$ from which the Arrhenius parameter A and the activation energy E_a can be determined (Figure 2.2.2).

The activation energy is required to initiate the reaction. In order to form a product from the reactants it is not only necessary for the reacting molecules to collide but also that the total kinetic energy be sufficient to overcome the barrier of the activation energy. Figure 2.2.3 illustrates the situation for the reaction A + B → C. As the reaction enthalpy ΔH^0 is negative the reaction is exothermic, i.e. during the overall reaction heat is released to the surroundings. Initially, however, the activation energy E_a has to be supplied in order to form the activated complex or transition state AB* (cf. Section 3.1.3, Chapter 3). The mean kinetic energy of the molecules of e.g. a gas results from the motion of each molecule with its particular velocity. At a given temperature the molecules exhibit a certain velocity distribution called Maxwell-Boltzmann distribution. There are a

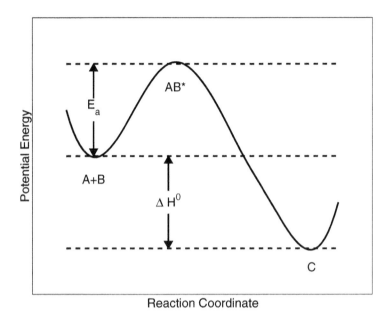

Figure 2.2.3: Illustration of energies for a reaction $A+B \rightarrow C$. As the reaction proceeds along the reaction coordinate, the activation energy E_a has to be supplied initially to form the transition state AB^*. The overall reaction is exothermic for the reaction enthalpy ΔH^0 is negative.

large number of molecules with velocities (absolute values) close to a certain mean velocity, some with higher and some with a smaller velocity than the mean. Only those molecules with sufficient energy to overcome E_a have the potential to react (Figure 2.2.4). Since the mean kinetic energy of the molecules increase with temperature it is now clear why the reaction rate increases with temperature for there are more molecules with sufficient energy to react.

Regarding enzyme-catalyzed reactions, it is interesting to note that the key function of enzymes is to lower the energy barrier for the transition state. As a consequence, more molecules are allowed to surmount the energy barrier, thus accelerating the reaction. Even when present in trace amounts, enzymes can speed up chemical reactions by up to several orders of magnitude. Without enzymes, which are active in the vast majority of metabolic pathways in living organisms, chemical reactions would take place much too slowly to support life.

Exercise 2.3 (**)

The temperature dependence of a rate constant has been measured: $k \;=$

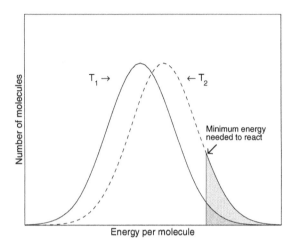

Figure 2.2.4: Schematic illustration of the energy content per molecule at different temperatures. At a higher temperature T_2, more molecules have the energy necessary to undergo reaction (light shaded area) than at a lower temperature T_1 (dark shaded area). This leads to an increase of the reaction rate.

$[5.0, 6.8, 8.6, 10.7, 12.5] \times 10^6$ kg mol^{-1} s^{-1} at $T = [4, 6, 20, 25, 31]°$C. Fit an Arrhenius equation to the data and determine the pre-exponential factor and the activation energy of the reaction.

2.3 Reactions and rate constants of the carbonate system

The kinetic rate constants of the carbonate system play a similar role in kinetics as the dissociation constants in thermodynamics. The rate constants are therefore of major importance for any quantitative description of the reaction kinetics of the carbonate system. The aim of this section is to provide a summary of the rate constants reported in the literature and their discussion. In contrast to the extensive studies of the dissociation constants of the carbonate system in seawater (see Appendix A), little work has hitherto been concerned with the kinetic rate constants in seawater. Some of the constants are controversial (see Section 2.3.2 and Appendix C.6) and the rate constants of the boric acid - borate equilibrium are widely unknown in marine chemistry and chemical oceanography. Thus, the current section includes the development of working knowledge rather than only reporting good working knowledge. The values of the rate constants are the fundamental quantities for the discussions to follow in subsequent sections of this chapter.

As discussed in Chapter 1, the following relations may be used to describe the thermodynamic equilibrium of the carbonate system:

$$CO_2 + H_2O \;\rightleftharpoons\; H^+ + HCO_3^- \qquad (2.3.8)$$
$$CO_3^{2-} + H^+ \;\rightleftharpoons\; HCO_3^- \qquad (2.3.9)$$
$$H_2O \;\rightleftharpoons\; H^+ + OH^- \qquad (2.3.10)$$
$$B(OH)_3 + H_2O \;\rightleftharpoons\; B(OH)_4^- + H^+ \; . \qquad (2.3.11)$$

Using the law of mass action, information on the concentration of each of the chemical species in equilibrium can be obtained. Unfortunately, the equilibria (2.3.8)-(2.3.11) do not provide information on the reactions or on the reaction mechanisms involved in the establishment of the equilibrium. The reaction mechanisms and the rate constants associated with the equilibria (2.3.8)-(2.3.11) will be introduced step by step in the following sections.

2.3.1   The hydration of carbon dioxide

First of all, we will discuss the hydration of carbon dioxide (for review see Kern (1960) and Edsall (1969)). The reaction scheme can be formulated as shown in Eq. (2.3.12) (cf. Eigen et al., 1961). Note that carbonic acid, H_2CO_3, always occurs in very small concentrations compared to CO_2. Thus state (II) is barely occupied.[2]

A very important feature of the CO_2 kinetics is that the hydration step (transition from state (I) to e.g. state (II)) is a 'slow' reaction. By slow

---

[2] According to Stumm and Morgan (1996) the ratio of $[CO_2]$ to $[H_2CO_3]$ at $25°C$ is in the range of 350 to 990. Thus the concentration of true H_2CO_3 is less than 0.3% of $[CO_2]$.

we mean that the reaction rate of this reaction is slow compared to a number of other reactions which are much faster. The rate constant of the dehydration of H_2CO_3 is on the order of 10^4 kg mol^{-1} s^{-1}, whereas rate constants of an e.g. diffusion-controlled reaction (see below) is on the order of 10^{10} kg mol^{-1} s^{-1}. The reaction between carbonic acid and bicarbonate (II) $\rightleftharpoons$ (III) is of that nature and is practically instantaneous. It can therefore be assumed that the equilibrium relation holds:

$$[H^+][HCO_3^-] = K^*_{H_2CO_3}[H_2CO_3] \quad . \tag{2.3.13}$$

It is impossible to decide whether the hydration reaction occurs directly, i.e., (I) $\rightarrow$ (III) or via (I) $\rightarrow$ (II) $\rightarrow$ (III) (Eigen et al., 1961). The only measurable quantity is a rate constant corresponding to an overall reaction (carbonic acid can be eliminated using Eq. (2.3.13)):

$$CO_2 + H_2O \quad \underset{k_{-1}}{\overset{k_{+1}}{\rightleftharpoons}} \quad H^+ + HCO_3^- \tag{2.3.14}$$

with

$$k_{+1} \;:=\; k^*_{+1} + k_{+2} \quad \text{(often named } k_{CO_2} \text{ in the literature)} \tag{2.3.15}$$

$$k_{-1} \;:=\; k^*_{-1} + \frac{k_{-2}}{K^*_{H_2CO_3}} \quad . \tag{2.3.16}$$

These relationships can be derived by writing down the rate laws for the reactions in the triangle in Eq. (2.3.12) (see Exercise). The concentration of the solvent, H_2O, is so large compared to the concentrations of the solutes that it can safely be assumed as constant.

The hydration of CO_2 via reaction (2.3.14) is predominant at low pH whereas at high pH the presence of an increasing number of OH$^-$ ions favors the following hydroxylation reaction:

$$CO_2 + OH^- \quad \underset{k_{-4}}{\overset{k_{+4}}{\rightleftharpoons}} \quad HCO_3^- \tag{2.3.17}$$

(compare Fig. 2.3.5). It is noted that the chemical equilibrium between CO_2 and HCO_3^- (that is the value of the first dissociation constant of carbonic acid, K^*_1) does not depend on the reaction mechanism. The thermodynamic equilibrium is given by the energy of the molecules (cf. Section 3.5) and therefore cannot depend on the chemical pathway by which the equilibrium is achieved.

It is important to note that the hydration and hydroxylation of CO_2 (Eqs. (2.3.14) and (2.3.17)) are elementary reactions. Thus, the relationship between equilibrium constants and rate constants can be written (cf.

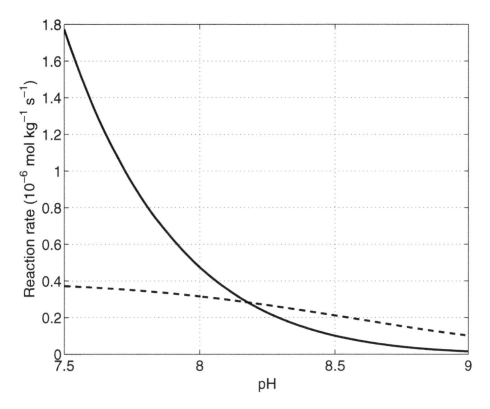

Figure 2.3.5: The forward reaction rates k_{+1} [CO$_2$] (solid line) and k_{+4} [CO$_2$][OH$^-$] (dashed line) of reactions (2.3.14) and (2.3.17) at $T_c = 25°$C, $S = 35$, and TA = 2300 μmol kg^{-1}. Reaction (2.3.14) is most important for conversion of CO$_2$ to HCO$_3^-$ at pH < 8.2. With increasing pH (decreasing [CO$_2$]) its rate decreases faster than that of reaction (2.3.17) which dominates at higher pH.

Eq. (2.1.4) in Section 2.1):

$$K_1^* = \frac{[\text{H}^+][\text{HCO}_3^-]}{[\text{CO}_2]} = \frac{k_{+1}}{k_{-1}} = \frac{k_{+4}}{k_{-4}} K_W^* \qquad (2.3.18)$$

with $K_W^* = [\text{H}^+][\text{OH}^-]$ being the ion product of water.

In summary, the set of reactions and rate constants which is considered to describe the hydration of CO$_2$ in aqueous solution reads:

$$CO_2 + H_2O \quad \overset{k_{+1}}{\underset{k_{-1}}{\rightleftharpoons}} \quad H^+ + HCO_3^- \tag{2.3.19}$$

$$CO_2 + OH^- \quad \overset{k_{+4}}{\underset{k_{-4}}{\rightleftharpoons}} \quad HCO_3^- \tag{2.3.20}$$

Interestingly, both reactions are equally important in seawater at typical surface ocean pH of 8.2, which can be demonstrated by considering the reaction rates:

$$
\begin{aligned}
r_{H_2O} &= k_{+1}[CO_2] \\
&\simeq (0.037)\,(1 \times 10^{-5}) \\
&= 3.7 \times 10^{-7} \text{ mol kg}^{-1} \text{ s}^{-1}
\end{aligned}
$$

and

$$
\begin{aligned}
r_{OH^-} &= k_{+4}[CO_2][OH^-] \\
&\simeq (4.0 \times 10^3)\,(1 \times 10^{-5})\,(1 \times 10^{-5}) \\
&= 4.0 \times 10^{-7} \text{ mol kg}^{-1} \text{ s}^{-1}
\end{aligned}
$$

where approximate values of $[CO_2]$ and $[OH^-]$ in seawater at $\Sigma CO_2 = 2$ mmol kg^{-1}, $pH = 8.2$, $S = 35$, and $T = 25°C$ were used. The values used for the reaction rates k_{+1} and k_{+4} will be discussed in detail in the following section.

2.3.2   Temperature dependence of the rate constants $k_{\pm1}$ and $k_{\pm4}$

The values of the rate constants of the 'slow' reactions ($k_{\pm1}$ and $k_{\pm4}$) and their temperature dependence as determined for seawater are displayed in Figure 2.3.6, whereas Table 2.3.1 on page 110 summarizes information on the complete set of rate constants. The values for the rate constants of the forward reactions, k_{+1} (Figure 2.3.6a, solid line) and k_{+4} (Figure 2.3.6b, dashed line), were obtained from the work of Johnson (1982) (diamonds). Johnson measured the rate constants of the hydration and hydroxylation reaction of CO_2 at 5, 15, 25, and 35°C using natural seawater which was aged for two months and then filtered. Also shown in Figure 2.3.6 are the values determined by Miller et al. (1971) in artificial seawater (squares) which are in good agreement with those of Johnson (see discussion below). The values of the rate constants of the backward reactions, k_{-1} (Figure 2.3.6b, solid

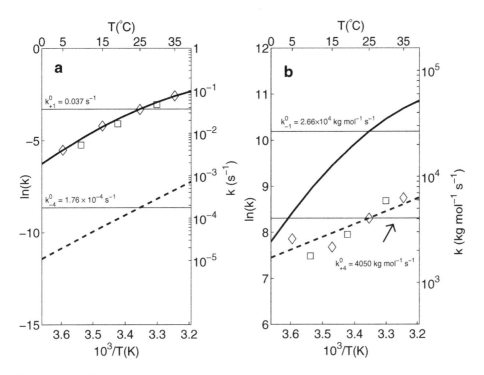

Figure 2.3.6: Rate constants of the 'slow' reactions of the carbonate system: $CO_2 + H_2O \underset{k_{-1}}{\overset{k_{+1}}{\rightleftharpoons}} H^+ + HCO_3^-$ and $CO_2 + OH^- \underset{k_{-4}}{\overset{k_{+4}}{\rightleftharpoons}} HCO_3^-$ as a function of temperature. Experimental data on (a) k_{+1} and (b) k_{+4} are from Johnson (1982) (diamonds) and Miller et al. (1971) (squares). The solid line (a) and the dashed line (b) are curves that were fitted to Johnson's data (see text). The curves for k_{-4} (a, dashed line) and k_{-1} (b, solid line) were calculated using equilibrium constants (Eq. 2.3.18); k_i^0 refers to the value of the rate constant k_i at 25°C.

line) and k_{-4} (Figure 2.3.6a, dashed line) were calculated from the forward reaction rates and the equilibrium constants using Eq. (2.3.18).

As observed in a number of reactions between solute, here CO_2, and solvent, here H_2O, the temperature dependence of the rate constant shows deviations from the simple 'Arrhenius behavior', i.e., a plot of $\ln(k)$ vs. $1/T$ shows considerable curvature (Figure 2.3.6a). It is then useful to fit the data to a curve of the form:

$$\ln k = A + B/T + C \ln T . \qquad (2.3.21)$$

The rate constant k_{+1} for the hydration of CO_2 shows this behavior in both fresh water (cf. Kern, 1960) and seawater (Johnson, 1982; Miller et al., 1971). The values for A, B, and C determined by Johnson (1982) are given in Table 2.3.1. The apparent activation energy of the reaction, E_a, is not

constant in this case but varies as a function of temperature. It is given
by the slope of $\ln(k)$ vs. $1/T$ which drops from $E_a = 90$ to 45 kJ mol^{-1}
for k_{+1} between $5°$ and $35°C$ in seawater. This apparent activation energy
is in good agreement with the values for fresh water (E_a drops from 80 to
45 kJ mol^{-1} between $0°$ and $40°C$, Kern (1960)).

Because only uncharged species (CO_2 and H_2O) are involved in the
hydration reaction, k_{+1} should not depend on the ionic strength of the
solution. Indeed, neither Johnson (1982) nor Miller et al. (1971) observed a
systematic change of k_{+1} when the salinity was varied from $3.4 - 37.1$ and
$30.7 - 39.7$, respectively.[3] The seawater value $k_{+1} = 0.037$ s^{-1} at $25°C$ is
virtually identical to that measured in fresh water, $k_{+1} = 0.040$ s^{-1} (Knoche,
1980).

The data for k_{+4} shown in Figure 2.3.6b (diamonds) was obtained from
the work of Johnson (1982), see Appendix C.6 for details. It is noted that
the data for k_{+4} show substantially more scatter than the data for k_{+1}.
The dashed line in Figure 2.3.6b is a fit to the data of Johnson (see also
Table 2.3.1). An Arrhenius equation was used because the deviations of the
temperature dependence of the measured values from the simple Arrhenius
behavior appear to be non-systematic and hence do not support a fit of
the form of Eq. (2.3.21). The value of k_{+4} at $25°C$ in seawater is about a
factor of 2 smaller than the value for fresh water (8500 kg mol^{-1} s^{-1}, Kern,
1960) which may be explained by the effect of the ionic strength on the
rate constants (see below). It is also noted that the activation energy of
~ 20 kJ mol^{-1}, given by the slope of the linear regression of $\ln(k)$ vs. $1/T$,
is about a factor of 3 smaller than in fresh water. Due to the non-systematic
scatter in the measured values of k_{+4}, however, this statement should not
be taken too rigorously. If the measured value of k_{+4} at $5°C$ was removed
from the data set, the calculated slope would increase dramatically. We
conclude that further experimental work on the rate constant k_{+4} of the
hydroxylation of CO_2 in seawater is desirable, particularly because of the
scatter of the data and because of the differences in the results of Miller et
al. (1971) and Johnson (1982). No attempt was made to include salinity
effects on k_{+4} (see Table 2.3.1).

Effect of ionic strength on reaction rates

One remark on the effect of the ionic strength and thus salinity on reaction
rates seems to be in order. Using transition-state theory it can be shown

---

[3] Miller et al. (1971) studied solutions of 17, 19, and 22‰ chlorinity (Cl) which is
defined as the mass in grams of Ag necessary to precipitate the halogens (Cl^- and Br^-)
in 328.5233 g of seawater and is related to Salinity by $S = 1.80655 \, Cl$.

that the reaction rate of ions in dilute solutions may be written as (Steinfeld et al., 1999):

$$\log k = \log k_0 + 1.02 \; z_A z_B \sqrt{I} \tag{2.3.22}$$

in water at 25°C, where k_0 is the reaction rate at infinite dilution, z_A and z_B are the charges of ion A and B, respectively; $I = 1/2 \; \Sigma \; c_i z_i^2$ is the ionic strength, where c's refer to the concentrations of all ions present in solution. It follows that the reaction rate should increase with ionic strength for similarly charged ions and decrease for oppositely charged ions, respectively. This behavior is often referred to as the 'primary salt effect' and is observed in a number of reactions (see also Moore and Pearson, 1981). The ionic atmosphere of similarly charged ions in solution tends to reduce the electrostatic repulsion of the ions which leads to enhanced collision rates between the ions. On the other hand, for oppositely charged ions the ionic atmosphere tends to reduce the electrostatic attraction of the ions therefore leading to reduced collision rates.

If one of the reactants is a neutral molecule, i.e. $z = 0$, Eq. (2.3.22) predicts no effect of the ionic strength on the reaction rate. However, a slight effect of the ionic strength on e.g. the reaction $CO_2 + OH^-$ in seawater was indeed observed (Johnson, 1982). At higher ionic strength, activity coefficients of ions show deviations from the Debye-Hückel theory on which Eq. (2.3.22) is based (see Section 1.1.2). Also, activity coefficients of neutral molecules are affected at higher ionic strength. Consequently, Eq. (2.3.22) cannot strictly be applied to solutions such as seawater because it holds exactly only for very dilute solutions.

An increase of the ionic strength may also alter the reaction mechanism as a result of ion interactions in solution and at mineral surfaces. One example is the precipitation of calcite from seawater at different ionic strengths. Zuddas and Mucci (1998) proposed that an increase in ionic strength leads to a change of the precipitation mechanism which may significantly alter processes at the surface of the growing calcite crystal.

2.3.3 Rate constants of the diffusion-controlled reactions

As mentioned earlier, the rate constants of the diffusion-controlled reactions are much larger than e.g. the rate constants of the hydration reaction of CO_2. The characteristic time scale or decay time of diffusion-controlled reactions is typically on the order of microseconds, whereas the decay time of the hydration of CO_2 is on the order of a minute. Thus, if time scales much larger than 10^{-6} s are considered, it can safely be assumed that the fast reactions are already at equilibrium. Considering time scales on the order of 10^{-6} s, however, this approximation is not valid.

The rate limiting step for diffusion-controlled reactions is not the reaction mechanism itself, i.e. the molecular rearrangement of the chemical bonds which might be associated with high activation energies. It is rather the diffusion driven encounter rate that limits the reaction rate (see box on page 103). One might say: once the molecules meet they do react. Typical activation energies of diffusion-controlled reactions are $8 - 16$ kJ mol^{-1} (Eigen and De Maeyer, 1963) which results in a small temperature dependence of the reaction rate (cf. Section 2.2).

One of the fastest reaction known in aqueous solution is the recombination of H$^+$ and OH$^-$:

$$\text{H}_2\text{O} \underset{k_{-6}}{\overset{k_{+6}}{\rightleftharpoons}} \text{H}^+ + \text{OH}^- . \qquad\qquad (2.3.23)$$

The rate constant k_{-6} measured in dilute solution is 1.4×10^{11} kg mol^{-1} s^{-1} at 25°C (Eigen, 1964). Using a fresh water value of $pK_W^* = 14.01$ one obtains a value of 1.4×10^{-3} mol kg^{-1} s^{-1} for k_{+6}. The dissociation of H$_2$O should be largely independent of the ionic strength of the medium because H$_2$O is uncharged. It is therefore probably safe to use the fresh water value of k_{+6} also for seawater, whereas k_{-6} for seawater can be determined from:

$$k_{-6} = \frac{k_{+6}}{K_W^*(T, S)}$$

which gives a value of $k_{-6} = 2.31 \times 10^{10}$ kg mol^{-1} s^{-1} at 25°C (see Table 2.3.1). As described in Section 2.1, this large rate constant leads to a reaction rate ($r = k \times [\text{H}^+] \times [\text{OH}^-]$) of about 6.3×10^{-3} mol kg^{-1} s^{-1} at typical surface seawater $pH = 8.2$ and 25°C or a turnover time of OH$^-$ for this reaction of about $1/630$ s.

Diffusion-controlled reactions

Eigen and Hammes (1963) write: "The maximum value of a bimolecular rate constant is determined by the rate at which two molecules can diffuse together. The rate constant for a diffusion-controlled association between molecules A and B can be written as (Debye, 1942)

$$k_D = \frac{4\pi N_A (D_A + D_B)}{\int_a^\infty e^{-U/kT} \, dr/r^2}. \tag{2.3.24}$$

Here N_A is Avogadro's number, D_A and D_B are the diffusion coefficients of A and B, a is their distance of closest approach, and U is the potential energy of interaction between the two molecules. The above equation was derived for spherical molecules ... If the reaction is between ions at very low concentrations, the potential energy U can be written as

$$U = \frac{z_A z_B e_0^2}{\epsilon a} \tag{2.3.25}$$

yielding

$$k_D = \frac{4\pi N_A a (D_A + D_B)(z_A z_B e_0^2/\epsilon a k T)}{\exp\left[z_A z_B e_0^2/\epsilon a k T\right] - 1}. \tag{2.3.26}$$

The z's designate the ionic valencies, e_0 is the electronic charge, and ϵ is the dielectric constant of the medium. ... the approximate range of values (for k_D) is from 10^{11} to 10^9 M^{-1} s^{-1}."

The dielectric constant ϵ is given by $\epsilon = \epsilon_r \epsilon_0$ where $\epsilon_0 = 8.8542 \times 10^{-12}$ F m^{-1} is the vacuum dielectric constant and $\epsilon_r = 80$ is the relative dielectric constant of water at 20°C; M = mol l^{-1}.

The rate constants of very fast reactions such as (2.3.23) may be measured by electric field-pulse or temperature-jump methods (for review, see Eigen and De Maeyer (1963)). These so-called relaxation methods allow measurements of reactions that are effectively complete in less than 10 μs. As already discussed, the equilibrium constant of a chemical reaction depends on the temperature and the pressure of the system. In a temperature-jump experiment, for instance, the temperature of a system which is in equilibrium is changed abruptly (temperature jumps of about 5 to 10 K can be achieved in ca. 1 μs). The system will then immediately start to adjust (or relax) towards the new equilibrium according to the new temperature, i.e., the concentrations of reactants and products will change. The velocity of the relaxation towards the new equilibrium depends on the rate constants of the reactions involved. The measurement of the concentrations of the reactants and products during the course of the relaxation of the system therefore provides a means of measuring the rate constants of very fast reactions. Manfred Eigen and his colleagues invented those powerful techniques in the 1950s and 1960s (cf. Eigen and De Maeyer (1963); Eigen (1964)).

2.3.4 Protolysis and hydrolysis

In order to complete our description of the kinetics of the carbonate system the equilibria between the acid-base pairs bicarbonate and carbonate, and boric acid and borate will be discussed:

$$CO_3^{2-} + H^+ \quad \rightleftharpoons \quad HCO_3^- \tag{2.3.27}$$
$$B(OH)_3 + H_2O \quad \rightleftharpoons \quad B(OH)_4^- + H^+ \ . \tag{2.3.28}$$

In general, acid-base equilibria in aqueous solution can be formulated according to a universal reaction scheme given by Eigen (1964). We will examine this scheme here using HCO_3^- and CO_3^{2-} as an example of an acid and its conjugate base (Eq. (2.3.29)).

$$
\text{(I)} \quad HCO_3^- + H_2O \quad
\underset{k_{-6}}{\overset{k_{+6}}{\rightleftharpoons}}
\quad H^+ + OH^- + HCO_3^- \quad \text{(II)}
$$

$$
\text{Protolysis } k_{+5}^{H^+} \ \diagdown \ k_{-5}^{H^+}
\qquad
k_{+5}^{OH^-} \diagup k_{-5}^{OH^-} \text{ Hydrolysis}
$$

$$H^+ + CO_3^{2-} + H_2O$$

$$\text{(III)} \tag{2.3.29}$$

There are two reaction paths to achieve equilibrium in this acid-base system. The first path (I) $\rightleftharpoons$ (III) is called protolysis in which the acid HCO_3^- dissociates and provides an excess hydrogen ion in the forward reaction or the base molecule CO_3^{2-} combines with a hydrogen ion in the backward reaction. The second path (II) $\rightleftharpoons$ (III) is called hydrolysis in which the acid HCO_3^- combines with OH^- in the forward reaction or the base molecule CO_3^{2-} reacts with a water molecule forming a defect proton in the backward reaction. The connection between the two reaction paths is provided by the neutralization and dissociation of water (I) $\rightleftharpoons$ (II).

The rate constant for the hydrolysis reaction

$$HCO_3^- + OH^- \quad \underset{k_{-5}^{OH^-}}{\overset{k_{+5}^{OH^-}}{\rightleftharpoons}} \quad CO_3^{2-} + H_2O \tag{2.3.30}$$

was determined by Eigen and co-workers to be $k_{+5}^{OH^-} \sim 6 \times 10^9$ kg mol^{-1} s^{-1} at ionic strength of 1.0 (see Eigen (1964) for review). Assuming that this value also applies to seawater and recognizing that $k_{-5}^{OH^-} = k_{+5}^{OH^-} \times K_W^*/K_2^*$, it follows that $k_{-5}^{OH^-} \sim 3 \times 10^5$ s^{-1} at 25°C in seawater.

We are not aware of any data on the reaction rates of the protolysis reaction

$$CO_3^{2-} + H^+ \quad \underset{k_{+5}^{H^+}}{\overset{k_{+5}^{H^+}}{\rightleftharpoons}} \quad HCO_3^- . \tag{2.3.31}$$

It is, however, probably safe to assume that the reaction rate of this reaction is similar to that measured for

$$HCO_3^- + H^+ \quad \underset{k_-^{H^+}}{\overset{k_+^{H^+}}{\rightleftharpoons}} \quad H_2CO_3 \tag{2.3.32}$$

which is $k_+^{H^+} = 4.7 \times 10^{10}$ kg mol^{-1} s^{-1} at ionic strength of 0 (cf. Eigen (1964)). Again, it is assumed that this value also applies to seawater. Using $k_{+5}^{H^+} \sim 5 \times 10^{10}$ kg mol^{-1} s^{-1}, it follows from $k_{-5}^{H^+} = k_{+5}^{H^+} \times K_2^*$ that $k_{+5}^{H^+} \sim$ 59 s^{-1} at 25°C in seawater. A summary of the reaction rates discussed in the last paragraph is given in Table 2.3.1.

It is noted that the equilibration time for the equilibrium $HCO_3^- \rightleftharpoons CO_3^{2-} + H^+$ (which is on the order of 10^{-7} s, see below) is a function of the concentrations of the reactants and therefore depends on the pH of the solution. At typical surface seawater pH of 8.2 (total scale, pH $= -\log[H^+]$), the hydrolysis reaction (2.3.30) is the dominant reaction which can be shown by comparison of the reaction rates

$$
\begin{aligned}
r_{OH^-} &= k_{+5}^{OH^-}[HCO_3^-][OH^-] \\
&= (6 \times 10^9)(1.7 \times 10^{-3})(1 \times 10^{-5}) \\
&\approx 100 \text{ mol kg}^{-1} \text{ s}^{-1}
\end{aligned}
$$

and

$$
\begin{aligned}
r_{H^+} &= k_{+5}^{H^+}[CO_3^{2-}][H^+] \\
&= (5 \times 10^{10})(3 \times 10^{-4})(6 \times 10^{-9}) \\
&\approx 0.1 \text{ mol kg}^{-1} \text{ s}^{-1}
\end{aligned}
$$

where approximate values of $[HCO_3^-]$ and $[CO_3^{2-}]$ in seawater at $\Sigma CO_2 = 2$ mmol kg^{-1}, $S = 35$, and $T = 25$°C were used.

2.3.5   Kinetics of the boric acid - borate equilibrium

In this section, we present reaction pathways and rate constants of the boric acid - borate equilibrium in seawater. The kinetics of the boric acid system were hitherto widely unknown in marine chemistry and chemical oceanography. We have obtained information on this subject from physico-chemical studies which investigated sound absorption in seawater (Mallo et al., 1984; Waton et al., 1984; see Zeebe et al. (2001) for more details).

Following Mellen et al. (1981) and Waton et al. (1984), the kinetics of the boric acid - borate equilibrium may be described by the reaction:

$$B(OH)_3 + OH^- \underset{k_{-7}}{\overset{k_{+7}}{\rightleftharpoons}} B(OH)_4^- . \tag{2.3.33}$$

The temperature dependence of k_{+7} determined by Waton et al. (1984) at ionic strength $I = 0.1$ is shown in Figure 2.3.7a. They observed only a marginal effect of the ionic strength on k_{+7} when I varied between 0.1 and 0.7. Thus, the values of k_{+7} and the temperature dependence at $I = 0.1$ reported by Waton et al. (1984) may also be applicable to seawater ($I \approx 0.7$).

A fit of an Arrhenius equation to the data yields $k_{+7} = 1.04 \times 10^7$ kg mol^{-1} s^{-1} at 25°C. It is emphasized that, in contrast to many other acid-base equilibria, reaction (2.3.33) is not diffusion-controlled. The rate constant is three to four orders of magnitude smaller than typical rate constants of diffusion-controlled reactions. This is probably due to the substantial structural change that is involved in the conversion from planar $B(OH)_3$ to tetrahedral $B(OH)_4^-$ (cf. Figure 3.4.34, page 231). The activation energy of this reaction, calculated from the linear regression of $\ln(k_{+7})$ vs. $1/T$, is $E_a = 20.8$ kJ mol^{-1}.

Introducing the equilibrium constant for reaction (2.3.33):

$$K_{B'}^* = \frac{k_{+7}}{k_{-7}} = \frac{[B(OH)_4^-]}{[B(OH)_3][OH^-]} ,$$

and using the dissociation constant of boric acid and the ion product of water:

$$K_B^* = \frac{[B(OH)_4^-][H^+]}{[B(OH)_3]} ; \qquad K_W^* = [H^+][OH^-]$$

the rate constant of the backward reaction, k_{-7}, can be determined from:

$$k_{-7} = \frac{k_{+7}}{K_{B'}^*} = k_{+7} \times \frac{K_W^*}{K_B^*} . \tag{2.3.34}$$

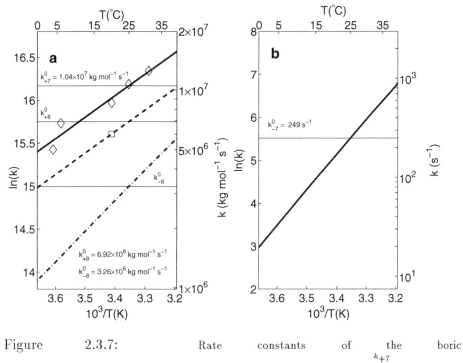

Figure   2.3.7:   Rate   constants   of   the   boric acid - borate reactions as a function of temperature: $B(OH)_3 + OH^- \underset{k_{-7}}{\overset{k_{+7}}{\rightleftharpoons}} B(OH)_4^-$ and $CO_3^{2-} + B(OH)_3 + H_2O \underset{k_{-8}}{\overset{k_{+8}}{\rightleftharpoons}} B(OH)_4^- + HCO_3^-$. (a) Experimental data on k_{+7} and k_{+8} are from Waton et al. (1984) (diamonds) and Mallo et al. (1984) (square), respectively. The solid line has been fitted to Watons's data (see text). The temperature dependence of k_{+8} (dashed line) was assumed to be equal to that of k_{+7}. The curve for k_{-8} (a, dot-dashed line) and k_{-7} (b, solid line) was calculated using equilibrium constants; k_i^0 refers to the value of the rate constant k_i at 25°C.

The temperature dependence of k_{-7} is shown in Figure 2.3.7b.

In seawater, i.e. in the presence of bicarbonate and carbonate ions, the acid-base exchange between $B(OH)_3/B(OH)_4^-$ and HCO_3^-/CO_3^{2-} equilibria complicates the situation. It has been shown that the bicarbonate concentration has a marked effect on the boron relaxation times, indicating interactions of the boron compounds with the HCO_3^- and CO_3^{2-} ions (see Mellen et al. (1981) and Mallo et al. (1984) and references therein). The overall reaction which is thought to be responsible for the coupling between boric acid - borate and bicarbonate - carbonate may be written as follows:

$$CO_3^{2-} + B(OH)_3 + H_2O \underset{k_{-8}}{\overset{k_{+8}}{\rightleftharpoons}} B(OH)_4^- + HCO_3^- . \qquad (2.3.35)$$

It is important to note that this mechanism may approximate a more complex process that involves intermediate steps and a sequence of coupled reactions (compare Mellen et al. (1983) and discussion in Zeebe et al. (2001)).

The forward rate constant of this reaction was determined to be $k_{+8} = 1.5 \times 10^7$ kg mol^{-1} s^{-1} at 20°C (Mallo et al., 1984). In order to obtain this value, Mallo et al. (1984) fitted their experimental data using fresh water equilibrium constants. Using seawater constants, one obtains $k_{+8} = 6 \times 10^6$ kg mol^{-1} s^{-1} at 20°C. The temperature dependence of k_{+8} has (to the best of our knowledge) not been studied so far. It is very likely that the activation energy of this reaction is similar to that of reaction (2.3.33) since the rate limiting step is the change of the coordination of the boron compounds in both cases. This assumption is in line with the observation that the rate constants of both reactions are similar at 20°C. It follows that the temperature dependence of k_{+8} is equal to that of k_{+7} (Figure 2.3.7, solid and dashed lines).

The rate constant of the backward reaction is determined from equilibrium relations. One sees from reaction (2.3.35) that:

$$\frac{k_{+8}}{k_{-8}} = \frac{[\text{B(OH)}_4^-][\text{HCO}_3^-]}{[\text{B(OH)}_3][\text{CO}_3^{2-}]} = \frac{K_B^*}{K_2^*}$$

from which follows:

$$k_{-8} = k_{+8} \frac{K_2^*}{K_B^*} \, .$$

The rate constant k_{-8} as a function of temperature is shown in Figure 2.3.7a (dot-dashed line).

2.3.6 Summary

The description of the reactions and the rate constants involved in the kinetics of the carbonate system can be summarized as follows. The set of reactions reads:

$$CO_2 + H_2O \; \underset{k_{-1}}{\overset{k_{+1}}{\rightleftharpoons}} \; HCO_3^- + H^+ \tag{2.3.36}$$

$$CO_2 + OH^- \; \underset{k_{-4}}{\overset{k_{+4}}{\rightleftharpoons}} \; HCO_3^- \tag{2.3.37}$$

$$CO_3^{2-} + H^+ \; \underset{k_{-5}^{H^+}}{\overset{k_{+5}^{H^+}}{\rightleftharpoons}} \; HCO_3^- \tag{2.3.38}$$

$$HCO_3^- + OH^- \; \underset{k_{-5}^{OH^-}}{\overset{k_{+5}^{OH^-}}{\rightleftharpoons}} \; CO_3^{2-} + H_2O \tag{2.3.39}$$

$$H_2O \; \underset{k_{-6}}{\overset{k_{+6}}{\rightleftharpoons}} \; H^+ + OH^- \tag{2.3.40}$$

$$B(OH)_3 + OH^- \; \underset{k_{-7}}{\overset{k_{+7}}{\rightleftharpoons}} \; B(OH)_4^- \tag{2.3.41}$$

$$CO_3^{2-} + B(OH)_3 + H_2O \; \underset{k_{-8}}{\overset{k_{+8}}{\rightleftharpoons}} \; B(OH)_4^- + HCO_3^- \tag{2.3.42}$$

for which the rate constants are given in Table 2.3.1.

Exercise 2.4 (*)

The activation energy for the reaction $B(OH)_3 + OH^- \xrightarrow{k_{+7}} B(OH)_4^-$ is about 20 kJ mol^{-1}. How does this compare to typical activation energies of diffusion-controlled reactions?

Exercise 2.5 (**)

Consider Eq. (2.3.22) which describes the primary salt effect. Plotting $\log k$ vs. $\sqrt{I}$, what slope (approximately) is to be expected for reactions between (a) similarly charged ions (b) oppositely charged ions (examine the cases $z_A = -2, -1, +1, +2$ and $z_B = -2, -1, +1, +2$)?

Exercise 2.6 (***)

The equilibrium between CO_2 and HCO_3^- is mediated by reactions (2.3.14) and (2.3.17). Calculate the pH at which the reaction rates of both reactions are equal: at $T_c = 25°C$ and $S = 30, 35,$ and 40.

Exercise 2.7 (**)

Derive Eqs. (2.3.14) to (2.3.16) (write down the rate law for CO_2 for the reactions in the triangle in Eq. (2.3.12)).

Table 2.3.1: Rate constants of the carbonate system in seawater (note that equilibrium constants used to calculate backward reaction rates refer to DOE (1994), total pH scale).

Rate constant	Value at $T_c = 25°C$, $S = 35$	Dependence on T, S	Remarks
k_{+1}	0.037 s^{-1}	$\ln k_{+1} = 1246.98 - 6.19 \times 10^4/T - 183.0\ln(T)$	Johnson (1982)
k_{-1}	2.66×10^4 kg mol^{-1} s^{-1}	$k_{-1} = k_{+1}(T)/K_1^*(T,S)$	calculated
k_{+4}	4.05×10^3 kg mol^{-1} s^{-1}	$k_{+4} = A_4 \exp(-E_4/RT)$ [a]	after Johnson (1982)
k_{-4}	1.76×10^{-4} s^{-1}	$k_{-4} = k_{+4}(T) \times K_W^*(T,S)/K_1^*(T,S)$	calculated
$k_{+5}^{H^+}$	5.0×10^{10} kg mol^{-1} s^{-1}	constant	Eigen (1964), see text
$k_{-5}^{H^+}$	59.4 s^{-1}	$k_{-5}^{H^+} = k_{+5}^{H^+} \times K_2^*(T,S)$	calculated
$k_{+5}^{OH^-}$	6.0×10^9 kg mol^{-1} s^{-1}	constant	Eigen (1964)
$k_{-5}^{OH^-}$	3.06×10^5 s^{-1}	$k_{-5}^{OH^-} = k_{+5}^{OH^-} \times K_W^*(T,S)/K_2^*(T,S)$	calculated
k_{+6}	1.40×10^{-3} mol kg^{-1} s^{-1}	constant	Eigen (1964)
k_{-6}	2.31×10^{10} kg mol^{-1} s^{-1}	$k_{-6} = k_{+6}/K_W^*(T,S)$	calculated
k_{+7}	1.04×10^7 kg mol^{-1} s^{-1}	$k_{+7} = A_7 \exp(-E_7/RT)$ [a]	after Waton et al. (1984)
k_{-7}	249 s^{-1}	$k_{-7} = k_{+7} \times K_W^*(T,S)/K_B^*(T,S)$	calculated
k_{+8}	6.92×10^6 kg mol^{-1} s^{-1}	$k_{+8} = A_8 \exp(-E_8/RT)$ [a]	after Mallo et al. (1984)
k_{-8}	3.26×10^6 kg mol^{-1} s^{-1}	$k_{-8} = k_{+8} \times K_2^*(T,S)/K_B^*(T,S)$	calculated

[a] $A_4 = 4.70 \times 10^7$ kg mol^{-1} s^{-1}; $A_7 = 4.58 \times 10^{10}$ kg mol^{-1} s^{-1}; $A_8 = 3.05 \times 10^{10}$ kg mol^{-1} s^{-1}; $E_4 = 23.2$ kJ mol^{-1}; $E_7 = E_8 = 20.8$ kJ mol^{-1}.

2.4 Approaching equilibrium: the carbonate system

The introduction to the mathematical analysis of reaction kinetics and the summary of the rate constants of the carbonate system given in the preceding sections form a good basis for the understanding of the disequilibrium properties of the carbonate system. This section provides a description of the response of the carbonate system to perturbations and the relaxation towards equilibrium, including a discussion of the carbon isotopes ^{12}C, ^{13}C, and ^{14}C; for oxygen isotopes see Section 3.3.6. It is noted that, in contrast to the equilibrium behavior of the carbonate system, which has been studied in detail (cf. Chapter 1), the kinetic properties of the carbonate system have hitherto received less attention. Lehman (1978) and Usdowski (1982) calculated the time required for the establishment of the chemical equilibrium of the carbon dioxide system. However, their results are restricted to a simplified model that assumes instantaneous equilibrium for the protonation of CO_3^{2-} and the equilibrium of H_2O. Using this approximation, time scales smaller than seconds cannot be described adequately. In addition, boron compounds were not included in both analyses and the work by Usdowski (1982) considers a fresh-water system.

In order to describe the disequilibrium properties of the carbonate system in seawater, we use the set of kinetic rate constants given in Section 2.3 (cf. also Wolf-Gladrow and Riebesell (1997) and Zeebe et al. (1999b)). This approach enables us to analyze the system on time scales ranging from microseconds to minutes. The obtained results may be utilized for, e.g., identification of inorganic carbon sources in marine diatoms (isotope disequilibrium technique, see e.g. Korb et al. (1997)) or the determination of isotope equilibration times in measurements of primary production.

It is important to note that a closed seawater system is considered in the following. In other words, the relaxation times calculated for chemical and isotopic equilibrium refer to a system in which no mass transport or gas exchange is allowed. Relaxation times of open systems may differ significantly from relaxation times of closed systems, particularly when isotopes are concerned (e.g. Broecker and Peng, 1974; Lynch-Stieglitz et al., 1995).

2.4.1 Equilibration time for CO_2

As discussed in Section 2.3.1, the slowest process of the relaxation of the carbonate system is the equilibration between CO_2 and the other chemical

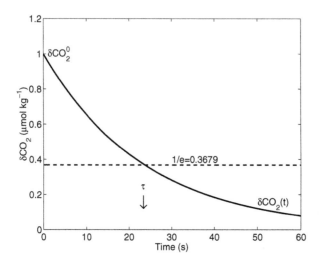

Figure 2.4.8: The relaxation of a perturbation in CO_2. After τ seconds, the perturbation has decreased to $\sim 37\%$ of its initial value $(\delta CO_2{}^0)$.

species. Because the recombination of H^+ and OH^- and the bicarbonate-carbonate equilibrium are very fast, it can be assumed that these reactions are in equilibrium on the time scale of the CO_2 equilibration. In this section, the kinetics of the carbonate system on time scales longer ~ 1 s are described, whereas all time scales (including μs) are discussed in Section 2.4.2.

In the following, the relaxation time τ as introduced in Section 2.1 is used to characterize the temporal behavior of the system. In this context, τ refers to the time after which a perturbation has reached about 37% ($\simeq 1/e$) of its initial value. For example, on time scales longer ~ 1 s, the rate law for a perturbation in CO_2 $(= \delta CO_2)$ can be written as (see Appendix C.7):

$$\frac{d(\delta CO_2)}{dt} = -\frac{1}{\tau}\, \delta CO_2 \; . \tag{2.4.43}$$

This equation is of the same form as the equation for the radioactive decay, where the decay rate dN/dt (number of decays per time) is proportional to the number of atoms (N): $dN/dt = -\lambda N$. Because same equations have the same solutions, the solution is an exponential function:

$$\delta CO_2 = \delta CO_2{}^0 \, \exp\left(-t/\tau\right) \tag{2.4.44}$$

where $\delta CO_2{}^0$ is the perturbation in CO_2 at $t = 0$. In other words, a perturbation in CO_2 is decreasing exponentially, reaching $\sim 37\%$ of its initial value after τ seconds (Figure 2.4.8).

A formula from which the CO_2 equilibration time as a function of pH can be calculated is given in Appendix C.7. The formula is only valid on time scales longer ~ 1 s on which the acid-base equilibria of H^+/OH^- and

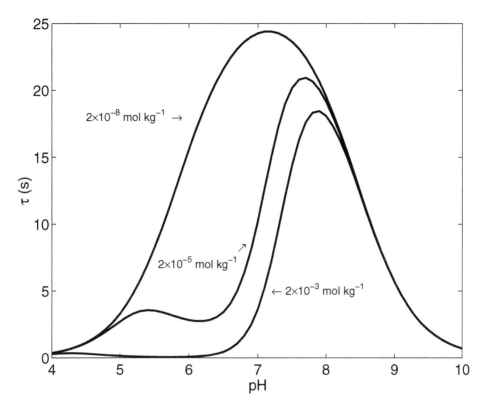

Figure 2.4.9: The relaxation time τ for the equilibration of CO_2 in a closed seawater system ($T = 25°C$, $S = 35$). Numbers indicate different values of ΣCO_2. Note that the small values of ΣCO_2 are not typical for ΣCO_2 in seawater - they are included here because these values are useful in the discussion of carbon isotopes ^{14}C and ^{13}C (see Section 2.5).

HCO_3^-/CO_3^{2-} have already been established (for the kinetics of those processes, cf. Section 2.4.2). In addition to the hydration of CO_2, the hydroxylation is also taken into account. The relaxation time for the equilibration of CO_2 according to Eq. (C.7.25) is displayed in Figure 2.4.9 as a function of pH. At high pH, the large concentration of OH^- is responsible for the small equilibration time (hydroxylation/dehydroxylation), whereas at low pH it is the large concentration of H^+ which speeds up the equilibration (hydration/dehydration). At intermediate pH, where the reactants have similar concentrations, the relaxation time is maximum. For typical surface seawater conditions, i.e. pH $= 8.2$ and $\Sigma CO_2 = 2 \times 10^{-3}$ mol kg^{-1}, a relaxation time of 13.5 s is calculated.

The influence of the concentration of the total dissolved inorganic carbon, ΣCO_2, on the relaxation time is also shown in Figure 2.4.9. Higher values of ΣCO_2 lead to slightly smaller relaxation times. For small concentrations of ΣCO_2 ($\lesssim 10^{-7}$ mol kg^{-1}), the relaxation time can be approximated by:

$$\frac{1}{\tau} = \left(k_{+1} + k_{-1} \, [\mathrm{H^+}]^* + \, k_{-4} + k_{+4} \, [\mathrm{OH^-}]^* \right) . \qquad (2.4.45)$$

For example, at $pH = 7$, $T = 25°C$, and $S = 35$ we have

$$\begin{aligned}
\tau &= \left(0.037 + (2.66 \times 10^4)(10^{-7}) + 1.76 \times 10^{-4} \right. \\
&\quad \left. + (4.05 \times 10^3)(10^{-13.22})/(10^{-7}) \right)^{-1} \\
&= 23.7 \text{ s} .
\end{aligned}$$

Exercise 2.8 (*)
Natural seawater is manipulated by the addition of 1 mg NaHCO$_3$ per kg. How long does it take (approximately) until chemical equilibrium is established ($T = 25°C$, $S = 35$)? Give a value for τ ($\sim$63% equilibration) and a value for 99% equilibration.

Exercise 2.9 (**)
Why is the relaxation time for the CO$_2$ equilibration (Eq. (C.7.25)) independent of the concentrations of the carbonate species at low ΣCO_2?

2.4.2 The complete chemical system

In this section, all time scales involved in the relaxation of the carbonate system are considered. These time scales range from μs to minutes. The complete chemical system (including boron) is given by the reactions (2.3.36)-(2.3.42) on page 109. The total dissolved carbon, the alkalinity, and the total dissolved boron for the reactions considered are defined by:

$$\begin{aligned}
\Sigma CO_2 &= [CO_2] + [HCO_3^-] + [CO_3^{2-}] \\
TA &= [HCO_3^-] + 2[CO_3^{2-}] + [B(OH)_4^-] + [OH^-] - [H^+] \\
B_T &= [B(OH)_3] + [B(OH)_4^-] .
\end{aligned}$$

Note that minor contributions of species such as PO$_4^{3-}$ and SiO(OH)$_3^-$ to total alkalinity have been neglected. The corresponding kinetic rate laws of the complete carbonate system are given in Appendix C.8.

A mathematical analysis of the system using the theory of dynamical systems (see e.g. Zeebe et al. (1999b)) yields 4 characteristic time scales of

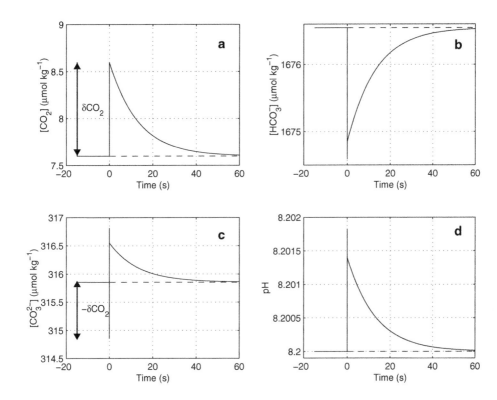

Figure 2.4.10: The relaxation of the carbonate species (linear system) for a perturbation in CO_2 (CO_3^{2-} at $t = 0$ was reduced by δCO_2 in order to keep ΣCO_2 constant, whereas OH^- was increased by $2\,\delta CO_2$ in order to keep alkalinity constant, see also Figure 2.4.11). All chemical species including boron are considered. The dashed lines indicate the equilibrium values of each species which were calculated from $\Sigma CO_2 = 2000\ \mu mol\ kg^{-1}$ and $TA = 2437\ \mu mol\ kg^{-1}$ ($T = 25°C$, $S = 35$).

the system. These time scales are associated with the following equilibria (1) $CO_2 + H_2O \rightleftharpoons HCO_3^- + H^+$ (2) $HCO_3^- \rightleftharpoons CO_3^{2-} + H^+$ (3) $H_2O \rightleftharpoons H^+ + OH^-$ and (4) $B(OH)_3 + H_2O \rightleftharpoons B(OH)_4^- + H^+$ (see Table 2.4.2). It is important to note that all the kinetic pathways, i.e. reactions (2.3.36)-(2.3.42), are involved in the relaxation towards equilibrium. The mathematical analysis - which shall not be elaborated here - yields the time development of the chemical compounds. The results are presented in Figures 2.4.10 and 2.4.11.

Starting from equilibrium at pH 8.2 and $[CO_2] = 7.6\ \mu mol\ kg^{-1}$, CO_2 is increased by 1 $\mu mol\ kg^{-1}$ at $t = 0$. In addition, CO_3^{2-} and OH^- are decreased and increased, respectively, at $t = 0$ by δCO_2 and $2\,\delta CO_2$ in order to keep ΣCO_2 and the alkalinity constant. As a consequence, HCO_3^-, CO_3^{2-}, and the pH respond to this perturbation within less than a microsecond.

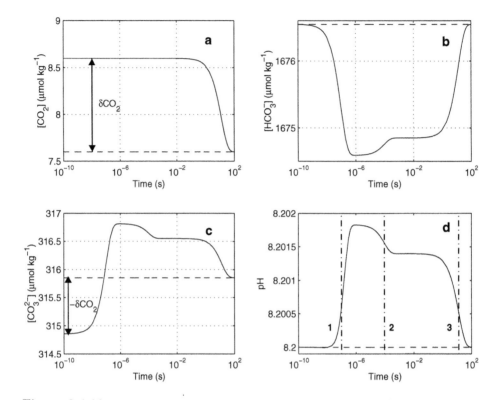

Figure 2.4.11: Same as Figure 2.4.10 with logarithmic time axes. (a) CO_2 decreases on a time scale of 10 s. (b) HCO_3^- and (c) CO_3^{2-} first decrease and increase, respectively, on a time scale of 10^{-7} s, then respond to the boric acid - borate equilibrium on a time scale of 10^{-4} s, and finally relax to the equilibrium value on a time scale of 10 s. (d) Since the recombination of OH^- and H^+ is a fast process (10^{-7} s) the pH exhibits changes on all time scales involved. The vertical dot-dashed lines in (d) indicate the different time scales (10^{-7} s, 10^{-4} s, and 10 s).

On a time scale of $\sim 10^{-4}$ s all components (except CO_2) respond to the boric acid - borate equilibrium. Subsequently, CO_2 is converted to HCO_3^-, while H^+ is released (~ 10 s). The pH decreases and eventually reaches the equilibrium value of 8.2. The time constant for the e-folding time of the calculated decrease in CO_2 is 13.1 seconds. This value is in excellent agreement with the value of 13.5 seconds calculated for the system discussed in Section 2.4.1, where it was assumed that the acid-base equilibria of H^+/OH^- and HCO_3^-/CO_3^{2-} were already established.

The same results as in Figure 2.4.10 are displayed in Figure 2.4.11 with logarithmic time axes. The temporal evolution of the carbon compounds can be followed by examining the pH (Figure 2.4.11d). After ca. 10^{-7} s (see

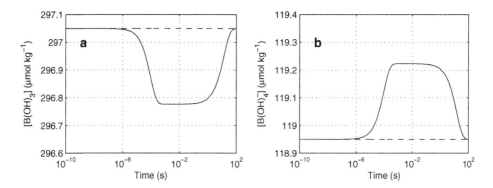

Figure 2.4.12: Relaxation of the boron compounds. The boric acid - borate equilibrium responds to a perturbation on a time scale of about 10^{-4} s.

Table 2.4.2: Equilibria and time scales of the carbonate system.

Process	Equilibrium	τ [a]
Recombination	$H_2O \rightleftharpoons H^+ + OH^-$	10^{-7} s
Protolysis/Hydrolysis	$CO_3^{2-} + H^+ \rightleftharpoons HCO_3^-$	10^{-7} s
Boron compounds	$B(OH)_3 + H_2O \rightleftharpoons B(OH)_4^- + H^+$	10^{-4} s
Hydration/Hydroxylation	$CO_2 + H_2O \rightleftharpoons H^+ + HCO_3^-$	10 s

[a] The indicated time scales are characteristic for the respective equilibria (note that these equilibria are established via several kinetic pathways, see text).

vertical dot-dashed line No 1 in Figure 2.4.11d) the pH increases due to the recombination of $OH^- + H^+ \rightarrow H_2O$ (OH^- was initially raised and thus H^+ is consumed) and the HCO_3^-/CO_3^{2-} equilibrium. This transient state (it exists only for about 10^{-3} s and is different from the final equilibrium) is in accordance with lower HCO_3^- and higher CO_3^{2-} concentrations (Figure 2.4.11b and c). The relaxation of the boron compounds leads to the small shift in HCO_3^-, CO_3^{2-}, and pH around $t = 10^{-4}$ s (vertical dot-dashed line No 2 in Figure 2.4.11d). Finally, CO_2 is converted to HCO_3^- (on a time scale of 10 s) and all components relax to their respective equilibrium values (vertical dot-dashed line No 3). The outlined analysis of the complete chemical system allows the identification of the processes involved and their characteristic time scales (see Table 2.4.2).

The time development of the concentrations of the boron compounds is

shown in Figure 2.4.12. As the boric acid - borate equilibration is slower
than e.g. the equilibration of the diffusion-controlled reactions, the concentrations of $B(OH)_3$ and $B(OH)_4^-$ do not exhibit any change until $\sim 10^{-4}$ s.
According to the transient state at a higher pH, $B(OH)_3$ decreases whereas
$B(OH)_4^-$ increases. Finally, the boron compounds relax to their ultimate
equilibrium values on the CO_2/HCO_3^- equilibration time scale.

2.5   Approaching isotopic equilibrium: ^{12}C, ^{13}C, and ^{14}C

We are now turning to the equilibration time of carbon isotopes in a closed
carbonate system (for open systems, see e.g. Broecker and Peng, 1974;
Lynch-Stieglitz et al., 1995). Although stable isotope fractionation[4] has
not been discussed in detail yet (this is subject of Chapter 3), it is very
convenient to include the calculation of carbon isotope equilibration at this
stage. This is because the analysis of the system is analogous to the analysis of the chemical equilibration of the carbonate system presented in the
previous sections. The reader who is not familiar with stable isotope fractionation may consider Chapter 3 first.

In the previous sections, a carbon dioxide system was considered that
contained only one species of carbon isotopes. Consequently, the calculated
relaxation time reflects the chemical equilibrium of the system. When different isotopes of variable concentrations are present in the solution, the
slower reaction rates of the heavier isotopes lead to fractionation effects between the carbon species. For example, the hydration step of CO_2 to HCO_3^-
(reaction (2.3.36) forward) results in isotopically 'lighter' HCO_3^- of about
13‰. On the other hand, the dehydration step (reaction (2.3.36) backward) results in 'lighter' CO_2 of about 22‰ (O'Leary et al., 1992). The
difference between the forward and backward reaction equals the equilibrium fractionation between CO_2 and HCO_3^- of about 9‰ (the equilibrium
value given by Mook (1986) is 8.97‰ at 25°C).

The rate constants for reactions involving ^{12}C, ^{13}C, and ^{14}C will be denoted by k_i^{12}, k_i^{13}, and k_i^{14}. Because all relevant fractionation effects are in
the range of several 10‰ at most, k_i^{14} and k_i^{13} differ only slightly from the
corresponding k_i^{12}. It is emphasized that the slightly different rate constants
have almost no effect on the estimates of the relaxation time constants - one

---

[4]On the time scales considered here (minutes), ^{14}C can be treated as a stable isotope
since the half life of ^{14}C is about 6,000 y.

may actually set $k_i^{14} = k_i^{13} = k_i^{12}$. When considering different isotopes, the essential new feature is the fact that the system contains carbon isotope species with very different concentrations. For instance, the natural ratio of ^{13}C to ^{12}C atoms is approximately 1:99. The coupling of the compounds containing different isotopes is brought about by the reactions of those compounds with H^+ and OH^-. This introduces additional time constants.

Isotopic compartments

Including ^{12}C, ^{13}C, and ^{14}C components, nine carbon species have to be considered which can be grouped into three compartments (cf. Section 3.2.2):

| $^{12}CO_2 \longleftrightarrow H^{12}CO_3^- \longleftrightarrow {}^{12}CO_3^{2-}$ | COMPARTMENT 1 |

| $^{13}CO_2 \longleftrightarrow H^{13}CO_3^- \longleftrightarrow {}^{13}CO_3^{2-}$ | COMPARTMENT 2 |

| $^{14}CO_2 \longleftrightarrow H^{14}CO_3^- \longleftrightarrow {}^{14}CO_3^{2-}$ | COMPARTMENT 3 |

There are no direct reactions between each of these compartments since each isotope is conserved during reaction. Thus, there are three *independent* systems with respect to carbon isotopes. The systems are, however, coupled via H^+ and OH^-. Consequently, isotopic equilibrium is achieved when chemical equilibrium is achieved in each compartment.

Before getting into the analysis of the time development of the system, some introductory remarks on ^{14}C are in order.

Radiocarbon (^{14}C)

Due to the multiple applications of radiocarbon, the radioactivity of ^{14}C that occurs in the respective investigations may range from 10^1 to 10^{13} dpm $(g\ C)^{-1}$ (disintegrations per minute per gram of carbon). This range includes the natural activity of for e.g., atmospheric CO_2 and the measurements of primary production (Steemann Nielsen, 1952). The natural activity of atmospheric CO_2, oceanic bicarbonate and living plants and animals is (Mook, 1980):

$A_0 \simeq 14$ dpm $(g\ C)^{-1}$.

In order to obtain insight into the behavior of the carbonate system under different circumstances, the relaxation time for radiocarbon components in seawater will be investigated for various concentrations and pH in this section. The calculations are analogous to the treatment of the stable isotope ^{13}C. Because of the long half-life of ^{14}C, changes of its concentration due to radioactive decay can be neglected on the time scales considered here. For the purpose of numerical calculations, the amount of radiocarbon in solution (which is often expressed in terms of radioactivity) has to be converted to a concentration. Radioactivity is usually given in units of Curie (Ci) or Becquerel (Bq). One Becquerel is the amount of radioactive material in which the average number of disintegrations is one per second.

$$1 \text{ Ci} = 3.7 \times 10^{10} \text{ Bq (disint. s}^{-1})$$

From a given activity of a solution (commonly expressed in $\mu Ci \text{ kg}^{-1}$) and the definition of Ci and Bq the concentration can be calculated. Using the law of radioactive decay the following formula is derived:

$$[c] = A \frac{3.7 \times 10^{10}}{N_A \ln 2} t_{1/2} \approx \frac{A}{62.43} \tag{2.5.46}$$

where $[c]$ is the concentration of the radiocarbon (in $\mu mol \text{ kg}^{-1}$), A is the activity ($\mu Ci \text{ kg}^{-1}$), $t_{1/2} = 5730$ years (1.807×10^{11} s) is the half-life of ^{14}C (Mook, 1980), and $N_A = 6.022136 \times 10^{23} \text{ mol}^{-1}$ is Avogadro's constant. Hence, activities typical for e.g. measurements of primary production (120-1200 μCi) correspond to total concentrations of ^{14}C ($[\Sigma^{14}CO_2]$) ranging from 2 to 20 $\mu mol \text{ kg}^{-1}$.

Relaxation times

To determine the time for the relaxation of ^{12}C, ^{13}C, and ^{14}C compounds the following reactions are considered:

$$^{\nu}CO_2 + H_2O \underset{k^{\nu}_{-1}}{\overset{k^{\nu}_{+1}}{\rightleftharpoons}} H^+ + H^{\nu}CO_3^- \tag{2.5.47}$$

$$^{\nu}CO_2 + OH^- \underset{k^{\nu}_{-4}}{\overset{k^{\nu}_{+4}}{\rightleftharpoons}} H^{\nu}CO_3^- \tag{2.5.48}$$

$$^{\nu}CO_3^{2-} + H^+ \rightleftharpoons H^{\nu}CO_3^-;$$

$$^{\nu}K_2^* = [^{\nu}CO_3^{2-}][H^+]/[H^{\nu}CO_3^-] \tag{2.5.49}$$

$$H_2O \rightleftharpoons H^+ + OH^-; \quad K_W^* = [H^+][OH^-] \tag{2.5.50}$$

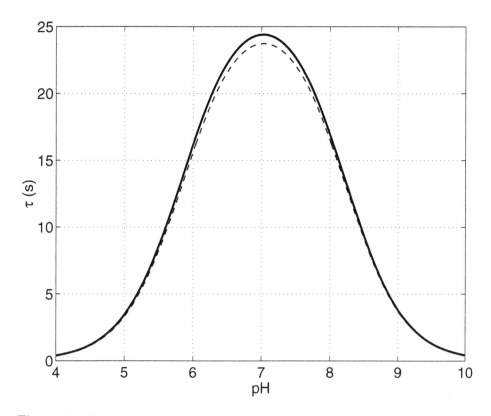

Figure 2.5.13: Relaxation time for chemical and isotopic equilibrium in a closed carbon dioxide system in seawater ($\Sigma CO_2 \sim 2000 \ \mu mol \ kg^{-1}$, $T = 25°C$, $S = 35$) with respect to the carbon isotopes ^{12}C, ^{13}C, and ^{14}C as a function of pH (solid line). The equilibrium is achieved more slowly when the concentrations of the chemical species have similar concentrations (around pH 7). Also shown is the curve according to Eq. (2.5.51) which may be used as an approximation (dashed line).

where ν is 12, 13, and 14.

Values for the ratios of the reaction rates for the hydration and hydroxylation of CO_2 with respect to ^{13}C and ^{12}C used in the calculations are given in Table 3.2.5, Section 3.2.6. The reaction rates for compounds containing ^{14}C can be calculated using Eq. (3.2.34) and the reaction rates of ^{13}C and ^{12}C. As already said, the slightly different rate constants for ^{14}C, ^{13}C, and ^{12}C have only a minor influence on the calculated time constants (less than 0.5 s in this case, $\sim 4\%$). They are mentioned here, however, in order to complete the picture.

The system can again be analyzed by the mathematical theory of dynamical systems (see Zeebe et al. (1999b)). The relaxation times of the system at $pH = 8.2$, $[\Sigma^{14}CO_2] = 2$ and 20 μmol kg^{-1} are 13.849 and 13.848 s, respectively. Thus, the relaxation time is virtually independent of the concentration of ^{14}C over this range. The relaxation time τ as a function of pH is presented in Figure 2.5.13. The maximum relaxation time was calculated at pH 7.14 to be 24.4 s. The dependence on pH results from the different pathways of the conversion between CO_2 and HCO_3^-. At low pH the concentrations of CO_2 and H^+ are high, thus equilibrium with HCO_3^- and OH^- (reaction (2.5.47)) is achieved quickly. At high pH the concentrations of HCO_3^- and OH^- are high, thus equilibrium with CO_2 and H^+ (reaction (2.5.48)) is achieved quickly. In the pH range where all chemical species have similar concentrations, equilibrium is achieved more slowly ($[H^+]^* \approx [OH^-]^*$ around pH 7).

The solid curve shown in Figure 2.5.13 was obtained by the analysis of the complete system. It can, however, be approximated by the simple formula:

$$\frac{1}{\tau} = \left(k_{+1} + k_{-1} \, [H^+]^* + k_{-4} + k_{+4} \, [OH^-]^* \right) . \qquad (2.5.51)$$

as indicated by the dashed line in Figure 2.5.13. For seawater at $T = 25°C$ and $S = 35$, τ can be written as:

$$\tau \;=\; \Big(0.037 + (2.66 \times 10^4)(10^{-pH}) + 1.76 \times 10^{-4}$$
$$+ (4.05 \times 10^3)(10^{-13.22})/(10^{-pH}) \Big)^{-1}$$

yielding $\tau = 13.2$ s at $pH = 8.2$ (for temperature dependence of the k's, see Table 2.3.1). Equation (2.5.51) was already derived in Section 2.4.1 for the equilibration time of CO_2 for small concentrations of ΣCO_2. The fact that Eq. (2.5.51) also describes the equilibration time of the system considered here is therefore not surprising because it contains ^{14}C in small concentrations.

In summary, the time required to establish chemical and isotopic equilibrium in the considered carbon dioxide system is on the order of minutes. The relaxation time is virtually independent of the amount of the radiocarbon present in the solution, when the concentrations range from 2 to 20 μmol kg^{-1} (activity ca. 120-1200 μCi).

Exercise 2.10 (**)

Derive the formula for the conversion between radioactivity and concentration (Eq. (2.5.46)). Note that the activity, A, is related to the total number of radioactive atoms, N, by $A = -dN/dt = \lambda N$.

2.6 Diffusion and Reaction

So far the kinetics of the carbonate system have been discussed in terms of a closed, homogeneous chemical system. In nature, however, we usually have to deal with heterogeneous systems. Consider, for example, the oceanic surface layer. Spatial concentration gradients are created by 'forcing' at the air-sea interface by, for instance, precipitation/evaporation or through up-take of nutrients by algae. Consequently, the system exhibits heterogeneities with respect to concentrations. As a response, chemical reactions and mass transfer by e.g. diffusion will tend to reduce those concentration gradients. A description of these processes has to take into account both reaction and diffusion which leads to so-called diffusion-reaction equations.

Here we will first discuss one-dimensional diffusion in plane and in spherical geometry. The stationary diffusion equations possess analytical solutions. One important result is that diffusion in spherical geometry provides its own spatial scale (i.e. the thickness of the diffusive boundary layer), whereas the spatial scale in plane geometry is set by other processes (such as turbulence). We will then add the reactions of the carbonate system. The coupling of reactions and diffusion yields the reacto-diffusive length scale, a_k, which has a descriptive interpretation and can be used to estimate the relative importance of reactions and diffusion. The mathematical equations derived in spherical geometry are used to discuss concentration profiles of e.g. carbon dioxide within the vicinity (microenvironment) of organisms such as microalgae and foraminifera.

The diffusion equation, which was first given by Fourier[5] (1768-1830) in 1807 and published in 1822 reads:

$$\frac{\partial c}{\partial t} = D \, \nabla^2 c \tag{2.6.52}$$

where D (m^2 s^{-1}) is the diffusion coefficient and c (mol kg^{-1}) is the concentration of the chemical compound under consideration. The symbol ∇ refers to the so-called nabla operator which represents the derivative with respect to space. In plane geometry and in one dimension, Eq. (2.6.52) reads:

$$\frac{\partial c}{\partial t} = D \, \frac{\partial^2 c}{\partial x^2}$$

---

[5]Publication of his paper (1807) *Théorie de la Propagation de la Chaleur dans les Solides* was rejected by the French Academy. Remember this case when your next paper is rejected.

which is also known as Fick's second law. Furthermore,

$$\nabla c = \frac{\partial c}{\partial x} \quad \text{and} \quad \nabla^2 c = \frac{\partial^2 c}{\partial x^2}$$

where ∇^2 is called the Laplace operator. The diffusion equation is related to the conservation of mass (of compound c), which is expressed by the continuity equation:

$$\frac{\partial c}{\partial t} = -\nabla \cdot \mathbf{F} . \tag{2.6.53}$$

which states that the mass of c (within a given volume) only changes if the net flux of c (denoted by $\mathbf{F}$) into or out of that volume changes. On the other hand, the flux obeys a relationship which is known as Fick's first law:

$$\mathbf{F} = -D\,\nabla c , \tag{2.6.54}$$

i.e. the flux is proportional to the concentration gradient ∇c. Combining Eqs. (2.6.54) and (2.6.53) yields the diffusion equation:

$$\frac{\partial c}{\partial t} = -\nabla(-D\,\nabla c) = D\,\nabla^2 c .$$

It is interesting to note that the diffusion equation, as formulated here for the diffusion of particles, is in complete analogy to the heat conduction equation:

$$\frac{\partial T}{\partial t} = \alpha\nabla^2 T \tag{2.6.55}$$

where T is temperature and α is the thermal diffusivity. For a couple of simple cases, the diffusion equation can be solved analytically. As an example, consider an isolated infinitely long rod at a constant temperature that is heated at $x = 0$ for a short moment. The solution of Eq. (2.6.55) is given by (see e.g Cussler, 1984):

$$T(x,t) = \frac{1}{2\sqrt{\pi\alpha t}} \exp\left(-\frac{x^2}{4\alpha t}\right) \tag{2.6.56}$$

which is graphically shown in Figure 2.6.14. When objects are considered which are not infinitely long, i.e. when boundaries come into play, things get more complicated. In those cases, analytical solutions can often be given (if at all possible) only in the form of infinite series (see Carslaw and Jaeger, 1959).

We will skip the time-dependent problems and discuss some simple stationary boundary value problems, i.e. systems which are in steady state ($\partial c/\partial t = 0$). Consider two well-mixed reservoirs of different concentrations c_1 and c_2 (of the same compound) which are separated by a thin film of thickness L in z-direction (Figure 2.6.15).

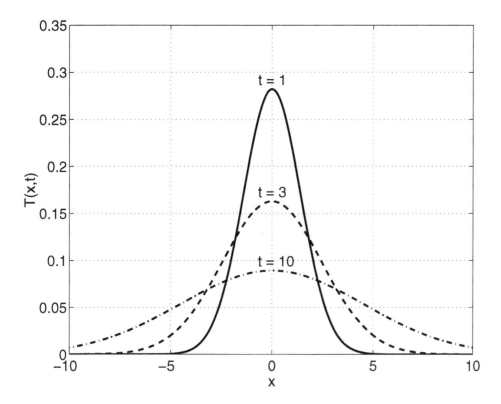

Figure 2.6.14: Solution (Eq. (2.6.56)) of the time-dependent one-dimensional diffusion equation for three different times ($\alpha = 1$).

The concentration is homogeneous in the $x - y$ - plane and therefore only depends on z. The diffusion equation reads:

$$D\frac{d^2c}{dz^2} = 0 \qquad\qquad (2.6.57)$$

The general solution is

$$c(z) = a\,z + b$$

where a and b are constants which are determined from the boundary conditions, i.e. the concentrations or the fluxes at z_1 and z_2. Using $c(z_1 = 0) = c_1$ and $c(z_2 = L) = c_2$, it follows ($0 \leq z \leq L$):

$$c(z) = c_1 + (c_2 - c_1)\,\frac{z}{L}$$

(see Figure 2.6.15). Thus, the concentration profile between z_1 and z_2 is independent of the diffusion coefficient. However, the flux of c which is

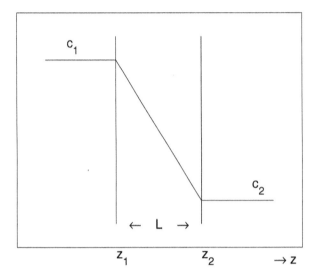

Figure 2.6.15: Diffusion between two large reservoirs across a thin film of thickness L (plane geometry).

given by Fick's first law:

$$F = -D\frac{dc}{dz} = (c_1 - c_2)\frac{D}{L}$$

does depend on the diffusion coefficient. For certain combinations of given boundary conditions, the diffusion equation (2.6.57) has unique solutions of which some are listed in Table 2.6.3.

Table 2.6.3: Some unique solutions of the diffusion equation in plane geometry[a].

Given boundary conditions		Solution, $c(z) =$
$c_1 \equiv c(z_1)$	$c_2 \equiv c(z_2)$	$c_1 + (c_2 - c_1)\frac{z}{L}$
$c_1 \equiv c(z_1)$	$F_2 \equiv -D\,(dc/dz)_{z_2} = -D\,a$	$c_1 - \frac{F_2}{D}\,z$
$c_1 \equiv c(z_1)$	$F_1 \equiv -D\,(dc/dz)_{z_1} = -D\,a$	$c_1 - \frac{F_1}{D}\,z$

[a] Solutions are given in the interval $0 = z_1 \leq z \leq z_2$; $L := z_2 - z_1$.

Exercise 2.11 (**)
Give boundary conditions that yield unique solutions of the diffusion equation (2.6.57) in plane geometry. Given $(dc/dz)_{z_1}$ and $(dc/dz)_{z_2}$, does this yield a unique solution?

Next we will briefly discuss diffusion in spherical coordinates. Consider two concentric spheres (the centers of the spheres coincide). When the outer

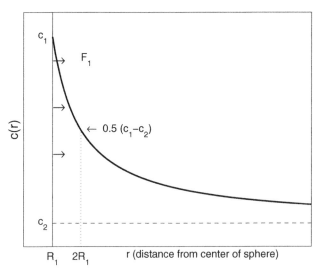

Figure 2.6.16: Diffusion in spherical geometry. The flux at the inner boundary, R_1, is denoted by F_1, whereas the concentration at $R_2 \to \infty$ is c_2. The concentration profile decreases proportional to $1/r$. The thickness of the diffusive boundary layer in spherical geometry is on the order of R_1.

sphere is at infinity we consider diffusion outside a sphere that is embedded in an unbounded medium. For the sake of simplicity, we assume that the concentration c is independent of an angle and thus depends on the radial distance r only. As in the case of plane geometry the diffusion equation (here in spherical coordinates) reduces to an ordinary differential equation:

$$\frac{D}{r} \frac{d^2 (r \cdot c)}{dr^2} = 0 \tag{2.6.58}$$

The general solution reads

$$c(r) = \frac{A}{r} + B$$

where A and B are constants which will be determined by the boundary conditions. For example, assume that the flux at R_1 (inner boundary) is given by:

$$F_1 \equiv -4\pi \; R_1^2 \; D \; \left(\frac{dc}{dr} \right)_{R_1} = 4\pi \; D \; A$$

and that the concentration at $R_2 \to \infty$ is given by c_2. It follows that $B = c_2$ and the solution is ($R_1 \le r < R_2$):

$$c(r) = \frac{F_1}{4\pi D} \frac{1}{r} + c_2$$

which is graphically shown in Figure 2.6.16. Some unique solutions of the diffusion equation in spherical geometry for given boundary conditions are listed in Table 2.6.16.

Table 2.6.4: Some unique solutions of the diffusion equation in spherical geometry[a].

Given boundary conditions[b]		Solution, $c(r) =$
c_1	c_2	$c_1 + \dfrac{R_2(c_1 - c_2)}{(R_2 - R_1)}\left(\dfrac{R_1}{r} - 1\right)$
	$R_2 \to \infty \Rightarrow$	$c_2 - (c_2 - c_1)\dfrac{R_1}{r}$
c_1	F_2	$c_1 + \dfrac{F_2}{4\pi D R_1}\left(\dfrac{R_1}{r} - 1\right)$
c_1	F_1	$c_1 + \dfrac{F_1}{4\pi D R_1}\left(\dfrac{R_1}{r} - 1\right)$
c_2	F_1	$c_2 + \dfrac{F_1}{4\pi D R_2}\left(\dfrac{R_2}{r} - 1\right)$
	$R_2 \to \infty \Rightarrow$	$c_2 + \dfrac{F_1}{4\pi D}\dfrac{1}{r}$

[a] Solutions are given in the interval $R_1 \leq r \leq R_2$.
[b] $c_i := c(R_i)$, $F_i := -4\pi R_i^2\, D\,(dc/dr)_{R_i} = 4\pi\, D\, A$.

In spherical geometry, the size of the region in which the concentration profile shows large gradients is on the order of the radius of the sphere. This region is called the diffusive boundary layer (DBL). For example, considering the profile shown in Figure 2.6.16, the concentration decreases to 50% of $(c_1 - c_2)$ at $r = 2R_1$. It is interesting to note that the thickness of the DBL in spherical geometry is given by the geometry itself. This is not the case in plane geometry.

Exercise 2.12 (*)

Marine organisms such as foraminifera take up oxygen for respiration. If the rate of O_2 consumption is 1 nmol h^{-1}, and the radius of the approximately spherical shell of the foraminifer is 300 μm, what is the O_2 concentration at the surface of the shell? Assume $D = 2 \times 10^{-9}$ m^2 s^{-1} and $[O_2] = 200$ μmol kg^{-1} in seawater.

Exercise 2.13 (**)

What is the maximum theoretical calcium uptake of this organism, if $D_{Ca^{2+}} = 1 \times 10^{-9}$ m^2 s^{-1} and $[Ca^{2+}] = 10.33$ mmol kg^{-1} in seawater?

2.6.1 Diffusion-reaction equations

Diffusion-reaction equations (DRE) are of the form

$$\frac{\partial c_i}{\partial t} = \text{Diffusion}(c_i) + \text{Reactions}(c_i, c_j) \,. \tag{2.6.59}$$

In general, the DRE for different compounds c_i are coupled by nonlinear reaction terms (c_i, c_j) and therefore have to be solved by numerical methods even in the case of steady state (compare, for example, Wolf-Gladrow and Riebesell, 1997, Wolf-Gladrow et al., 1999a, and Zeebe et al., 1999a). Here we will discuss a more simple case: the stationary DRE in one spatial dimension with linear reaction kinetics. In plane geometry it reads:

$$\underbrace{D\frac{d^2 c(z)}{dz^2}}_{\text{diffusion}} + \underbrace{k\,(c_b - c(z))}_{\text{reaction}} = 0 \tag{2.6.60}$$

where c_b refers to the concentration at the outer boundary (e.g. the bulk concentration). In spherical geometry it reads:

$$\underbrace{\frac{D}{r}\frac{d^2\,(r \cdot c(r))}{dr^2}}_{\text{diffusion}} + \underbrace{k\,(c_b - c(r))}_{\text{reaction}} = 0 \tag{2.6.61}$$

where spherical symmetry has been assumed, i.e the concentration is a function of the radial distance only. Equations of that form can be derived, for example, as an approximation to the coupled DRE of the carbonate system (Gavis and Ferguson (1975); Wolf-Gladrow and Riebesell (1997)):

$$\frac{D_{CO_2}}{r}\frac{d^2\,(r \cdot [CO_2](r))}{dr^2} + k_{CO_2}\,([CO_2]_b - [CO_2](r)) = 0$$

where D_{CO_2} is the diffusion coefficient of CO_2 and k_{CO_2} is a rate constant (the term $k_{CO_2}([CO_2]_b - [CO_2](r))$ gives the rate of chemical conversion between HCO_3^- and CO_2). The CO_2 concentration in the bulk medium is denoted by $[CO_2]_b$.

The diffusion-reaction equation in plane geometry as well as in spherical geometry (Eqs. (2.6.60) and (2.6.61)) both possess analytical solutions for given concentrations at the respective boundaries.

Plane geometry

In plane geometry the boundary conditions are $(-\infty < z_1 \leq z \leq z_2 < \infty)$:

$$c(z_1) = c_1, \quad c(z_2) = c_2 = c_b$$

and the solution of the diffusion-reaction equation reads

$$c(z) = c_b + A e^{-z/a_k} + B e^{z/a_k} \tag{2.6.62}$$

with

$$a_k = \sqrt{\frac{D}{k}} \quad \text{(reacto-diffusive length scale)} \tag{2.6.63}$$

and

$$A = -\frac{(c_b - c_1)\, e^{z_2/a_k}}{2\sinh [L/a_k]}, \quad B = \frac{(c_b - c_1)\, e^{-z_2/a_k}}{2\sinh [L/a_k]}$$

$$L := z_2 - z_1 \; .$$

The reacto-diffusive length scale, a_k, is central to the description of diffusion-reaction systems. It is given by the square root of the ratio of the diffusion coefficient, D, and the reaction rate constant, k, and measures the relative importance of diffusion and reaction. As an example, consider the carbonate system and in particular the diffusion of CO_2 and the chemical conversion between HCO_3^- and CO_2. At 25°C and $pH = 8.2$, we have (see Section 2.3.6 for values of rate constants):

$$\begin{aligned}
a_k &= \left(\frac{D_{CO_2}}{k_{+1} + k_{+4}[OH^-]}\right)^{\frac{1}{2}} = \left(\frac{1.8 \times 10^{-9}}{0.037 + 3820 \cdot 9.6 \times 10^{-6}}\right)^{\frac{1}{2}} \\
&= 156\ \mu\text{m} \; .
\end{aligned}$$

Figure 2.6.17 shows a_k as a function of temperature and pH. Typical values for a_k in seawater are in the range of a few hundred μm. As will be discussed below, this has interesting consequences for e.g. the CO_2 concentration in the boundary layer of microalgae (Riebesell et al., 1993).

In the limit $a_k \to \infty$ the reaction can be neglected and the appropriate solution of the diffusion-reaction equation in plane geometry is given by:

$$c(z)^{\text{diffusion}} = c_1 + (c_b - c_1)\, \frac{z - z_1}{z_2 - z_1} \tag{2.6.64}$$

which is identical to the solution given in the previous section if $z_1 = 0$ (Table 2.6.3). Figure 2.6.18 shows examples of the solutions of the diffusion-reaction equation (2.6.62) for two different values of the reacto-diffusive length scale, a_k, and the solution for pure diffusion (2.6.64). For the sake of simplicity, concentrations and distances are dimensionless. Note also that

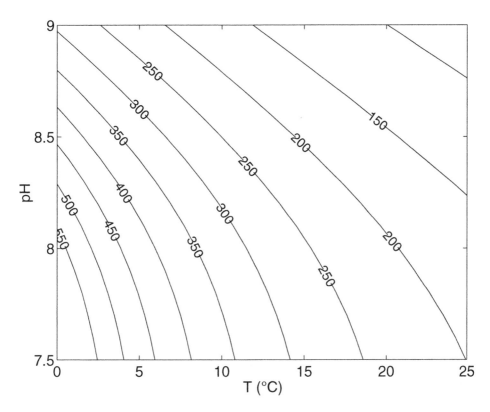

Figure 2.6.17: The reacto-diffusive length scale, a_k (in μm), for the conversion between HCO_3^- and CO_2 as a function of temperature and pH (see Wolf-Gladrow and Riebesell, 1997).

the boundaries ($z_1 = 0$ and $z_2 = 1$) and the concentrations at the boundaries ($c_1 = 0$ and $c_2 = 1$) are chosen arbitrarily.

For the example considered here in plane geometry, the diffusive boundary layer (DBL) is the region attached to the plane at $z = z_1$ where transport is dominated by diffusion. As already noted, the thickness of the diffusive boundary layer in plane geometry has to be set by external processes because the mathematical problem lacks any characteristic length scale. One example of those external processes is turbulence within the bulk medium. When reactions can be neglected, the concentration of chemical species in the DBL varies linearly in z between the left boundary (at $z = z_1 = 0$) and the bulk medium (at $z = z_2 = 1$; Figure 2.6.18, dot-dashed line). However, when the contribution of chemical reactions are taken into account the concentration profile deviates from a straight line. (Considering the carbonate system, and in particular a CO_2 profile, the contribution of chemical reac-

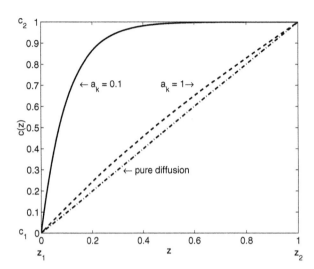

Figure 2.6.18: Solution of the diffusion-reaction equation in plane geometry. The reacto-diffusive length scale, a_k, determines the shape of the curves (solid line: $a_k = 0.1$, dashed line: $a_k = 1$). In the limit $a_k \to \infty$ (reaction term is very small), the curves approach the solution for pure diffusion (dot-dashed line).

tion refers to the conversion between HCO_3^- and CO_2.) These deviations are small provided that the reacto-diffusive length scale, a_k, is larger than or equal to L (Figure 2.6.18, dashed line). The differences with respect to the linear profile become appreciable when $a_k \ll L$ (Figure 2.6.18, solid line).

Spherical geometry

In spherical geometry the boundary conditions are $(0 < R_1 \leq r \leq R_2 \leq \infty)$:

$$c(R_1) = c_1, \quad c(R_2) = c_2 = c_b.$$

and the solution of the diffusion-reaction equation reads

$$c(r) = c_b + \frac{A}{r}e^{-r/a_k} + \frac{B}{r}e^{r/a_k} \tag{2.6.65}$$

with

$$A = -\frac{R_1\,(c_b - c_1)\,e^{R_2/a_k}}{2\sinh\left[(R_2 - R_1)/a_k\right]}, \quad B = \frac{R_1\,(c_b - c_1)\,e^{-R_2/a_k}}{2\sinh\left[(R_2 - R_1)/a_k\right]}.$$

Please note that the differential equation for $\tilde{c}(r) = r \cdot c(r)$ reads

$$\frac{d^2\tilde{c}(r)}{dr^2} + \frac{1}{a_k^2}\left(\tilde{c}_b - \tilde{c}(r)\right) = 0$$

and thus is of the same form as Eq. (2.6.60) after division by D. Same equations have the same solutions. Therefore the general solution $[c(z) - c_b]$ (Eq. (2.6.62)) is identical to $[r\,(c(r) - c_b)]$ (Eq. (2.6.65)).

In the limit $R_2 \to \infty$ (sphere in an unbounded medium), the solution simplifies to

$$c(r) = c_b - \frac{R_1}{r}(c_b - c_1) e^{-(r-R_1)/a_k}. \tag{2.6.66}$$

On the other hand, in the limit $a_k \to \infty$ (reaction terms can be neglected) the appropriate solution for the diffusion equation is given by

$$c(r)^{\text{diffusion}} = c_1 + \frac{R_2(c_1 - c_b)}{(R_2 - R_1)}\left(\frac{R_1}{r} - 1\right)$$

which was already derived for pure diffusion in the previous section (compare Table 2.6.4). For $a_k \to \infty$ and $R_2 \to \infty$ the solution in spherical geometry simplifies to

$$c(r)^{\text{diffusion}, \infty} = c_b - (c_b - c_1)\frac{R_1}{r}.$$

The solution of the diffusion-reaction equation (2.6.65) for different values of a_k and R_2 is shown in Figure 2.6.19.

The situation in spherical geometry is quite similar to the situation in plane geometry when the 'total' thickness of the DBL is set by external processes (Figure 2.6.19a, $R_2 = 2$). Here the thickness of the DBL may be defined as the radial distance from the surface of the sphere to the point where the concentration is almost equal to the bulk concentration (say 99%). Analogous to the concentration profiles in plane geometry (Figure 2.6.18) the differences between pure diffusion (dot-dashed line) and diffusion+reaction (dashed and solid line) become appreciable when the reacto-diffusive length scale, a_k, is equal to the 'effective' thickness of the DBL $(= 1)$.

When the thickness of the DBL is not set by external processes it is set by the radius of the inner sphere (Figure 2.6.19b and c, $R_2 \to \infty$). Again, a_k determines the shape of the curves. For $a_k = 0.1$, the bulk value is already approached at $r = 1.5$, i.e. the thickness of the DBL is smaller than R_1. On the contrary, the concentration for a nonreactive compound (pure diffusion, dot-dashed line) does not approach the bulk value at $r = 2R_1$, i.e. the thickness of the DBL is actually larger than R_1. Strictly speaking the concentration in this case does approach the bulk concentration only at $r \to \infty$. We will argue later on, however, that the 'effective' thickness of the DBL (with respect to the diffusion flux) is indeed equal to the radius of the inner sphere, R_1.

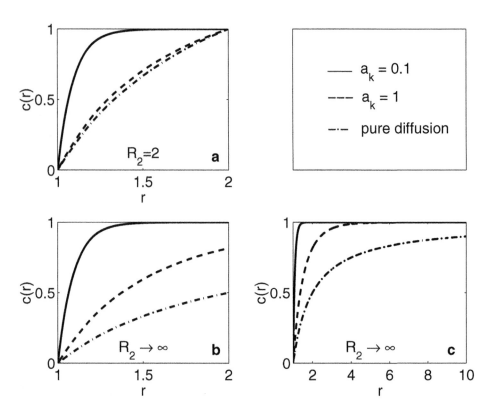

Figure 2.6.19: Solution of the diffusion-reaction equation (2.6.65) in spherical geometry (for the sake of simplicity, concentrations and distances are dimensionless). The curves refer to different values of the reacto-diffusive length scale: $a_k = 0.1$ (solid line), $a_k = 1$ (dashed line), and pure diffusion, $a_k \to \infty$ (dot-dashed line). (a) The outer boundary is at $r = R_2 = 2$, i.e. the 'total' thickness of the diffusive boundary layer, L, is 1. (b) The outer boundary is at infinity ($R_2 \to \infty$). (c) Same as (b) with extended horizontal axis. Note that the 'effective' thickness of the DBL is finite (even if $R_2 \to \infty$) and on the order of the radius of the inner sphere.

The boundary layer of tiny marine organisms

What do the results presented here tell us about diffusion and reaction in the boundary layer of particles and organisms in the ocean? As an approximation, plankton species such as microalgae, radiolaria, and foraminifera may be modeled as spherical particles. Thus, the concentration profiles calculated in spherical geometry may be used to determine concentration profiles of oxygen, carbon dioxide, and other nutrients within the vicinity (microenvironment) of these organisms. Because life processes such as respiration and photosynthesis disturb the chemical equilibrium within the microenvironment, the concentration that the organism actually 'sees' may be significantly different from the concentration within the bulk medium.

One result which follows directly from our mathematical description concerns the thickness of the diffusive boundary layer. The DBL for particles in the ocean is set by external processes such as turbulence. Consequently, the thickness of the DBL for particles with radii similar to the size of the smallest eddies ($\lambda \sim 1$ mm) is on the order of λ. On the other hand, for spherical particles with radii much smaller than the size of the smallest eddies, the thickness of the DBL is set by the particle radius, R_1 (Figure 2.6.19b and c).

Another interesting result that has to do with primary production in the ocean is the supply of carbon dioxide to the surface of photosynthesizing cells such as marine diatoms. Let us assume that the profiles shown in Figure 2.6.19 represent the concentration of CO_2 in the vicinity of microalgae. As can be seen from e.g. Figure 2.6.19c, the concentration profiles for diffusion+reaction deviate from the profiles based on pure diffusion. In terms of carbon dioxide, this is because the conversion between HCO_3^- and CO_2 is taken into account when the chemistry is included (solid line), whereas it is ignored in the case of pure diffusion (dot-dashed line). The additional CO_2 supply becomes appreciable when the reacto-diffusive length scale, a_k, is equal to the 'effective' thickness of the DBL, R_1. Typical values for a_k in seawater are about several hundred μm (see Figure 2.6.17). Thus, for marine diatoms of radius ~ 10 μm the ratio R_1/a_k is much smaller than 1 and therefore conversion between HCO_3^- and CO_2 cannot contribute to the CO_2 supply (Riebesell et al., 1993; Wolf-Gladrow and Riebesell, 1997). On the other hand, considering photosynthesis of symbiotic algae in foraminifera (foraminifera are much larger, typical radius ~ 300 μm) the conversion between HCO_3^- and CO_2 plays an important role in their DBL (Wolf-Gladrow et al., 1999a).

Exercise 2.14 (**)

Given the fact that there is a large pool of dissolved inorganic carbon in seawater, why can the growth rates of marine microalgae be limited by the CO_2 supply? Under which circumstances, in terms of carbon uptake and cell size, might this happen?

Mass transport, Damköhler number, and enhancement factor

The arguments developed in the preceding paragraph can be formulated more elegantly when mass transport is considered. The concentration gradient in plane as well as in spherical geometry (Figures 2.6.18 and 2.6.19) increases with increasing reaction rate (decreasing a_k). So does the flux, F, at $z = z_1$ and $r = R_1$, respectively.

In plane geometry the flux per unit area (absolute value) at $z = z_1$ is given by

$$
\begin{aligned}
f(z_1) \quad &:= \quad D\left(\nabla c\right)_{z=z_1} \cdot \mathbf{e_z} \\
&\underbrace{=}_{\text{Eq. (2.6.62)}} \quad D\left(-Ae^{-z_1/a_k} + Be^{z_1/a_k}\right)/a_k \\
&= \quad \frac{D}{a_k}(c_b - c_1)\frac{e^{L/a_k} + e^{-L/a_k}}{2\sinh[L/a_k]} \\
&= \quad \frac{D}{a_k}(c_b - c_1)\coth\left(L/a_k\right) \qquad\qquad (2.6.67)
\end{aligned}
$$

where $\mathbf{e_z}$ is the unit vector in z-direction. The flux consists of two parts. The flux due to pure diffusion reads

$$
f(z_1)^{\text{diffusion}} = D\frac{c_b - c_1}{L}, \qquad\qquad (2.6.68)
$$

whereas the contribution to the flux by reaction is given by

$$
\begin{aligned}
f(z_1)^{\text{reaction}} &= \quad f(z_1) - f(z_1)^{\text{diffusion}} \\
&= \quad D(c_b - c_1)\left(\frac{\coth\left(L/a_k\right)}{a_k} - \frac{1}{L}\right). \qquad\qquad (2.6.69)
\end{aligned}
$$

In spherical geometry, the expression for the flux per unit area (absolute value, $f = |F/4\pi R_1^2|$) for $R_2 \to \infty$ is:

$$
\begin{aligned}
f(R_1) \quad &:= \quad D\left(\nabla c\right)_{r=R_1} \cdot \mathbf{e_r} \\
&= \quad D\left(\frac{\partial c(r)}{\partial r}\right)_{r=R_1} \\
&\underbrace{=}_{\text{Eq. (2.6.66)}} \quad D\left(\left[\frac{1}{r} + \frac{1}{a_k}\right]\frac{R_1}{r}(c_b - c_1)\,e^{-(r-R_1)/a_k}\right)_{r=R_1} \\
&= \quad D\frac{c_b - c_1}{R_1}\left(1 + \frac{R_1}{a_k}\right) \qquad\qquad (2.6.70)
\end{aligned}
$$

where $\mathbf{e_r}$ is the unit vector in radial direction. The flux due to pure diffusion is given by

$$
f(R_1)^{\text{diffusion}} = D\frac{c_b - c_1}{R_1}, \qquad\qquad (2.6.71)
$$

whereas the contribution to the flux by reaction (e.g. through conversion between HCO_3^- and CO_2) is:

$$
f(R_1)^{\text{reaction}} = D\frac{c_b - c_1}{R_1}\frac{R_1}{a_k} = f(R_1)^{\text{diffusion}} \times \frac{R_1}{a_k}. \qquad\qquad (2.6.72)
$$

The expressions for the diffusive fluxes in spherical and in plane geometry (Eqs. (2.6.68) and (2.6.71)) are very similar: f is proportional to the diffusion coefficient, D, and to the concentration difference over the DBL, $c_b - c_1$, and in inverse proportion to a length scale. This length scale is the thickness of the DBL, L, in plane geometry and the radius of the sphere, R_1, in spherical geometry, respectively. It follows that the diffusion fluxes per surface area are equal in both cases when $L = R_1$. This is the reason why the 'effective' boundary layer thickness in spherical geometry (with respect to the diffusive fluxes) is equal to the radius of the sphere (cf. discussion to Figure 2.6.19).

The contribution of chemical reaction to the total flux (spherical geometry, Eq. (2.6.72)) is given by the ratio R_1/a_k. If $a_k \gg R_1$ the ratio R_1/a_k is small compared to 1 and therefore the contribution of reactions to the total flux is negligible. On the other hand, if $a_k \ll R_1$ the contribution by diffusion can be neglected. This relationship is expressed by the dimensionless quantity

$$D_a := \left(\frac{R_1}{a_k}\right)^2 = \frac{k \cdot R_1^2}{D} \qquad (2.6.73)$$

which is called the Damköhler number (Boucher and Alves, 1959). Diffusion dominates at small Damköhler numbers ($D_a \ll 1$), whereas reaction dominates at large Damköhler numbers ($D_a \gg 1$). The Damköhler number can be interpreted as the ratio of the diffusion time scale, $\tau_d = R_1^2/D$, and the reaction time scale, $\tau_r = 1/k$. The Damköhler number is small when reactions are slow compared to diffusion.

An alternative description of the contribution of reactions to the total flux uses the enhancement factor, EF, which is defined as the ratio between the true flux (diffusion plus reaction) and the flux predicted if there was no reaction (Emerson, 1975). The enhancement factors in plane and in spherical geometry read:

$$\mathrm{EF}^{\mathrm{plane}} = \frac{L}{a_k} \coth(L/a_k) \qquad (2.6.74)$$

$$\mathrm{EF}^{\mathrm{spherical}} = 1 + \frac{R_1}{a_k} = 1 + \sqrt{D_a}. \qquad (2.6.75)$$

The two factors are both a function of the ratio of the (effective) thickness of the boundary layer and the reacto-diffusive length scale (Figure 2.6.20).

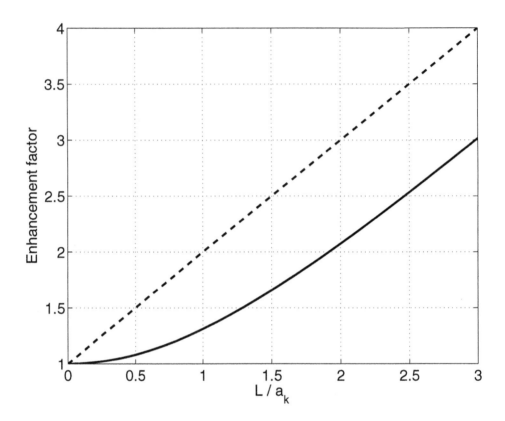

Figure 2.6.20: The enhancement factor, EF, as a function of the (effective) thickness of the diffusive boundary layer, L divided by the reacto-diffusive length scale, a_k: in plane geometry (solid line) and in spherical geometry (dashed line; $L = R_1$, $R_2 \to \infty$).

The enhancement factor and the Damköhler number are related by

$$\mathrm{EF}^{\text{plane}} = \sqrt{D_a}\ \coth\left(\sqrt{D_a}\right); \quad (D_a := (L/a_k)^2)$$

$$\mathrm{EF}^{\text{spherical}} = 1 + \sqrt{D_a}$$

and thus contain the same information. It is merely a matter of taste whether the enhancement factor (in the context of air-sea gas exchange) or the Damköhler number is used (in the context of nutrient uptake by microalgae).

As an example of the use of the enhancement factor, the $CO_2(g)$ exchange between the ocean and the atmosphere is considered. The thickness of the diffusive boundary layer at the ocean surface has been estimated as ~ 50 μm (Broecker, 1974). Given a typical value of $a_k = 300$ μm leads to an enhancement factor of 1.01. If the boundary layer is thicker, say 200 μm and a_k is smaller, say also 200 μm, the enhancement factor is 1.31, i.e. $\sim 30\%$.

2.7 Summary

The kinetics of the carbon dioxide system in seawater have been discussed in this chapter. The presented analysis of the disequilibrium properties has led to a detailed understanding of the mechanisms which determine the chemical behavior of the carbonate system on time scales ranging from microseconds to minutes. In general, these results shall serve to decide whether equilibrium or disequilibrium properties of the carbonate system should be considered.

Regarding measurements of primary production via addition of radio-carbon (Steemann Nielsen, 1952), an equilibration time of ca. 2 minutes has to be taken into account in order to achieve isotopic equilibrium. Otherwise, the carbon uptake by phytoplankton could be underestimated because the added $H^{14}CO_3^-$ or $^{14}CO_3^{2-}$ has not been converted to $^{14}CO_2$. This very property can, on the other hand, be utilized to identify carbon sources for photosynthesis of marine phytoplankton (e.g. Korb et al., 1997).

The mathematical analysis of diffusion-reaction equations in plane and in spherical geometry led to the reacto-diffusive length scale. This quantity is central to the description of diffusion-reaction systems and measures the relative importance of diffusion and reaction. The application of diffusion-reaction equations to the microenvironment of plankton such as marine microalgae and foraminifera leads to interesting results regarding the carbon supply, chemical gradients, and the pH in the vicinity of those organisms (e.g. Riebesell et al., 1993; Wolf-Gladrow et al., 1999a).

There are many more subjects in which the kinetics of the carbonate chemistry are of importance. A few examples are the $CO_2(g)$ exchange between the ocean and the atmosphere (e.g. Emerson, 1995; Wanninkhof, 1996) and the effects of the carbonate system kinetics on the precipitation of calcium carbonate (e.g. Dreybrodt et al., 1997). The kinetics of $CaCO_3$ precipitation/dissolution itself are another important aspect of carbonate system kinetics (e.g. Mucci et al., 1989; Wollast, 1990, Morse and Mackenzie, 1990). This subject is not discussed here because we confine ourselves to the description of the kinetics of the dissolved carbonate species in solution. The biochemical and physiological aspects of the CO_2 kinetics in living organisms (see e.g. Forster et al., 1969) have not been touched upon at all since a detailed discussion this subject is, of course, beyond the scope of this book.

Chapter 3

Stable Isotope Fractionation

In 1913, Sir J.J. Thomson found that the element neon has two different kinds of atoms with atomic weights of 20 and 22, respectively. This was the first experimental proof of a hypothesis by F. Soddy that different atoms might occupy the same place in the Periodic Table. Soddy named these atoms isotopes, which in Greek means equal places (Soddy, 1913). For his experiments, Thomson used a so-called 'positive ray apparatus', a predecessor of today's mass spectrometers (Thomson, 1914). Around 1920, F.W. Aston greatly improved the apparatus proposed by Thomson and subsequently discovered 212 of the 287 naturally occurring isotopes. In 1931, Harold C. Urey studied hydrogen by spectroscopic methods and detected a substance that had the same chemical properties as hydrogen but exhibited a larger mass than hydrogen known at this time (Urey et al., 1932). Because the mass was about twice the mass of hydrogen, Urey named it deuterium. The reason for atoms of the same element having different weights was, however, unclear until the discovery of the neutron by Chadwick in 1932 (Chadwick, 1932).

The discoveries summarized above are milestones at the beginning of a whole branch of science dealing with stable isotopes. Among the different properties of isotopic substances, one is of particular importance for earth sciences: their slightly different physico-chemical behaviors that lead to so-called isotopic fractionation effects. In 1947, Urey published a paper on the thermodynamic properties of isotopic substances (Urey, 1947). This provided the basis for the utilization of stable isotopes in modern disciplines such as stable isotope geochemistry, isotope geology, biogeochemistry, paleoceanography and others. For instance, the analysis of the ratio of stable oxygen isotopes in calcium carbonate, secreted by organisms like belemnites, mollusca, and foraminifera and buried in deep-sea sediments, has permitted the reconstruction of paleotemperatures for the last 150 million years or so (McCrea, 1950; Epstein et al., 1953; Emiliani, 1966).

This chapter is dedicated primarily to the stable isotopes of the elements of the carbonate system, and thus focuses on carbon, oxygen, and boron. Overviews on these and other elements can be found in textbooks by Faure (1986), Bowen (1988), Mook (1994), Clark and Fritz (1997), Hoefs (1997), and Criss (1999).

3.1   Notation, abundances, standards

The nuclei of isotopes of a certain element all contain the same number of protons (Z) but different numbers of neutrons (N). The notation $^A_Z E$ is used, where E is the element and $A = Z+N$ is the mass number that equals the sum of protons and neutrons in the nucleus. Often the subscript Z is omitted as in the discussion of isotopes in this book (e.g. $^{12}_6 C = {}^{12}C$).

An important observation is that most of the stable nuclides have even numbers of protons and neutrons, i.e. they are more abundant than nuclides that have, e.g. even numbers of protons and odd numbers of neutrons (Table 3.1.1). In addition, the ratio of the natural abundances of isotopes of the same element often follow the rule that the even-even nucleus is most abundant (Table 3.1.2). For example, the ratio of abundances of the stable carbon isotopes ^{13}C (even-odd) to ^{12}C (even-even) is about 1:99. Thus, the light isotope ^{12}C is much more abundant. On the contrary, the heavy stable isotope of boron ^{11}B (odd-even) is more abundant than the light isotope ^{10}B (odd-odd); the ratio of ^{11}B:^{10}B is approximately 80:20. However, these simple rules cannot explain, for instance, the differences between the abundances of ^{16}O and ^{18}O (both even-even). A sound understanding of abundances of nuclides requires the discussion of concepts of nuclear physics such as nuclear binding energy and 'magic numbers'. Magic numbers refer to nuclei in which N or Z is equal to 2, 8, 20, 28, 50, 82, and 126. These nuclei have greater binding energies than their neighbors in the chart of the nuclides. If Z and N is magic, the nuclei are referred to as 'doubly magic'; examples of doubly magic nuclei are: ^4He, ^{16}O, and ^{40}Ca (for a more detailed discussion of this subject, see e.g. Williams (1991)). A further interesting property of nuclides is the relative abundances of the elements found in nature which is related to the production of matter during the early stages of the universe and to element synthesis in stars (see e.g. Fowler (1984); Broecker (1985)).

Table 3.1.1: Number of stable nuclides with even and odd numbers of protons and neutrons (after Faure (1986)).

Z	N	Number of stable nuclides
Even	Even	157
Even	Odd	53
Odd	Even	50
Odd	Odd	4

Table 3.1.2: Abundances of stable isotopes with different combinations of even and odd numbers of protons and neutrons.

Isotope	Z	N	Abundance
^{10}B	Odd	Odd	~20%
^{11}B	Odd	Even	~80%
^{12}C	Even	Even	~99%
^{13}C	Even	Odd	~1%
^{16}O	Even	Even	~99.76%
^{17}O	Even	Odd	~0.04%
^{18}O	Even	Even	~0.2%

Atomic weight. Given the masses and the abundances of the stable isotopes of an element, the atomic weight as given in the Periodic Table can be calculated. By definition, the mass of ^{12}C is 12 amu (atomic mass unit), whereas the mass of ^{13}C is 13.0034 amu. The abundance of ^{12}C is ca. 98.89%, while the abundance of ^{13}C is ca. 1.11%. Thus, the atomic weight of carbon is $12 \times 0.9889 + 13.0034 \times 0.0111 \approx 12.01$. The International Union of Pure and Applied Chemistry gives 12.0107 for the atomic weight of carbon (IUPAC, 1999). Note that this value is not an integral multiple of the amu. In other words, the atomic weight of an element does not have to be an even number when expressed in terms of the amu - it does not even have to be close to it. One example is boron which has an atomic weight of $10.0129 \times 0.1982 + 11.0093 \times 0.8018 \approx 10.81$.

At this stage a warning should be given about calculations dealing with stable isotopes. In the context of the carbonate system it is convenient to start the discussion of isotopes with carbon (as it is done in this book). When stable carbon isotopes are concerned, it is almost always valid to assume that the concentration of the sum of the stable isotopes can be approximated by the concentration of ^{12}C because the concentration of ^{13}C is generally very small. The same assumption applied to boron leads to completely erroneous results because the ratio of ^{11}B to ^{10}B is about 4 to 1. To a student who is asking what to begin with, when studying calculations of isotopic fractionation, one should actually recommend starting with boron (this could avoid a lot of pitfalls). However, since carbon compounds are of major interest here, we will start with carbon (Section 3.2).

Exercise 3.1 (*)

The sum of the weight of eight protons, neutrons and electrons is 16.1379 amu. On the other hand, the weight of an oxygen atom (^{16}O) is only 15.9949 amu. Where does the mass difference come from?

Exercise 3.2 (*)

Calculate the corresponding mass difference for 1 mol O_2 and multiply it by the square of the speed of light. Compare the result to the typical energy of a covalent bond (400 kJ mol^{-1}).

3.1.1   Notation

The ratio of the numbers of atoms of two isotopes within a chemical compound is denoted by R. For example, the ratio of stable carbon isotopes in CO_2 is given by[1]:

$$^{13}R_{CO_2} = \frac{[^{13}CO_2]}{[^{12}CO_2]} \qquad\qquad (3.1.1)$$

where the left superscript indicates that ^{13}R refers to the stable carbon isotope ratio $^{13}C/^{12}C$. Analogously, ^{18}R, for example, refers to the stable oxygen isotope ratio $^{18}O/^{16}O$. We use R as the ratio of the heavy isotope to the light isotope (not vice versa). A quantity which is very useful in e.g. mass-balance calculations is the fractional abundance (see Hayes (1982) and Section 3.1.5):

$$^{13}r_{CO_2} = \frac{[^{13}CO_2]}{[^{13}CO_2 + {}^{12}CO_2]} \qquad\qquad (3.1.2)$$

---

[1] If not stated otherwise, the abbreviation CO_2 is used for dissolved CO_2, whereas gaseous CO_2 is denoted as $CO_2(g)$.

Table 3.1.3: Example of the values of various fractionation factors commonly used in calculations.

δ_A	δ_B	$(\delta_A - \delta_B)$	$\varepsilon_{(A-B)}$	$10^3 \ln(\alpha_{(A-B)})$	$\alpha_{(A-B)}$
-30	10	-40	-39.60	-40.41	0.9604
-20	10	-30	-29.70	-30.15	0.9703
-10	10	-20	-19.80	-20.00	0.9802
0	10	-10	-9.90	-9.95	0.9901
10	10	0	0	0	1
20	10	10	9.90	9.85	1.0099
30	10	20	19.80	19.61	1.0198
40	10	30	29.70	29.27	1.0297
50	10	40	39.60	38.84	1.0396

which relates to the isotope ratio ^{13}R by:

$$^{13}r = \frac{^{13}R}{1 + ^{13}R} \quad ; \qquad ^{13}R = \frac{^{13}r}{1 - ^{13}r} \; .$$

The fractionation factor α is defined as the ratio of the numbers of atoms of two isotopes within one chemical compound divided by the corresponding ratio in another compound. For instance,

$$^{13}\alpha_{(CO_2-HCO_3^-)} = \frac{^{13}R_{CO_2}}{^{13}R_{HCO_3^-}} \; . \tag{3.1.3}$$

Since α values are mostly very close to 1.0, the numbers $10^3 \ln(\alpha)$ or ε (see Exercise 3.3) are commonly used to express isotopic fractionations in per mil (‰):

$$^{13}\varepsilon_{(CO_2-HCO_3^-)} = (^{13}\alpha_{(CO_2-HCO_3^-)} - 1) \times 10^3 \; . \tag{3.1.4}$$

The isotopic composition of a substance determined by mass spectrometric methods is measured with respect to a standard (standards for certain elements will be discussed below). The value obtained is expressed as the δ value of the sample. Analyzing CO_2 gas, for instance, results in:

$$\delta^{13}C_{CO_2(g)} = \left(\frac{^{13}R_{CO_2(g)}}{^{13}R_{Stand.}} - 1 \right) \times 10^3 \tag{3.1.5}$$

where the factor 10^3 converts the δ value to per mil. The fractionation factor α between a sample A and a sample B is related to δ values by:

$$\alpha_{(A-B)} = \frac{\delta_A + 10^3}{\delta_B + 10^3} \; . \tag{3.1.6}$$

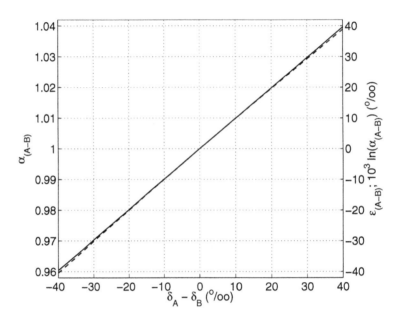

Figure 3.1.1: Fractionation factor $\alpha_{(A-B)}$ (solid line, left vertical axis), $\varepsilon_{(A-B)}$ (solid line, right vertical axis), and $10^3 \ln(\alpha_{(A-B)})$ (dashed line, right vertical axis) corresponding to the values given in Table 3.1.3.

Provided that α is close to unity, the fractionation (ε) between two substances can be approximated by the difference of their δ values (see Exercise 3.3). For example,

$$\varepsilon_{(A-B)} = \frac{\delta_A - \delta_B}{1 + \delta_B/10^3} \approx \delta_A - \delta_B \ .$$

It is emphasized that this approximation generally does not apply to δ values involving hydrogen isotope ratios D/H (^{2}H/^{1}H = deuterium/protium), where α can be very different from unity. For instance, if $\delta D_A = +200\%_0$ and $\delta D_B = -200\%_0$, i.e. $\alpha_{(A-B)} = 1.5$, it follows that $\varepsilon_{(A-B)} = 500\%_0$, whereas $\delta D_A - \delta D_B = 400\%_0$! Table 3.1.3 and Figure 3.1.1 display examples of ε, $10^3 \ln(\alpha)$, and α values for given δ values of two samples A and B.

Exercise 3.3 (**)

Calculate the values given in Table 3.1.3. Show that $\varepsilon_{(A-B)} = (\delta_A - \delta_B)/(1 + \delta_B/10^3)$, and $10^3 \ln(\alpha) \approx \varepsilon$.

3.1.2 Isotopic fractionation: Beans and peas

A German Professor, who has been working on geochemistry for decades, illustrated the phenomenon of isotopic fractionation by a person eating beans and peas. For some reason, the person might have a slightly higher affinity for beans. Thus, almost equal portions of beans and peas would be removed from the plate, however, since beans are slightly preferred, they would accumulate within the stomach of the person, leaving a slight deficit of beans on the plate (compared to the initial mixture). In the following, an example is presented to illustrate the basic mechanism and associated calculations of an isotopic fractionation process. Instead of beans and peas, white and black balls will serve as the model isotopes.

Consider a box (A) filled with a lot of white balls and fewer black balls (Figure 3.1.2). The white balls are a little bit lighter than the black balls, corresponding to the mass difference between e.g. ^{12}C and ^{13}C. The ratio of black to white balls in A should be $102/10^3$ and we define the standard as $100/10^3$. The δ value of box A is readily calculated.

$$\delta_A = \left(\frac{102/10^3}{100/10^3} - 1 \right) \times 10^3$$

$$= +20\,\%_0$$

Now the balls are transferred successively into a second box (B) (see Figure 3.1.2) which was initially empty. Assuming that the transfer of the heavier black balls is a little more energy-consuming, this may result in a slightly lower probability for the black balls to be carried to B. This feature corresponds to a statistical process like a chemical reaction where the activation energy is greater for the heavy isotope. After a certain time (reservoir A is assumed to be infinitely large) we count the number of white and black balls in B and find that the ratio is $98/10^3$. The δ value of B therefore is

$$\delta_B = \left(\frac{98/10^3}{100/10^3} - 1 \right) \times 10^3$$

$$= -20\,\%_0$$

The fractionation factor α which describes the statistical process is:

$$\alpha_{(A-B)} = \left(\frac{102/10^3}{98/10^3} \right) = 1.0408\;,$$

whereas ε is given by:

$$\varepsilon_{(A-B)} = \left(\frac{102/10^3}{98/10^3} - 1 \right) \times 10^3$$

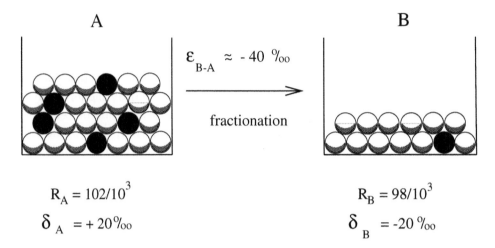

$$R_A = 102/10^3$$

$$\delta_A = +20\%_{oo}$$

$$R_B = 98/10^3$$

$$\delta_B = -20\%_{oo}$$

Figure 3.1.2: An example of fractionation: The transfer of balls from one box to another.

$$= 40.8\%_{oo}$$
$$\approx 40\%_{oo} = \delta_A - \delta_B$$

or

$$\varepsilon_{(B-A)} = \left(\frac{98/10^3}{102/10^3} - 1\right) \times 10^3$$
$$= -39.2\%_{oo}$$
$$\approx -40\%_{oo} = \delta_B - \delta_A$$

This example illustrates the essential quantities used to describe isotopic fractionation (α or ε) and isotopic composition (δ).

- The fractionation factor α (or ε) describes the fractionation which is characteristic for the process concerned. The isotopic composition of reservoir B depends on that of A and on the fractionation, which is an intrinsic feature of the process.

- The δ value describes the isotopic composition of the substance relative to a certain standard. Thus, the δ value is the result of the history of the sample in the sense that the sum of accumulations or depletions of isotopes during processes such as genesis, metamorphosis or degradation are expressed in the δ value measured at a certain time. The standard can be chosen arbitrarily, similar to the temperature scale in degrees Celsius or Fahrenheit. A positive δ value corresponds to 'isotopically heavy' (heavier than the standard) while a negative δ value corresponds to 'isotopically light' (lighter than the standard).

3.1.3   Isotope effects and isotope fractionation in nature

Differences in the chemical and physical properties of atoms and molecules that contain different isotopes of the same element are called 'isotope effects'. Isotope effects can cause isotope fractionation among different substances that contain a common element, i.e., the isotopes are then not equally distributed or partitioned among the different substances. One example in which an isotope effect can cause isotope fractionation is the partitioning of isotopes between the reactants/products of a chemical reaction which has a sound theoretical basis (Urey, 1947; Bigeleisen and Mayer, 1947, Bigeleisen, 1965). Molecules containing either the heavy or the light isotope differ, for instance, in their zero-point energies of molecular vibration. This property is illustrated in Figure 3.1.3 by the different binding energies of a diatomic molecule (e.g. H_2 or O_2, cf. Figure 3.1.4) containing the heavy isotope ($E_{b,h}$) and the light isotope ($E_{b,l}$). In classical mechanics, the equilibrium distance between the atoms corresponds to the minimum of the potential energy curve. In quantum mechanics, however, this state is forbidden due to Heisenberg's uncertainty principle - even at a temperature of absolute zero there is vibration of the atoms of the molecule. The zero-point energy of a diatomic molecule (with respect to the bottom of the potential energy curve) is $h\nu/2$, where $h = 6.626 \times 10^{-34}$ J s is Planck's constant and ν is the fundamental frequency of vibration (see Eq. (3.1.7)).

Regarding different substances that contain a common element, the different vibrational energies can cause the heavy isotope to concentrate in a particular substance and to be depleted in another. Thus, isotope fractionation will occur between the two substances. In thermodynamic equilibrium, differences in the vibrational frequencies of different materials are the primary cause for isotope fractionation. In the following, it will be explained why the vibrational energy of a molecule is affected by isotopic substitution.

Molecular vibration

The effect of isotopic substitution on molecular vibration can be understood by considering a model of a diatomic molecule (see also Criss (1999)). In classical mechanics, a diatomic molecule can be treated as two masses connected by a spring, where each mass represents one atom of the molecule and the spring represents the chemical bond (Figure 3.1.4). The force F between the masses is proportional to a force constant κ and the displacement from the equilibrium positions of the masses, x. This is mathematically expressed by $F = -\kappa\, x$. The frequency of vibration, ν, is then given by

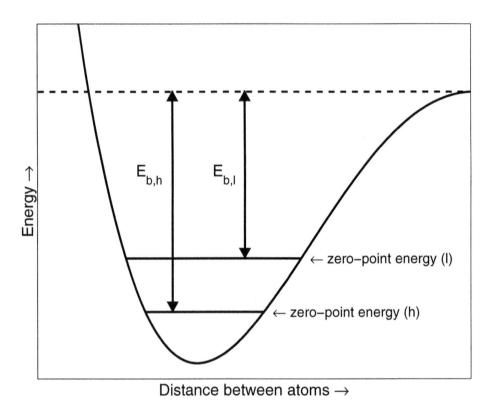

Figure 3.1.3: Schematic representation of binding energies of a diatomic molecule containing the heavy ($E_{b,h}$) and the light ($E_{b,l}$) isotope, respectively. The minimum of the potential energy curve corresponds to the equilibrium distance of the atoms of the molecule (classical consideration). The zero-point energy of the molecule containing the heavy isotope is lower (the frequency of vibration is smaller) than that of the molecule containing the light isotope.

(see textbooks on physics, e.g. Alonso and Finn (1992)):

$$\nu = \frac{1}{2\pi}\sqrt{\frac{\kappa}{\mu}} \qquad\qquad (3.1.7)$$

where $\mu = m_1 m_2/(m_1 + m_2)$ is the so-called reduced mass.

Equation (3.1.7) is a very useful expression because it provides a means of estimating the effect of isotopic substitution on the vibrational frequency of a diatomic molecule. The force in the molecule depends on the electronic structure, the nuclear charges, and the positions of the atoms in the molecule. For two molecules with the same chemical formula but of different isotopic species, these properties are virtually independent of the masses of the nuclei. Thus the force in the two molecules is the same. In other words,

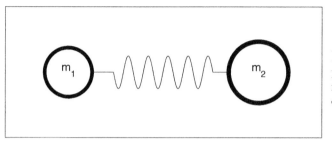

Figure 3.1.4: A simple model of a diatomic molecule: Two masses connected by a spring.

the force constant κ in Eq. (3.1.7) of the diatomic molecule of interest is equal to the force constant of the isotopically substituted diatomic molecule. The ratio of the frequencies of these molecules is then given by:

$$\frac{\nu'}{\nu} = \sqrt{\frac{\mu}{\mu'}} \qquad (3.1.8)$$

where the primed quantities refer to the molecule containing the heavy isotope.

As an example, let us consider the molecules $^{16}O^{16}O$ and $^{18}O^{16}O$. The reduced masses are $\mu \simeq 16 \times 16/(16+16) = 8$ and $\mu' \simeq 18 \times 16/(18+16) = 8.4706$. The vibrational frequency[2] of $^{16}O^{16}O$ is $\omega = 1580.36$ cm^{-1} (Richet et al., 1977), from which ω' can be calculated as 1535.83 cm^{-1}. Thus, the vibrational frequency of the molecule containing the heavy isotope ($^{16}O^{18}O$) is lower than that of the molecule containing the light isotope ($^{16}O^{16}O$), i.e. its zero-point energy is lower.

In summary, due to the higher mass, and thus lower vibrational frequency of the molecule that contains the heavy isotope, its ground state energy is generally lowered (Figure 3.1.3). The differences in the energies of the molecules lead to isotopic fractionation in, for example, chemical reactions which are usually on the order of per mil (except for hydrogen isotopes). The theoretical basis of these effects and the method to calculate the magnitude of the isotope fractionation associated with these effects are discussed in detail in Section 3.5.

Equilibrium isotope fractionation

Isotope fractionation between different phases or compounds of a system can occur in thermodynamic equilibrium. In terms of chemical reactions, this means that even if there is no net reaction (forward and backward reaction

---

[2]Vibrational frequencies are commonly reported in terms of wavenumbers, $\omega = \nu/c$, where $c = 299792458$ m s^{-1} is the speed of light.

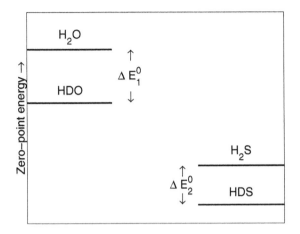

Figure 3.1.5: Relative zero-point energies of HDS, H_2S, HDO, and H_2O. The deuterium atoms concentrate preferentially in the water because the difference in the zero-point energy between H_2O and HDO is greater than the difference between H_2S and HDS.

rates are equal), different molecular species that contain a common element may have different isotope ratios. Equilibrium isotope fractionation can be most easily explained by considering isotopic exchange reactions. As an example, the exchange of protium and deuterium between liquid water and hydrogen sulfide gas will be discussed (cf. Bigeleisen, 1965). The isotopic exchange reaction can be written as:

$$H_2O(l) + HDS(g) \;\rightleftharpoons\; HDO(l) + H_2S(g) \tag{3.1.9}$$

In isotopic equilibrium, the protium and deuterium atoms will be distributed or partitioned among the reactants and products of reaction (3.1.9) in a way that is characteristic for this exchange reaction. In other words, the D/H ratio in the water, $(D/H)_{H_2O}$, and in the hydrogen sulfide, $(D/H)_{H_2S}$, will take on a particular value (which is a function of temperature and pressure). The equilibrium fractionation is expressed by the equilibrium constant of reaction (3.1.9), which is equal to the fractionation factor in this case[3]:

$$K \;=\; \alpha \;=\; \frac{[HDO][H_2S]}{[HDS][H_2O]} \;=\; \frac{(D/H)_{H_2O}}{(D/H)_{H_2S}} \;=\; 2.35$$

at 25°C (Bigeleisen, 1965). The deuterium atoms concentrate preferentially in the water which can be understood by considering the relative zero-point energies of the molecules involved (Figure 3.1.5). The zero-point energies increase sequentially from HDS, H_2S, HDO, to H_2O.

Using the energy change, ΔE, of reaction (3.1.9) and the relation $\ln K \simeq -\Delta E/RT$ (R is the gas constant and T is the absolute temperature in

---

[3]Note that the fractionation factors involving hydrogen isotopes are usually much larger than those involving stable carbon and oxygen isotopes. The latter are generally close to unity.

Kelvin), the logarithm of the equilibrium constant, K, is given by:

$$\ln K \simeq (\Delta E_1^0 - \Delta E_2^0)/RT \qquad (3.1.10)$$

where ΔE_1^0 and ΔE_2^0 are the differences between the zero-point energies of H_2O and HDO, and H_2S and HDS, respectively (Figure 3.1.5). Inspection of Eq. (3.1.10) shows that for $\Delta E_1^0 > \Delta E_2^0$, $\ln K > 0$ or $K > 1$, i.e. the D/H ratio in the water is greater than in the hydrogen sulfide. In other words, the deuterium atoms concentrate preferentially in the water because the difference in the zero-point energy between H_2O and HDO is greater than the difference between H_2S and HDS. The reason for this is that the chemical binding of hydrogen to oxygen is much stronger than that of hydrogen to sulfur. As a rule:

"The heavy isotope goes preferentially to the chemical compound in which the element is bound most strongly." (Bigeleisen, 1965).

Consider for example, the carbon isotope equilibrium between dissolved CO_2 and HCO_3^-:

$$^{12}CO_2 + H^{13}CO_3^- \ \rightleftharpoons \ ^{13}CO_2 + H^{12}CO_3^- .$$

Because the carbon atom is bound more strongly in the bicarbonate ion, HCO_3^- is enriched in ^{13}C relative to CO_2 by some 9‰ at 25°C. Another important example is the oxygen isotope equilibrium between liquid H_2O and gaseous CO_2, where the ^{18}O concentrates in the CO_2. The enrichment of ^{18}O in $CO_2(g)$ relative to H_2O is about 41‰ at 25°C (see Section 3.5.2).

Nonequilibrium effects

In the discussion of isotope partitioning, equilibrium and nonequilibrium effects should be distinguished. Nonequilibrium effects are associated with incomplete or unidirectional processes such as evaporation, kinetic isotope effects in chemical reactions, metabolic effects, and diffusion. Kinetic isotope effects in chemical reactions occur, when the reaction rates of the reaction involving the heavy and the light isotope are different. Consider for example, the hydration of CO_2:

$$CO_2 + H_2O \ \xrightarrow{k_+} \ HCO_3^- + H^+ \qquad (3.1.11)$$

and let us assume that no backward reaction is allowed (which could be accomplished by the immediate precipitation of the bicarbonate ion as carbonate). Considering also the stable carbon isotopes ^{13}C and ^{12}C, reaction (3.1.11) can be written as two reactions:

$$^{13}CO_2 + H_2O \ \xrightarrow{^{13}k_+} \ H^{13}CO_3^- + H^+$$

$$^{12}CO_2 + H_2O \ \xrightarrow{^{12}k_+} \ H^{12}CO_3^- + H^+ .$$

If the reaction rate involving the light isotope, $^{12}k_+$, is greater than that involving the heavy isotope, $^{13}k_+$, then ^{12}C will be preferentially incorporated in the product (HCO_3^-), while ^{13}C will be enriched in the unreacted residue (CO_2). Indeed, O'Leary et al. (1992) reported that reaction (3.1.11) produces HCO_3^- that is depleted in ^{13}C by about 13‰ at 25°C. The reverse reaction:

$$HCO_3^- + H^+ \quad \overset{k_-}{\longrightarrow} \quad CO_2 + H_2O$$

results in a depletion of ^{13}C in CO_2 of ca. 22‰ (O'Leary et al., 1992). The carbon isotope equilibrium fractionation between CO_2 and HCO_3^- is just equal to the difference of the kinetic fractionations - which is ca. 9‰ at 25°C.

In the case discussed here, an isotope effect causes isotope fractionation. It is noted, however, that even the largest isotope effect may not cause fractionation if the reaction occurs quantitatively. If the reactant is completely transformed into product, then the final isotope ratio of the product will be identical to the initial isotope ratio of the reactant, irrespective of whether the reaction rate is sensitive to the mass of the reacting species or not. This is a result of conservation of mass: just as in a pipeline, everything that goes in - including neutrons - will eventually have to come out (Hayes, 1982). Thus, for a kinetic isotope effect to be expressed, an incomplete reaction is required.

It is generally observed that the reaction rate of the reaction involving the light isotope is greater than that involving the heavy isotope. This is called a normal kinetic isotope effect. When the opposite is true, it is called an inverse isotope effect (Bigeleisen and Wolfsberg, 1958; Hayes, 1982). Kinetic isotope effects can quantitatively be understood on the basis of a theory called transition state theory (e.g. Bigeleisen and Wolfsberg, 1958; Melander and Saunders, 1980). A qualitative description of this theory can be given by considering the potential energy profile of a reaction as shown in Figure 3.1.6 (see also Section 2.2). The horizontal axis represents the reaction coordinate along which the reaction proceeds from reactants to products, while the vertical axis represents the potential energy. The intermediate state associated with the maximum of the energy profile is called transition state. Briefly, the transition state represents the configuration of the reacting species which is most difficult to attain on the potential energy profile along which the reaction takes place.

For the sake of simplicity, an approximation will be discussed in the following. Quantum-mechanical tunneling is neglected and only zero-point

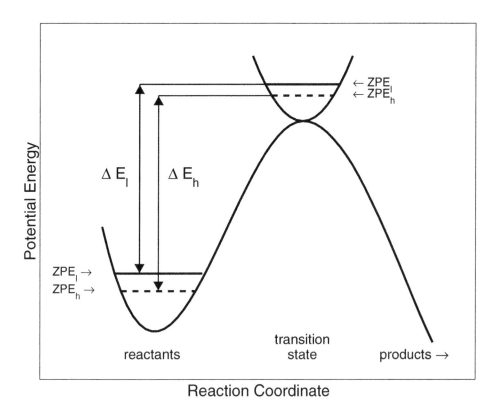

Figure 3.1.6: Schematic illustration of the potential energy profile of a reaction. The difference in the zero-point energy (ZPE) of the two isotopic molecules (light/heavy) is greater in the reactants than in the transition state. It follows that $\Delta E_l < \Delta E_h$, i.e. the molecule containing the light isotope has a smaller activation energy and reacts more readily (normal kinetic isotope effect). Note that this illustration is a simplification, cf. Melander and Saunders (1980).

energies are considered. The latter approximation is valid only at low temperatures (Melander and Saunders, 1980). In order to form products, an activation energy is necessary to surmount the energy barrier set by the potential energy surface (Figure 3.1.6). The binding energy in the reactants is generally greater than in the transition state because the force constants in the transition state are weaker. Thus, the difference in the zero-point energy of the two isotopic molecules (light/heavy) is greater in the reactants than in the transition state (compare the differences between the solid and dashed horizontal lines in Figure 3.1.6). In our approximation, the differences in the activation energies for the light and heavy species is simply given by the difference in the difference in the zero-point energies in the reactants and in the transition state. It follows that the activation energy of the molecule containing the light isotope is smaller than that of the molecule containing the heavy isotope and thus its reaction rate is generally greater (for a general approach, see Bigeleisen and Wolfsberg (1958)).

Diffusion

Another property that is affected by the different masses of isotopes is diffusion. The mobility of the molecule containing the heavy isotope is decreased which leads to a smaller diffusion coefficient. For gases it can be shown that the diffusion coefficient D is proportional to the mean relative velocity of the molecules $\bar{v}$ and obeys the following relationship:

$$D \propto \bar{v} \propto \left(\frac{k_B T}{M}\right)^{\frac{1}{2}}$$

where $k_B = 1.38 \times 10^{-23}$ J K^{-1} is Boltzmann's constant, T is the absolute temperature in Kelvin, and M is the mass of the molecule. If diffusion of a gas A through a gas B is considered, M has to be replaced by the reduced mass (e.g. Mook (1986)):

$$\mu = \frac{M_A M_B}{M_A + M_B}$$

Since the diffusion coefficient is in inverse proportion to the square root of the reduced mass, the ratio of diffusion coefficients of the isotopic molecules A and A' is given by:

$$\frac{D_A}{D_{A'}} = \left(\frac{M_A + M_B}{M_A M_B} \frac{M'_A M_B}{M'_A + M_B}\right)^{\frac{1}{2}} \tag{3.1.12}$$

where primes indicate the presence of the heavy isotope. As an example, the diffusion of gaseous CO_2 containing either ^{13}C or ^{12}C through air may be calculated as:

$$\frac{D_{^{12}CO_2(g)}}{D_{^{13}CO_2(g)}} = \left(\frac{44.00 + 28.92}{44.00 \times 28.92} \frac{44.99 \times 28.92}{44.99 + 28.92}\right)^{\frac{1}{2}} = 1.0044$$

showing that the diffusion of $^{12}CO_2(g)$ is about 4.4‰ greater than the diffusion of $^{13}CO_2(g)$. It is emphasized that Eq. (3.1.12) does not hold for the diffusion of molecules in condensed media. For instance, the ratio of the diffusion coefficients of dissolved $^{12}CO_2$ and $^{13}CO_2$ in water as predicted by Eq. (3.1.12) is 1.0032. However, measured values are 1.0007 (O'Leary, 1984) and 1.00087 (Jähne et al., 1987), demonstrating that interactions between the water molecules and dissolved substances significantly reduce the fractionation effect that is expected for gases.

Exercise 3.4 (*)

Figure 3.1.3 schematically shows the potential energy curve of a diatomic molecule. What would follow for the distance between the atoms if the zero-point energy was equal to the minimum of the potential energy curve? Why is this forbidden?

Exercise 3.5 (*)

Calculate the ratio of the diffusion coefficients of $^{12}C^{16}O_2$ and $^{13}C^{18}O_2$ in air (note that e.g. $^{12}C^{16}O_2 \equiv {}^{12}C^{16}O^{16}O$). Considering all combinations of ^{12}C, ^{13}C and ^{16}O, ^{18}O - how many different diffusion coefficients of CO_2 are there?

Exercise 3.6 (**)

The energy states of a diatomic molecule can be calculated (approximately) by quantum mechanical treatment of a harmonic oscillator. The quantum mechanical case shall not be elaborated here - instead, we discuss the classical case: The differential equation describing a classical harmonic oscillator (see Figure 3.1.4) is $\mu\ddot{x} \equiv \mu \cdot d^2x/dt^2 = -\kappa x$. Solve this equation for initial conditions $x(t = 0) = x_0$ and $\dot{x}(t = 0) \equiv dx/dt(t = 0) = 0$.

3.1.4 Natural abundances and standards

In this section the natural abundances and the standards used in isotope analysis of hydrogen, boron, carbon, and oxygen are briefly summarized.

Hydrogen has two stable isotopes, ^1H (protium) and ^2H (or simply D for deuterium). Their natural abundances in Standard Mean Ocean Water (SMOW) are:

^1H : 99.984426%

^2H :   0.015574%

(IUPAC, 1998). The relative mass difference between protium and deuterium is the largest of all stable isotopes of the same element and can lead to dramatic fractionation effects up to 700‰ in terrestrial samples. The value for the absolute ratio of D/H in SMOW as given by Hagemann et al. (1970) is 155.76×10^{-6}.

Boron has two stable isotopes

^{10}B : 19.82%

^{11}B : 80.18%

(IUPAC, 1998). The standard commonly used in oceanographic studies is the NBS SRM 951 boric acid standard (National Bureau of Standards): $^{11}B/^{10}B = 4.0014$ (Hemming and Hanson, 1992).

The stable isotopes of carbon are ^{12}C and ^{13}C, with abundances

^{12}C :   98.8922%

^{13}C :   1.1078%

(IUPAC, 1998). The standard widely used for carbon is the PDB (Pee-Dee Belemnite) standard: A fossil of *Belemnitella americana* (an 'ancient squid') from the Cretaceous Peedee formation in South Carolina, which has been long exhausted. In practice, a working standard is utilized which has a known value relative to PDB.

Oxygen has three stable isotopes with abundances

^{16}O :   99.7628%

^{17}O :   0.0372%

^{18}O :   0.20004%

(IUPAC, 1998). Due to the higher abundance of ^{18}O compared to ^{17}O, the ratio of ^{18}O to ^{16}O is usually reported. The standards used for stable oxygen isotopes are V-SMOW (Vienna-Standard Mean Ocean Water) and V-PDB (Vienna-Pee-Dee Belemnite). Whereas the V-SMOW scale refers to the isotopic composition of ocean water ('water-scale'), the V-PDB standard refers to the composition of a $CaCO_3$ formation ('$CaCO_3$-scale'). As will be discussed later, the stable oxygen isotopic compositions of water, and calcium carbonate formed in equilibrium with that water, are different. This fact is also reflected in the difference between the standards: A sample that has a $\delta^{18}O$ value of 0‰ with respect to the V-PDB standard has a $\delta^{18}O$ value of about 31‰ with respect to the V-SMOW standard. The relationship between the two standards is shown in Figure 3.1.7.

The V-SMOW is the 'natural' oxygen isotope standard for water samples since the primary source of H_2O in the hydrological cycle is the water of the oceans. Experimentally, the isotopic composition of water is not determined directly from the oxygen of the H_2O, but from oxygen of CO_2 gas that has been equilibrated with the water ($CO_2(g)$ can easily be run on mass spectrometers). The V-PDB standard is the 'natural' oxygen isotope standard for oxygen contained in carbon compounds such as $CaCO_3$. Just as for H_2O, the oxygen isotopic composition of $CaCO_3$ is routinely determined from CO_2 gas. Calcium carbonate is reacted with 100% phosphoric acid (H_3PO_4) and the liberated CO_2 is analyzed (but cf. Section 3.3.3). Only two-thirds of the oxygen of the $CaCO_3$ is contained in the CO_2 after reaction. Unfortunately, the isotopic composition of the CO_2 is not equal to the isotopic composition of the total carbonate because isotope partitioning occurs during the process. Thus, a characteristic acid fractionation factor has to be applied to calculate the isotopic composition of the

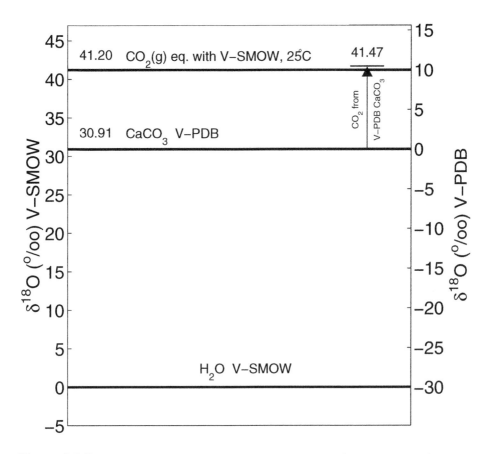

Figure 3.1.7: Relationship between the V-SMOW standard (left vertical axis) and the V-PDB standard (right vertical axis) used for stable oxygen isotopes (cf. Friedman and O'Neil, 1977; O'Neil, 1986; Hut, 1987). Note that gaseous CO_2 is analyzed on a mass spectrometer. Gaseous CO_2 which is equilibrated with V-SMOW at 25°C has an isotopic composition of 41.20‰ with respect to V-SMOW (average of results from 4 different laboratories). On the other hand, CO_2 released from V-PDB carbonate at 25°C has a $\delta^{18}O$ of 41.47‰ with respect to V-SMOW, therefore being 0.27‰ heavier than the CO_2 equilibrated with V-SMOW.

calcium carbonate. Values commonly used for calcite and aragonite are $\alpha_{(CO_2(g)-CaCO_3)} = 1.01025$ and 1.01034, respectively (Friedman and O'Neil, 1977)[4], whereas more recent work suggests larger values. Kim and O'Neil (1997) determined $\alpha_{(CO_2(g)-CaCO_3)}$ to be 1.01049 and 1.01107 for calcite and aragonite at 25°C, respectively. A compilation of acid fractionation factors

---

[4]Note that α is reported here, and not $10^3 \ln \alpha$, which give 10.20 and 10.29 for calcite and aragonite, respectively.

for other minerals can be found in Friedman and O'Neil (1977).

Also shown in Figure 3.1.7 is the δ^{18}O of $CO_2(g)$ released from V-PDB carbonate, and the δ^{18}O of $CO_2(g)$ equilibrated with V-SMOW at 25°C. The two values are roughly 10‰ heavier than the V-PDB standard. It is important to note that the V-PDB standard is the solid carbonate, not the acid-liberated CO_2. The latter is often referred to as the V-PDB standard which may cause confusion. Based on the ^{18}O fractionation between water and carbon dioxide at 25°C (1.0412), the acid fractionation factor (1.01025), and the difference between CO_2 released from V-PDB carbonate and CO_2 equilibrated with V-SMOW at 25°C (1.00027), the relationship between V-PDB and V-SMOW can be determined. The conversion between $\delta^{18}O_{V-PDB}$ and $\delta^{18}O_{V-SMOW}$ is (Coplen et al., 1983):

$$\delta^{18}O_{V-SMOW} = 1.03091 \times \delta^{18}O_{V-PDB} + 30.91‰ . \tag{3.1.13}$$

Exercise 3.7 (*)

Which general rule can be applied to explain that $CaCO_3$ is enriched in ^{18}O by about 30‰ with respect to water?

Exercise 3.8 (**)

Derive Eq. (3.1.13) using a fractionation factor α between V-PDB ($CaCO_3$) and V-SMOW (H_2O) of 1.03091.

3.1.5   Mass-balance calculations

Mass-balance calculations are of great importance in isotope studies. For example, if the stable isotopes of a certain element within a sample are distributed among different chemical species, a mass-balance calculation is very helpful to keep track of the molar quantities of the isotopes of the element of interest within that sample (see Section 3.2.5 for ^{13}C/^{12}C in the carbonate system). Another typical example of the use of a mass-balance calculation is the determination of the isotopic composition of a mixture of two or more samples of differing isotopic compositions. The fundamental equation that can be applied in those cases reads (e.g. Hayes (1982)):

$$r_T\, c_T = r_1\, c_1 + r_2\, c_2 + ... + r_n\, c_n \tag{3.1.14}$$

where r_i is the fractional abundance of the isotope of interest in sample i which reads e.g. for carbon:

$$^{13}r_i = \left. \frac{^{13}C}{^{13}C + {}^{12}C} \right|_{\text{in sample } i}$$

and $c_1, .., c_n$ refer to the molar quantities of the element of interest in sample (or chemical species) 1 to n of which the total sample consists (or sum of the chemical species, subscript T). Since c_T and c_i refer to the total number of carbon atoms in the sample, i.e. $^{13}C + {}^{12}C$, Eq. (3.1.14) is an exact mass-balance equation for ^{13}C.

Let us consider an example to evaluate Eq. (3.1.14). Two water samples containing 1 and 3 mmol total inorganic carbon are mixed, the $\delta^{13}C$ of sample 1 and 2 being $-20\%_0$ and $+80\%_0$ (assume $R_{\text{Stand.}} = 0.01$ for simplicity). What is the isotopic composition of the total carbon of the final mixture? The isotope ratios are $R_1 = 0.0098$ and $R_2 = 0.0108$ from which follows $r_1 \approx 0.0097049$ and $r_2 \approx 0.0106846$. The mass-balance equation therefore reads:

$$r_T \times 4 = 0.0097049 \times 1 + 0.0106846 \times 3 .$$

which gives $r_T = 0.0104397$ and a δ value of the total sample of $54.98\%_0$. Equation (3.1.14) could also have been formulated in terms of R or δ:

$$\delta_T \, c_T = \delta_1 \, c_1 + \delta_2 \, c_2 + ... + \delta_n \, c_n$$

which introduces an error in the calculation. In this case, the result for δ_T for the example considered above would be $\delta_T = 55.00\%_0$ which is very close to the exact result. It can be shown that the error introduced depends on the deviation of the following ratios from 1.0:

$$\frac{1 + (\delta_T/10^3 + 1) \, R_{\text{Stand.}}}{1 + (\delta_i/10^3 + 1) \, R_{\text{Stand.}}}$$

which approximately equals 1.0, provided that $R_{\text{Stand.}}$ and the δ values involved are small (i.e. $R_{\text{Stand.}} \ll 1$ and $\delta \ll 10^3$). One should therefore be cautious when elements are considered for which the isotope ratio (heavy over light isotope) is not small. For example, considering boron, $R_{\text{Stand.}} = {}^{11}B/^{10}B \approx 4$, $r \approx 0.8$ and the error in δ_T for the example considered above would be $1.5\%_0$.

3.1.6 Rayleigh process

A phenomenon that occurs in many processes associated with isotopic fractionation is the so-called Rayleigh distillation/condensation (see e.g. Bigeleisen and Wolfsberg (1958) or Mook (1994) for further reading). A well known example of a Rayleigh process is the isotopic fractionation of water vapor of a cloud and the raindrops released from the cloud. The water which is condensing from the cloud is enriched in ^{18}O by ca. $9\%_0$ at $25°C$. As more rain condenses from the cloud, the water of the cloud becomes progressively depleted in ^{18}O because with every raindrop a small surplus of

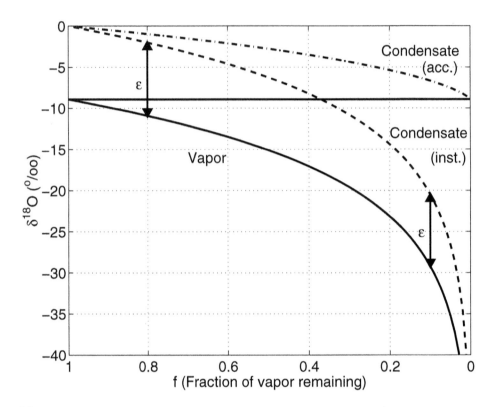

Figure 3.1.8: Rayleigh fractionation process of water condensing from vapor. The oxygen isotope fractionation between reservoir (vapor) and the instantaneous product (condensate, inst.) is constant, $\varepsilon \sim 9\permil$. When no vapor remains ($f = 0$), the accumulated product (condensate, acc.) has the same isotopic composition as the initial vapor since the vapor has completely condensed (horizontal line).

^{18}O is removed from the cloud. Since the fractionation between the water vapor and the liquid water is constant (at constant temperature), the raindrops condensing from the isotopically changing reservoir are progressively depleted in ^{18}O as the condensation proceeds (Figure 3.1.8).

The isotope ratio of the water of the cloud (the reservoir) during the condensation process is given by

$$\frac{R_r}{R_r^0} = f^{\,(\alpha-1)} \tag{3.1.15}$$

where R_r and R_r^0 are the isotopic ratios of the reservoir at time t and $t = 0$, respectively. The fraction of water remaining in the cloud is denoted by f ($0 \leq f \leq 1$), whereas the fractionation factor α refers to the fractionation characteristic for the process (e.g. ~ 1.009 for the condensation of liquid water from water vapor). The instantaneous isotope ratio of the product R_p^i is given by

$$\frac{R_p^i}{R_r^0} = \alpha \, f^{\,(\alpha-1)} \; . \tag{3.1.16}$$

Note that this ratio is the isotope ratio of a small amount of rain leaving the vapor at a particular time, whereas the isotope ratio of the accumulated product is:

$$\frac{R_p}{R_r^0} = \frac{f^\alpha - 1}{f - 1} \; . \tag{3.1.17}$$

Eqs. (3.1.15) - (3.1.17) are derived as follows. The reservoir may consist of a number of particles of the abundant isotope (N_1^0) and a number of particles of the rare isotope (N_2^0) at time $t = 0$, whereas at time t we have N_1 and N_2. On the other hand, the number of particles of the product are denoted by n_1 and n_2, respectively. It is assumed that $N_1 \gg N_2$ (N_1 is much more abundant than N_2), therefore the fraction of substance remaining in the reservoir f can be approximated by:

$$f = \frac{N_1}{N_1^0}$$

(for a more general approach, see Bigeleisen and Wolfsberg (1958)). The fractionation factor for the process considered is denoted by α. The fundamental equation describing the process is (dn refers to an infinitesimal small number of particles removed or added per time increment):

$$\frac{dn_2}{dn_1} = \alpha \frac{N_2}{N_1} \; . \tag{3.1.18}$$

In words: The ratio of isotopes accumulating in the product per time increment is given by the isotope ratio of the reservoir, times the fractionation factor. Since mass is conserved, we have:

$$\begin{aligned} N_1 + n_1 &= const. \\ N_2 + n_2 &= const. \end{aligned}$$

and thus

$$\begin{aligned} dN_1 &= -dn_1 \\ dN_2 &= -dn_2 \; . \end{aligned}$$

Eq. (3.1.18) can therefore be rewritten as:

$$\frac{dN_2}{dN_1} = \alpha \frac{N_2}{N_1} \; ,$$

which is solved by integration:

$$\int_{N_2^0}^{N_2} \frac{\mathrm{d}N_2'}{N_2'} = \alpha \int_{N_1^0}^{N_1} \frac{\mathrm{d}N_1'}{N_1'}$$

resulting in:

$$\ln\left(\frac{N_2}{N_2^0}\right) = \ln\left(\frac{N_1}{N_1^0}\right)^{\alpha}$$

or:

$$\left(\frac{N_2}{N_2^0}\right) = \left(\frac{N_1}{N_1^0}\right)\left(\frac{N_1}{N_1^0}\right)^{(\alpha-1)}$$

Since $(N_2/N_1)/(N_2^0/N_1^0) = R_r/R_r^0$ and $f = N_1/N_1^0$, we have:

$$\frac{R_r}{R_r^0} = f^{(\alpha-1)}.$$

The instantaneous isotope ratio of the product at any time is given by the isotope ratio of the reservoir, times the fractionation factor:

$$\frac{R_p^i}{R_r^0} = \alpha \, f^{(\alpha-1)} \tag{3.1.19}$$

whereas the isotope ratio of the accumulated product can be found by integration of (3.1.19):

$$
\begin{aligned}
\frac{R_p}{R_r^0} &= \frac{1}{(N_1 - N_1^0)} \int_{N_1^0}^{N_1} \alpha \left(\frac{N_1'}{N_1^0}\right)^{\alpha-1} \mathrm{d}N_1' \\
&= \frac{1}{(N_1 - N_1^0)} \frac{1}{(N_1^0)^{-1}} \left[\left(\frac{N_1'}{N_1^0}\right)^{\alpha}\right]_{N_1^0}^{N_1} \\
&= \frac{f^{\alpha} - 1}{f - 1}.
\end{aligned}
$$

Rayleigh processes with different numbers of sources and sinks are discussed in Mook (1994).

Exercise 3.9 (**)
H_2O in the form of ice on the Greenland icecap is heavily depleted in ^{18}O ($\sim -30\permil$) with respect to V-SMOW. Describe a mechanism that explains this feature.

Exercise 3.10 (***)
Consider a Rayleigh process for a reservoir with two sinks. Show that the isotopic ratio of the reservoir (analogous to Eq. (3.1.15)) can be written as $R_r/R_r^0 = f^{\,[(\alpha_1-1)x_1+(\alpha_2-1)x_2]}$, where x_1 and x_2 refer to the fractional contribution to the total sink ($x_1 + x_2 = 1$).

3.1.7 A Rayleigh process: Uptake of silicon by marine microalgae

An interesting example of a Rayleigh process which has recently received attention in chemical oceanography is the fractionation of silicon isotopes by marine microalgae (diatoms) in surface waters (De La Rocha et al., 1998). Marine diatoms produce siliceous skeletons or frustules ($SiO_2 \cdot nH_2O$) from the dissolved silicon in seawater which is mostly present as $Si(OH)_4$ ($\sim$95%) at pH 8.2. During the incorporation of silicon into the frustules of the organisms, the stable silicon isotope ^{28}Si is slightly preferred over ^{30}Si, resulting in a depletion of ^{30}Si of ca. 1.1‰ in the diatom silica relative to the $Si(OH)_4$ in seawater[5]. During algal growth, dissolved silicon is progressively removed from surface waters. The process is analogous to the removal of rain drops from water vapor as discussed in Section 3.1.6. In the case of silicon removal, however, the fractionation factor is smaller than one, $\varepsilon = 0.9989$ (De La Rocha et al., 1997), resulting in a slight enrichment of ^{30}Si in both the dissolved silicon in the water (reservoir) and in the biogenic silica of the diatoms as silicon uptake proceeds.

De La Rocha et al. (1998) suggested that the isotopic composition of the siliceous frustules during algal growth is in accordance with the laws of a Rayleigh process and is therefore a function of the utilization of dissolved silicon, i.e. diatom productivity. If so, changes in the utilization of dissolved silicon in the past should be documented in the sedimentary record where siliceous frustules have been preserved. Indeed, De La Rocha et al. (1998) investigated three sediment cores from the Southern Ocean and found a decrease of $\delta^{30}Si$ of ca. 0.5‰ $-$ 0.7‰ during the last glacial maximum (LGM), relative to the present interglacial in each of the three cores (Figure 3.1.9). Hence, their results imply a strongly diminished percentage utilization of silicic acid by diatoms in the Southern Ocean during the last glaciation.

Using equations derived in Section 3.1.6, one can calculate the isotopic ratios of the dissolved silicon remaining in seawater (diss. $Si(OH)_4$, reservoir), the instantaneous isotope ratio of the frustules (Silica, inst.), and the accumulated isotope ratio of the frustules (Silica (acc.), dot-dashed line in Figure 3.1.9). With a $\delta^{30}Si$ value of 1.6‰ for dissolved silicon supplied to the euphotic zone (De La Rocha et al., 1998), the accumulated product (diatom opal) would increase from 0.5‰ at 0% utilization to 1.6‰ at 100% utilization. De La Rocha et al. (1997) proposed that this range ($\sim$1.1‰)

---

[5]Natural abundances of stable silicon isotopes are ^{28}Si: 92.23%, ^{29}Si: 4.68%, ^{30}Si: 3.09% (IUPAC, 1998).

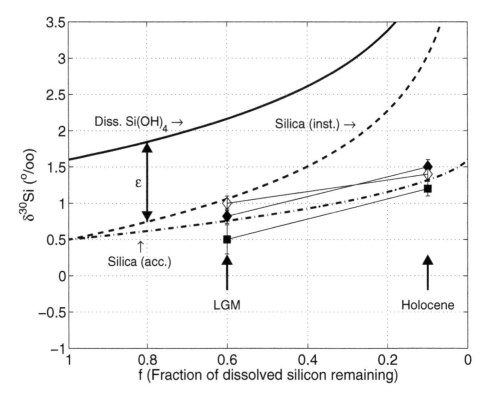

Figure 3.1.9: Rayleigh fractionation of silicon isotopes through silicon uptake by marine diatoms. Each pair of symbols represent δ^{30}Si values from a single sediment core from the Southern Ocean (squares: RC13-269, closed diamonds: E50-11, and open diamonds: RC11-94) for the Holocene (heavier values) and the last glacial maximum (LGM, lighter values) (De La Rocha et al., 1998). With the fractionation factor, $\varepsilon = 0.9989$ (De La Rocha et al., 1997), and a δ^{30}Si value of 1.6‰ for dissolved Si(OH)$_4$ supplied to the euphotic zone (De La Rocha et al., 1998), the δ^{30}Si values of dissolved Si(OH)$_4$ (solid line), instantaneous silica (dashed line), and accumulated silica (dot-dashed line) can be calculated.

should be close to the range expected for variations in sedimentary opal since the accumulation of diatoms in sediments reflect an integrated signal.

The δ^{30}Si record of each sediment core from the Southern Ocean studied by De La Rocha et al. (1998) shows a decrease from heavier to lighter values from the Holocene to the last glacial maximum (compare each pair of symbols in Figure 3.1.9). Provided that the downcore signal reflects changes in silicon utilization according to the laws of a Rayleigh process, approximate percentage values of 90% and 40% Si utilization can be assigned to the Holocene and the LGM, respectively.

It is noted that critical variables influencing the reliability of this pale-oproxy have to be examined in more detail. Those variables are, for example, the temperature dependence of the fractionation factor, vital effects, and the validity of the assumption of an underlying Rayleigh process. If silicon isotopes prove to be a reliable proxy of nutrient utilization, isotopic studies would become possible in areas such as the Southern Ocean where the abundance of $CaCO_3$ in sediments is small.

After having introduced some basic concepts of stable isotope fractionation (definitions, abundances, standards, mass-balance calculations, Rayleigh processes and more) we now turn to the stable isotopes of selected elements of the carbonate system in seawater (carbon, oxygen, and boron).

3.2 Carbon

Stable carbon isotopes are utilized in various branches of the marine sciences. To name only a few subjects in which this tool is used to understand processes on different time and length scales or is employed as a paleoindicator, one might consider the ^{13}C signal of organic matter as recorded in marine phytoplankton (e.g. Sackett, 1991; Hayes et al., 1999) which has been proposed as a proxy to reconstruct atmospheric CO_2 partial pressures during the past 150 Ma (so-called 'CO_2 paleobarometer'; Popp et al., 1989; Freeman and Hayes, 1992). Another example is the natural change of the $\delta^{13}C$ of atmospheric CO_2 on glacial-interglacial time scales which is preserved in the gas bubbles of ice cores and is believed to indicate changes of the carbon exchange fluxes between the carbon reservoirs of the atmosphere, the ocean, and the land biosphere (Indermühle et al., 1999). The current decrease of the $\delta^{13}C$ of atmospheric CO_2, which is a result of the invasion of isotopically light anthropogenic carbon dioxide (so-called ^{13}C Suess effect), can be used to study net global loss or gain of biospheric carbon (e.g. Keeling et al., 1980; Quay et al., 1992; Tans et al., 1993; Bacastow et al., 1996). A further example of the use of carbon isotopes as a paleoproxy is the difference of the $\delta^{13}C$ recorded in the calcium carbonate shells of surface-dwelling and deep ocean organisms such as foraminifera which might be used as a tracer for the ΣCO_2 of the ocean in the past (e.g. Broecker, 1982; Shackleton, 1985).

The comprehension of many of the subjects (and the application of many of the tools) mentioned in the last paragraph, requires a basic knowledge of the carbon isotopic composition of the chemical compounds involved - and of the stable carbon isotope fractionation between gaseous CO_2 and the dissolved carbon species in seawater. In order to provide the reader with that knowledge the natural variations of carbon isotope values are briefly surveyed first. Then the equilibrium fractionation effects between $CO_2(g)$, the dissolved forms of CO_2 in seawater, and $CaCO_3$ are summarized. Whereas the results of isotopic fractionation factors as determined by experimental work are presented at this stage, a summary of the theoretical, thermodynamic basis for isotopic fractionation is given in Section 3.5.

3.2.1 Natural variations

Figure 3.2.10 shows natural variations in the $\delta^{13}C$ of different carbon bearing compounds relative to the PDB calcium carbonate standard (see Section 3.1.4). The $\delta^{13}C$ of the total dissolved carbon in the ocean (ΣCO_2) shows a natural range of about $0\permil - 2\permil$. The surface ocean is usually enriched in the heavy isotope ^{13}C with respect to the deep ocean. This is because biological production in the euphotic zone preferentially removes the light isotope ^{12}C and hence slightly enriches the surface ocean in ^{13}C. It appears as if plants live on a diet of 'light' ^{12}C atoms - however, they are obviously rather lax about their diet because the difference of ^{12}C consumption over ^{13}C consumption is only in the order of per mil. The reason for the preferential uptake of ^{12}C is the fractionation occurring during photosynthetic fixation of carbon which will be discussed below. At depth, the organic matter is degraded and the isotopically light carbon is released, lowering the $\delta^{13}C$ of the ΣCO_2 of the deep ocean, with respect to the surface ocean.

Organic matter is generally depleted in ^{13}C. Depending on the pathway of carbon fixation, the isotopic composition of the organic matter varies roughly between $-32\permil$ and $-22\permil$ in so-called C_3 plants and between $-16\permil$ and $-10\permil$ in so-called C_4 plants (one might say that C_3 plants keep to their diet more strictly than C_4 plants do). The ^{13}C depletion in C_3 plants is mainly a result of the large fractionation associated with the enzyme Rubisco (ribulose bisphosphate carboxylase/oxygenase) which catalyzes the first step in the CO_2 fixation (e.g. O'Leary et al., 1992; Hayes, 1993). The name of the C_3 metabolism arises from the fact that the first intermediate compound which is formed during carbon fixation is a molecule containing 3 carbon atoms (analogously there are 4 carbon atoms in the C_4 metabolism). The majority of today's plants use the C_3 pathway, as is the case for marine phytoplankton (see, however, Reinfelder et al., 2000). During photosynthetic carbon fixation via the C_4 pathway, phosphoenolpyru-

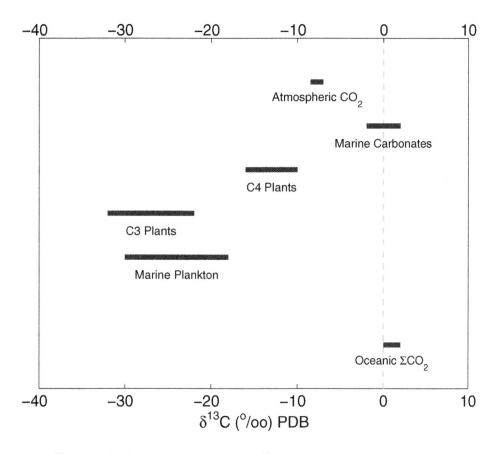

Figure 3.2.10: Natural variations in $\delta^{13}C$ of some terrestrial compounds.

vate carboxylase catalyzes the first carbon fixation step which is associated with a smaller carbon isotope fractionation compared to the fractionation associated with Rubisco. An example of a C_4 plant is maize.

The carbon isotopic composition of marine carbonates is usually close to the isotopic composition of ΣCO_2 (or dissolved HCO_3^-) in the ocean. The vast majority of the $\delta^{13}C$ values of $CaCO_3$ precipitated by marine organisms such as foraminifera, corals, and mollusca fall within the range from $-2\permil$ to $+2\permil$ (see Wefer and Berger, 1991). However, strong isotope depletions in ^{13}C by up to $-10\permil$ (often in conjunction with ^{18}O depletions) are observed in certain corals. This feature has been explained by kinetic isotope effects during the precipitation of calcium carbonate (McConnaughey, 1989a,b). Carbon isotope variations in foraminifera as a function of the seawater chemistry (on the order of $2-3\permil$) have been observed by Spero et al. (1997); for modeling of those effects, see Zeebe et al. (1999a) and Zeebe (2001a).

To a first order approximation, the carbon isotope composition of bulk atmospheric CO_2 is in equilibrium with the $\delta^{13}C$ of the dissolved HCO_3^- in the ocean. The difference between the preindustrial δ value of atmospheric CO_2 ($\approx -6.5‰$) and $\delta^{13}C_{HCO_3^-} \approx +2‰$ is $8.5‰$, which can be compared to the value of $9.1‰$ as determined in the laboratory at $15°C$ (Mook, 1986). On a regional scale, the $\delta^{13}C$ of atmospheric CO_2 is not uniform. For example, due to the combustion of fossil fuel (which is ultimately derived from organic matter which is isotopically light) the atmospheric CO_2 in industrial areas can be strongly depleted in ^{13}C. The continuous injection of 'light' anthropogenic CO_2 into the atmosphere since the beginning of the industrialization has led to a decrease of $\delta^{13}C$ of atmospheric carbon dioxide (^{13}C Suess effect) from a preindustrial value of $-6.3‰$ to about $-7.8‰$ in 1995 (e.g. Keeling et al., 1995; Ciais et al., 1995; Francey et al., 1999; Battle et al., 2000). Natural seasonal variations in $\delta^{13}C_{CO_2}$ on the order of $1‰$ modulate this long-term trend and occur simultaneously to changes in the atmospheric CO_2 concentrations of up to 20 ppmV. This effect is more pronounced in the northern hemisphere and is due to the 'breathing' of the biosphere, i.e. the periodic uptake and release of CO_2 by plants and soils during summer and winter.

A knowledge of the carbon isotope fractionation between gaseous carbon dioxide and the dissolved forms of carbon dioxide in the ocean is indispensable to gain understanding of the subjects discussed so far. In order to take a step in this direction, the carbon isotope fractionation within the carbonate system is studied in the following.

3.2.2 Equilibrium ^{13}C fractionation in the carbonate system

As demonstrated in Section 3.1.3, isotope fractionation between different phases or compounds of a system also occurs in thermodynamic equilibrium. A useful rule in this context is that the heavy isotope usually concentrates in the compound in which the element is most strongly bound. As it is to be expected, equilibrium isotope fractionation also occurs between the compounds of the carbonate system. This is the subject of this section.

The chemical structures of the dissolved forms of carbon dioxide are shown in Figure 3.2.11. As will be demonstrated later on, the structure of the molecules may help to understand carbon (and also oxygen) isotope partitioning in the system.

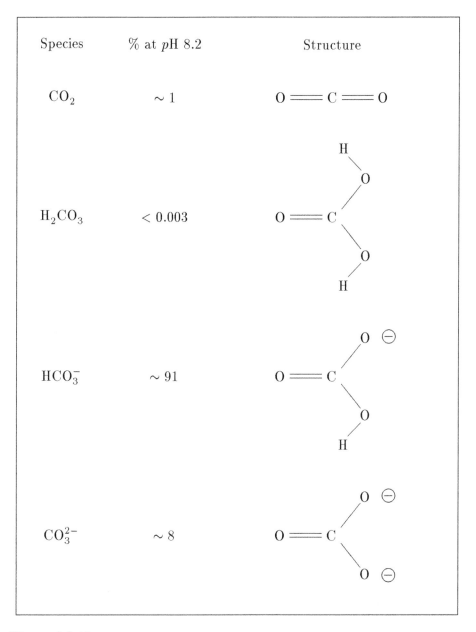

Species	% at pH 8.2	Structure
CO_2	~ 1	
H_2CO_3	< 0.003	
HCO_3^-	~ 91	
CO_3^{2-}	~ 8	

Figure 3.2.11: Chemical structures of the dissolved forms of CO_2. For structures of H_2CO_3 and HCO_3^-, see Nguyen and Ha (1984) and Davis and Oliver (1972). Note that bond angles, bond lengths and assignment of electrons is only schematic. Also given is the percentage (ca.) of each species with respect to the total dissolved carbon (ΣCO_2) in seawater at pH 8.2.

The chemical reactions of the carbonate species, which have been discussed in detail in Chapter 1 and Chapter 2, may be summarized by:

$$CO_2 + H_2O \underset{k_{-1}}{\overset{k_{+1}}{\rightleftharpoons}} H^+ + HCO_3^-$$

$$CO_2 + OH^- \underset{k_{-4}}{\overset{k_{+4}}{\rightleftharpoons}} HCO_3^- \tag{3.2.20}$$

$$CO_3^{2-} + H^+ \underset{k_{-5}}{\overset{k_{+5}}{\rightleftharpoons}} HCO_3^-$$

where the concentration of each chemical species represents the sum of the respective stable isotopes. The equilibrium constants of the reactions of the carbon compounds are:

$$K_1 = \frac{k_{+1}}{k_{-1}} = \frac{[H^+][HCO_3^-]}{[CO_2]}$$

$$\frac{K_1}{K_W} = \frac{k_{+4}}{k_{-4}} = \frac{[HCO_3^-]}{[CO_2][OH^-]} \tag{3.2.21}$$

$$K_2 = \frac{k_{-5}}{k_{+5}} = \frac{[CO_3^{2-}][H^+]}{[HCO_3^-]}$$

where K_1 and K_2 correspond to the first and second dissociation constant of carbonic acid (note that in this notation $K_2 = k_{-5}/k_{+5}$, not vice versa); K_W is the dissociation constant (or ion product) of water.

The concentrations of the carbonate species $[CO_2]$, $[HCO_3^-]$, and $[CO_3^{2-}]$ represent the sum of the stable carbon isotopes ^{13}C and ^{12}C within each species:

$$\begin{aligned} [CO_2] &= [^{12}CO_2] + [^{13}CO_2] \\ [HCO_3^-] &= [H^{12}CO_3^-] + [H^{13}CO_3^-] \\ [CO_3^{2-}] &= [^{12}CO_3^{2-}] + [^{13}CO_3^{2-}]. \end{aligned}$$

In order to describe the chemistry including stable isotopes, six carbonate species have to be considered which can be grouped into two compartments:

$$\boxed{\begin{array}{c} \text{I.} \\[4pt] {}^{12}\text{CO}_2 \quad \longleftrightarrow \quad \text{H}^{12}\text{CO}_3^- \quad \longleftrightarrow \quad {}^{12}\text{CO}_3^{2-} \end{array}}$$

$$\boxed{\begin{array}{c} \text{II.} \\[4pt] {}^{13}\text{CO}_2 \quad \longleftrightarrow \quad \text{H}^{13}\text{CO}_3^- \quad \longleftrightarrow \quad {}^{13}\text{CO}_3^{2-} \end{array}}$$

There are no direct reactions between the two compartments since each isotope is conserved during reaction. It is emphasized that there are two *independent* systems with regard to stable carbon isotopes. The chemical coupling of both systems is brought about by the chemical reactions of the carbon compounds with H^+ and OH^-. The chemical reactions including ${}^{13}\text{C}$ are:

$${}^{13}\text{CO}_2 + \text{H}_2\text{O} \underset{k'_{-1}}{\overset{k'_{+1}}{\rightleftharpoons}} \text{H}^+ + \text{H}^{13}\text{CO}_3^-$$

$${}^{13}\text{CO}_2 + \text{OH}^- \underset{k'_{-4}}{\overset{k'_{+4}}{\rightleftharpoons}} \text{H}^{13}\text{CO}_3^- \tag{3.2.22}$$

$${}^{13}\text{CO}_3^{2-} + \text{H}^+ \underset{k'_{-5}}{\overset{k'_{+5}}{\rightleftharpoons}} \text{H}^{13}\text{CO}_3^- .$$

Analogously to the equilibrium constants (3.2.21) the chemical equilibrium constants of the corresponding reactions of the ${}^{13}\text{C}$ species are:

$${}^{13}K_1 = \frac{k'_{+1}}{k'_{-1}} = \frac{[\text{H}^+][\text{H}^{13}\text{CO}_3^-]}{[{}^{13}\text{CO}_2]}$$

$$\frac{{}^{13}K_1}{K_{\text{W}}} = \frac{k'_{+4}}{k'_{-4}} = \frac{[\text{H}^{13}\text{CO}_3^-]}{[{}^{13}\text{CO}_2][\text{OH}^-]} \tag{3.2.23}$$

$${}^{13}K_2 = \frac{k'_{-5}}{k'_{+5}} = \frac{[{}^{13}\text{CO}_3^{2-}][\text{H}^+]}{[\text{H}^{13}\text{CO}_3^-]}$$

Isotopic fractionation factors (α) for the chemical reactions of the total carbon compounds (3.2.20) and the ${}^{13}\text{C}$ compounds (3.2.22) are given by the ratio of the equilibrium constants[6]:

$$\alpha_{(\text{CO}_2 - \text{HCO}_3^-)} = {}^{12}K_1 / {}^{13}K_1 = \left(\frac{[{}^{13}\text{CO}_2]/[{}^{12}\text{CO}_2]}{[\text{H}^{13}\text{CO}_3^-]/[\text{H}^{12}\text{CO}_3^-]} \right) \tag{3.2.24}$$

---

[6]The equilibrium constants for the sum of ${}^{12}\text{C}$ and ${}^{13}\text{C}$ (K_1) and for ${}^{12}\text{C}$ (${}^{12}K_1$) slightly

$$\alpha_{(CO_3^{2-}-HCO_3^-)} \;=\; {}^{13}K_2/{}^{12}K_2 = \left(\frac{[{}^{13}CO_3^{2-}]/[{}^{12}CO_3^{2-}]}{[H{}^{13}CO_3^-]/[H{}^{12}CO_3^-]} \right) \qquad (3.2.25)$$

The conversion between CO_2 and HCO_3^- occurs mainly via hydration (first reaction of (3.2.20)) at low pH and via hydroxylation (second reaction of (3.2.20)) at high pH. However, the isotopic ratio of CO_2 and HCO_3^- in equilibrium does not depend on the reaction path between the two species (thermodynamic constraint). Thus, the isotopic ratio of CO_2 and HCO_3^- in equilibrium is always given by (3.2.24).

The fractionation factor (3.2.24) can be interpreted as the equilibrium constant of the following reaction:

$$ {}^{12}CO_2 + H{}^{13}CO_3^- \;\;\rightleftharpoons\;\; {}^{13}CO_2 + H{}^{12}CO_3^- \;. $$

It should be kept in mind that this reaction actually represents the hydration/dehydration reaction of ${}^{13}CO_2$ and ${}^{12}CO_2$. In other words, carbon isotopes are not directly exchanged between the two molecules ${}^{12}CO_2$ and $H{}^{13}CO_3^-$ via chemical reaction. It is the equilibrium within each compartment, i.e. the reactions between ${}^{12}CO_2$, $H{}^{12}CO_3^-$, and ${}^{12}CO_3^{2-}$ on the one hand, and the reactions between ${}^{13}CO_2$, $H{}^{13}CO_3^-$, and ${}^{13}CO_3^{2-}$ on the other hand, which lead to isotopic fractionation.

In summary, the carbon isotope exchange between gaseous CO_2, the dissolved carbon species and $CaCO_3$ may be written as:

---

differ. The conversion between K_1 and ${}^{12}K_1$ for example is:

$$ K_1 = {}^{12}K_1 \, \alpha'_{(CO_2-HCO_3^-)} $$

with

$$ \alpha'_{(CO_2-HCO_3^-)} = \frac{1 - [{}^{13}CO_2]/[CO_2]}{1 - [H{}^{13}CO_3^-]/[HCO_3^-]} $$

The error using α instead of α' expressed in terms of $\delta^{13}C$ is in the order of 0.1‰.

$$^{12}CO_2(g) + H^{13}CO_3^- \;\rightleftharpoons\; ^{13}CO_2(g) + H^{12}CO_3^- \;;$$

$$\alpha_{gb} := \alpha_{(CO_2(g)-HCO_3^-)} \cdot$$

$$^{12}CO_2 + ^{13}CO_2(g) \;\rightleftharpoons\; ^{13}CO_2 + ^{12}CO_2(g) \;;$$

$$\alpha_{dg} := \alpha_{(CO_2-CO_2(g))} \cdot$$

$$^{12}CO_2 + H^{13}CO_3^- \;\rightleftharpoons\; ^{13}CO_2 + H^{12}CO_3^- \;;$$

$$\alpha_{db} := \alpha_{(CO_2-HCO_3^-)} \cdot$$

$$^{12}CO_3^{2-} + H^{13}CO_3^- \;\rightleftharpoons\; ^{13}CO_3^{2-} + H^{12}CO_3^- \;;$$

$$\alpha_{cb} := \alpha_{(CO_3^{2-}-HCO_3^-)} \cdot$$

$$Ca^{12}CO_3 + H^{13}CO_3^- \;\rightleftharpoons\; Ca^{13}CO_3 + H^{12}CO_3^- \;;$$

$$\alpha_{(CaCO_3-HCO_3^-)}$$

where g = gaseous CO_2, d = dissolved CO_2, b = bicarbonate ion (HCO_3^-), and c = carbonate ion (CO_3^{2-}).

3.2.3   Temperature dependence of fractionation factors

In the following, the values of the carbon isotope fractionation factors among the dissolved carbonate species are discussed. These values have been measured in fresh (or distilled) water rather than in seawater. As discussed by Zhang et al. (1995) the differences of the fractionation factors between $CO_2(g)$, CO_2, and HCO_3^- in fresh and in seawater are probably small. However, the fractionation between CO_3^{2-} and e.g $CO_2(g)$ might be significantly affected by ion complexes such as $MgCO_3^0$ and $CaCO_3^0$ (cf. also Thode et al. (1965)). This should be kept in mind when the fractionation factors summarized here are applied to seawater systems. A discussion of this subject and a theoretical calculation of the carbon isotope fractionation factor between CO_3^{2-} and $CO_2(g)$ are given in Section 3.5.3.

The fractionation factors between the carbonate species and gaseous CO_2 as given by Mook (1986), are ($\varepsilon = (\alpha - 1) \times 10^3$):

$$
\begin{aligned}
\varepsilon_{gb} &:= \varepsilon_{(CO_2(g)-HCO_3^-)} &&= -9483/T + 23.89\%_0 \\
\varepsilon_{dg} &:= \varepsilon_{(CO_2-CO_2(g))} &&= -373/T + 0.19\%_0 \\
\varepsilon_{db} &:= \varepsilon_{(CO_2-HCO_3^-)} &&= -9866/T + 24.12\%_0 \\
\varepsilon_{cb} &:= \varepsilon_{(CO_3^{2-}-HCO_3^-)} &&= -867/T + 2.52\%_0
\end{aligned}
\tag{3.2.26}
$$

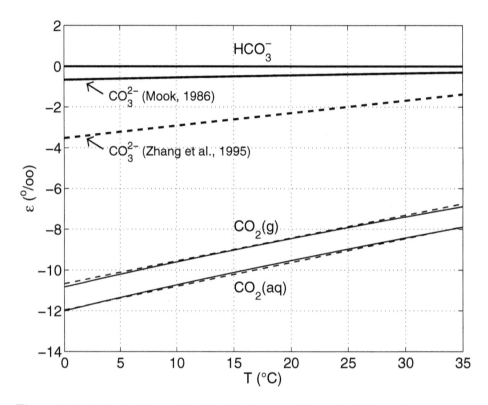

Figure 3.2.12: Carbon isotope fractionation between the species of the carbonate system as a function of temperature with respect to HCO_3^-. Values according to Mook (1986) and Zhang et al. (1995) are indicated by solid and dashed lines, respectively.

with T being the absolute temperature in Kelvin. The values of these fractionation factors as a function of temperature are displayed in Figure 3.2.12 (solid lines). It is interesting to note that $CO_2(g)$ is enriched in the heavy isotope ^{13}C relative to $CO_2(aq)$ as one might expect that the gaseous phase would be depleted in the heavy isotope relative to the dissolved phase.

The values of ε_{gb} and ε_{dg} have been measured by different authors, yielding quite similar results (ε_{gb}: Mook, 1974; Leśniak and Sakai (1989); Zhang et al., 1995; ε_{dg}: Vogel et al., 1970; Zhang et al., 1995; Szaran, 1998). For instance, the values and the temperature dependence of the fractionation factors given by Zhang et al. (1995):

$$\varepsilon_{bg} := \varepsilon_{(HCO_3^- - CO_2(g))} = -0.1141\,T_c + 10.78\text{‰} \tag{3.2.27}$$

$$\varepsilon_{dg} = \varepsilon_{(CO_2 - CO_2(g))} = +0.0049\,T_c - 1.31\text{‰}\,, \tag{3.2.28}$$

(where T_c is the temperature in °C) are very similar to those given by

Mook (1986) (Figure 3.2.12).[7] However, the value of the carbon isotope fractionation between carbonate ion and bicarbonate (ε_{cb}) as given by these authors differ by up to 2.5‰. The fractionation between CO_3^{2-} and $CO_2(g)$ given by Zhang et al. (1995) is:

$$\varepsilon_{cg} := \varepsilon_{(CO_3^{2-}-CO_2(g))} = -0.052\,T_c + 7.22‰ \, . \tag{3.2.29}$$

Using Eqs. (3.2.29) and (3.2.27), ε_{cb} according to Zhang et al. (1995) can be determined from:

$$\varepsilon_{cb} = \varepsilon_{(CO_3^{2-}-HCO_3^-)} = \frac{\varepsilon_{cg} - \varepsilon_{bg}}{1 + \varepsilon_{bg} \times 10^{-3}}$$

which is shown in Figure 3.2.12. It is obvious that compared to the good agreement between the values given for the fractionation between HCO_3^-, CO_2, and $CO_2(g)$, there is little agreement between the values given for the fractionation between e.g. HCO_3^- and CO_3^{2-}. The carbon isotope fractionation between carbonate ion and the other carbon species is discussed controversially in the literature which is elaborated in Section 3.5.3 (cf. Figure 3.5.38).

3.2.4 Fractionation between $CaCO_3$ and HCO_3^-

The carbon isotope fractionation between calcium carbonate and bicarbonate as reported by different authors is presented in Figure 3.2.13. For the analysis of the data it is useful to recall some features of the mineralogy of $CaCO_3$. Biogenic calcium carbonate in the ocean is mainly precipitated as calcite or aragonite. Whereas organisms such as planktonic foraminifera and coccolithophorids prefer to produce calcite, corals and pteropods prefer to build their shells of aragonite. The crystal structure of calcite is rhombohedral whereas the structure of aragonite is orthorhombic (for review cf. e.g. Hurlbut (1971) or Reeder (1983)). The different structures of both minerals lead to different physical and chemical properties such as density and solubility (Table 3.2.4) and also to different isotopic fractionation factors between the dissolved carbon in the water and the carbon in the calcium carbonate.

Rubinson and Clayton (1969) experimentally found an enrichment of ^{13}C in calcite and aragonite relative to bicarbonate of 0.9‰ and 2.7‰,

---

[7]Note that Zhang et al. report $\varepsilon_{(HCO_3^--CO_2(g))}$, whereas Mook reports $\varepsilon_{(CO_2(g)-HCO_3^-)}$. In general, the conversion is: $\varepsilon_{BA} = -\varepsilon_{AB}/(1 + \varepsilon_{AB}/10^3)$.

Table 3.2.4: Properties of calcite and aragonite.

Carbonate	Formula	Structure	Density $(g\ cm^{-3})$	$K_{sp}^* \times 10^7$ ^{a} $(mol^2\ kg^{-2})$
Calcite	$CaCO_3$	Rhombohedral	2.71	4.27
Aragonite	$CaCO_3$	Orthorhombic	2.95	6.48

^{a} K_{sp}^* is the stoichiometric solubility product in seawater $K_{sp}^* = [Ca^{2+}]_T[CO_3^{2-}]_T$, where e.g. $[CO_3^{2-}]_T$ refers to the equilibrium total (free + complexed) carbonate ion concentration. Values after Mucci (1983) at $T = 25°C$ and $S = 35$.

respectively (Figure 3.2.13). This finding has a theoretical basis (cf. Section 3.5). Since the vibrational frequencies of the carbonate ion in calcite are slightly different from the frequencies in aragonite, i.e. the energy of the carbonate ion within these minerals is different, carbon isotope fractionation between calcite and aragonite should be observed. Rubinson and Clayton (1969) calculated the aragonite-calcite fractionation to be 0.9‰ at 25°C.

Emrich et al. (1970) determined the carbon isotope fractionation between $CaCO_3$ and HCO_3^- at various temperatures (open squares in Figure 3.2.13). Unfortunately, since mineralogy was not controlled in their experiments it is not clear whether the observed temperature dependence is real or a result of the precipitation of different mixtures of aragonite and calcite.

Probably the most accurate study of carbon isotope fractionation in synthetic calcium carbonate was provided by Romanek et al. (1992). They examined the influence of precipitation rate and temperature on the ^{13}C fractionation in aragonite and calcite and checked mineralogy of the final precipitate by X-ray diffraction (triangles in Figure 3.2.13). Their reported difference in the carbon isotopic composition of calcite and aragonite is in good agreement with the findings of Rubinson and Clayton (1969) at 25°C. Romanek et al. (1992) concluded that the calcite-bicarbonate and aragonite-bicarbonate fractionation is essentially independent of temperature (over the range $10° - 40°C$) and precipitation rates ($\sim 10^{2.5} - 10^{4.5}$ $\mu mol\ m^{-2}\ h^{-1}$).

On the contrary, Turner (1982) reported a kinetic isotope fractionation effect during calcium carbonate precipitation as a function of the precipitation rate. Considering only those runs of Turner in which 100% calcite was precipitated, the effect as reported by Turner is on the order of 1.5‰ with isotopically lighter calcite produced at higher precipitation rates. Romanek et al. (1992) later recalculated Turner's values and argued against a kinetic

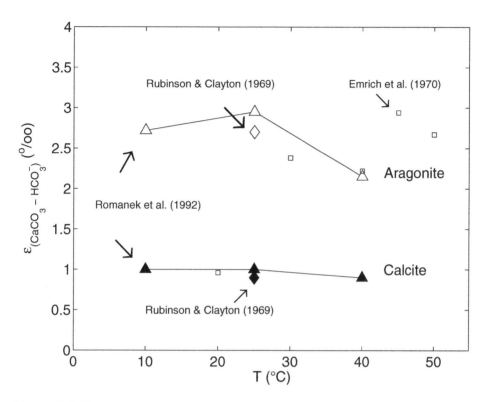

Figure 3.2.13: Carbon isotope fractionation between calcium carbonate and HCO_3^- as a function of temperature (solid symbols refer to calcite whereas open symbols refer to aragonite). The mineralogy of the samples of Emrich et al. (1970) (squares) is inconclusive since mineralogy was not controlled.

effect since the recalculated values did not correlate with precipitation rate.

The carbon isotope fractionation between aragonite and calcite as observed in synthetic $CaCO_3$ (Rubinson and Clayton, 1969; Romanek et al., 1992) is similar to fractionation effects observed in biogenic $CaCO_3$. Eisma et al. (1976) observed an enrichment of $1\%_0 - 2\%_0$ in the aragonitic part (relative to the calcitic part) of the shell of the euryhaline mollusc *Mytilus edulis* (euryhaline: tolerating a broad salinity range). Grossman and Ku (1986) showed that the aragonitic foraminifer *Hoeglundina elegans* and several aragonitic gastropods are heavier in ^{13}C by $0.5\%_0 - 2.5\%_0$ than the calcitic foraminifera *Uvigerina curticosta, U. flintii*, and *U. peregrina* (see also Grossman (1984)). It is emphasized, however, that $\delta^{13}C$ values of biogenic calcite and aragonite often show a wide range of scatter and also exhibit inconsistent fractionation relative to inorganically precipitated $CaCO_3$ (see Eisma et al. (1976); González and Lohmann (1985)). In addition, the

observed temperature dependence of the ^{13}C fractionation in biogenic aragonite of about $-2\permil$ per $20°C$ (Grossman and Ku, 1986) was not confirmed
in synthetic aragonite (Romanek et al., 1992).

The result that aragonite is heavier in ^{13}C than calcite is also in accordance with studies of inorganic carbonate cements. González and Lohmann
(1985) found a $1.0\permil$ to $1.4\permil$ enrichment in ^{13}C in inorganic marine aragonite cement compared to equilibrium calcite.

3.2.5 Carbon isotope partitioning as a function of *p*H

As shown in Section 3.2.3 there is considerable fractionation of carbon isotopes between the species of the carbonate system, i.e. for a given isotopic
composition of the total inorganic carbon (ΣCO_2), the isotopes are not distributed equally among CO_2, HCO_3^-, and CO_3^{2-}. Rather, the ratio of ^{13}C
to ^{12}C in HCO_3^- is, for instance, greater than in CO_2. It is obvious, however, that the sum of all the isotopes within the different chemical species
must be equal to the sum of the isotopes in ΣCO_2. Since the percentage of
CO_2, HCO_3^-, and CO_3^{2-} are a function of the *p*H of the solution, the carbon
isotopic composition ($\delta^{13}C$) of each carbonate species changes as a function
of *p*H. This behavior can easily be understood by considering the Bjerrum
plot (Figure 3.2.14a). At very low *p*H all dissolved inorganic carbon is es-
sentially CO_2 - thus the $\delta^{13}C$ of CO_2 is equal to $\delta^{13}C_{\Sigma CO_2}$. At very high *p*H
all dissolved inorganic carbon is essentially CO_3^{2-} - thus the $\delta^{13}C$ of CO_3^{2-} is
equal to $\delta^{13}C_{\Sigma CO_2}$. Since the fractionation between the different chemical
species is constant at a given temperature (i.e. the offset is constant, see
Figure 3.2.12), the $\delta^{13}C$ of each carbonate species changes as a function of
*p*H (Figure 3.2.14b).

In order to calculate the $\delta^{13}C$ of the carbonate species for a given value
of $\delta^{13}C_{\Sigma CO_2}$ we make use of a mass-balance relation. As was shown in
Section 3.1.5, the error introduced by replacing the fractional abundance r
by δ in the mass-balance equation for carbon is small. Thus, we can write:

$$\delta^{13}C_{\Sigma CO_2}\,[\Sigma CO_2]$$

$$= \delta^{13}C_{CO_2}[CO_2] + \delta^{13}C_{HCO_3^-}[HCO_3^-] + \delta^{13}C_{CO_3^{2-}}[CO_3^{2-}]\,.$$

Neglecting terms $\varepsilon \times 10^{-3}$, the $\delta^{13}C$ values of the carbonate species can be

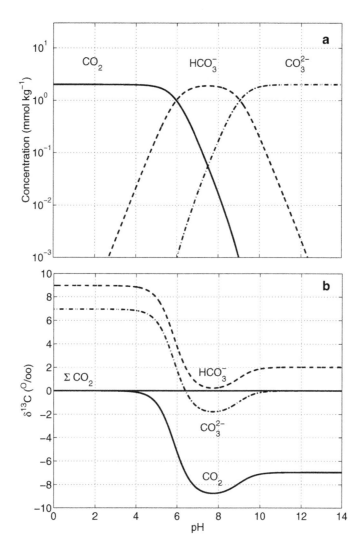

Figure 3.2.14: Carbon isotope partitioning in a closed seawater carbonate system as a function of pH. (a) Concentrations of the dissolved species at $\Sigma CO_2 = 2$ mmol kg^{-1}, $T = 25°C$, and $S = 35$. (b) $\delta^{13}C$ of the dissolved species for $\delta^{13}C_{\Sigma CO_2} = 0\%_0$. Fractionation factors as given by Zhang et al. (1995).

determined [8]:

$$\delta^{13}C_{HCO_3^-} = \delta^{13}C_{\Sigma CO_2} - \frac{\varepsilon_{db}[CO_2] + \varepsilon_{cb}[CO_3^{2-}]}{[\Sigma CO_2]} \qquad (3.2.30)$$

---

[8]The formula including terms $\varepsilon \times 10^{-3}$ is:

$$\delta^{13}C_{HCO_3^-} = \frac{\delta^{13}C_{\Sigma CO_2}[\Sigma CO_2] - (\varepsilon_{db}[CO_2] + \varepsilon_{cb}[CO_3^{2-}])}{(1 + \varepsilon_{db} \times 10^{-3})[CO_2] + [HCO_3^-] + (1 + \varepsilon_{cb} \times 10^{-3})[CO_3^{2-}]}$$

$$\delta^{13}C_{CO_2} = \delta^{13}C_{HCO_3^-}(1 + \varepsilon_{db} \times 10^{-3}) + \varepsilon_{db}$$

$$\delta^{13}C_{CO_2} = \delta^{13}C_{HCO_3^-} + \varepsilon_{db} \tag{3.2.31}$$

$$\delta^{13}C_{CO_3^{2-}} = \delta^{13}C_{HCO_3^-} + \varepsilon_{cb} \tag{3.2.32}$$

Figure 3.2.14b displays the $\delta^{13}C$ of the different carbonate species at a temperature of 25°C, $S = 35$ and a $\delta^{13}C_{\Sigma CO_2}$ of 0‰ using the fractionation factors as given by Zhang et al. (1995) (see Section 3.2.3). It is noted that the calculations presented here neglect the influence of ion complexes such as $MgCO_3^0$ on the fractionation among the dissolved carbonate species in seawater (cf. Section 3.5.3).

3.2.6   Kinetic ^{13}C fractionation in the carbonate system

As discussed in Section 3.1.3, a normal kinetic isotope effect occurs when the reaction rate of the reaction involving the light isotope is greater than that involving the heavy isotope. This is the case, for example, in the hydration of carbon dioxide:

$$CO_2 + H_2O \xrightarrow{k_{+1}} H^+ + HCO_3^- \ ,$$

leading to HCO_3^- that is lighter than CO_2 by about 13‰ at 24°C. Similarly, the dehydration step

$$H^+ + HCO_3^- \xrightarrow{k_{-1}} CO_2 + H_2O$$

results in lighter CO_2 of about 22‰ (O'Leary et al., 1992). These values were measured at 24°C (O'Leary, pers. comm. 1998). In the following they are, however, also used at 25°C because the temperature dependence has, to the best of our knowledge, not been determined so far. The difference between the forward and backward reaction equals the equilibrium fractionation between CO_2 and HCO_3^-, $\alpha_{(HCO_3^- - CO_2)}$, of about 9‰ (the equilibrium value given by Mook (1986) is 8.97‰ at 25°C). In general, the relationship between equilibrium fractionation and kinetic fractionation is mathematically expressed by:

$$\frac{K'}{K} = \alpha = \left(\frac{k'_+}{k'_-}\right)\left(\frac{k_-}{k_+}\right) \tag{3.2.33}$$

where K' and K are the equilibrium constants for the reaction involving ^{13}C and ^{12}C, respectively, α is the equilibrium fractionation factor, and $k'_\pm$

$$\delta^{13}C_{CO_3^{2-}} = \delta^{13}C_{HCO_3^-}(1 + \varepsilon_{cb} \times 10^{-3}) + \varepsilon_{cb}$$

The difference in e.g. $\delta^{13}C_{CO_2}$ using Eq. (3.2.30) instead of Eq. (3.2.31) is usually < 0.1‰.

and $k_{\pm}$ refer to the forward and backward reaction rate constant involving ^{13}C and ^{12}C, respectively. Considering the hydration of CO_2, for instance, the reaction involving ^{13}C and ^{12}C is:

$$^{13}CO_2 + H_2O \xrightarrow{k'_{+1}} H^+ + H^{13}CO_3^-$$

and

$$^{12}CO_2 + H_2O \xrightarrow{k_{+1}} H^+ + H^{12}CO_3^- \ ,$$

respectively. The kinetic isotope effect of this reaction is expressed by the ratio k_{+1}/k'_{+1} ($\equiv$ $^{12}k_{+1}/^{13}k_{+1}$):

$$\frac{k_{+1}}{k'_{+1}} \simeq 1.013 \qquad \text{at } 25°C$$

(O'Leary et al., 1992). The kinetic isotope effect of the backward reaction may be calculated from equilibrium fractionation:

$$\frac{k_{-1}}{k'_{-1}} = \alpha_{(HCO_3^- - CO_2)} \frac{k_{+1}}{k'_{+1}}$$

$$= 1.009 \times 1.013$$

$$\simeq 1.022 \qquad \text{at } 25°C \ .$$

Measured values of the kinetic isotope fractionation of ^{13}C in the carbonate system are summarized in Table 3.2.5. It is noted that the values for the hydroxylation of CO_2 as given by O'Leary (pers. comm. 1998), Siegenthaler and Münnich (1981), and Usdowski et al. (1982) differ by about 30‰. In order to clarify this obvious discrepancy, further work on the kinetic isotope effect of the hydroxylation of carbon dioxide in aqueous solution is desired.

Remark: ^{14}C

With regard to the isotopic fractionation of ^{14}C as opposed to the fractionation of ^{13}C, an interesting feature can be derived from first principles. The relative influence of the fractionation, whether equilibrium or kinetic, is twice as large for ^{14}C as it is for ^{13}C (see Bigeleisen (1952); Craig (1954); Mook (1980)). This feature may be expressed by:

$$\left(^{14}x - 1\right) \times 10^3 = 2 \left(^{13}x - 1\right) \times 10^3 \qquad (3.2.34)$$

where x represents the isotopic fractionation factor α in case of equilibrium or the ratio of e.g. reaction rate constants (with respect to $^{14}C/^{12}C$ and $^{13}C/^{12}C$) in case of kinetics. It follows that the reaction rates for compounds containing ^{14}C can immediately be calculated using Eq. (3.2.34) and the values given in Table 3.2.5.

Table 3.2.5: Kinetic ^{13}C fractionation in the carbonate system.

Reaction	Constants	$^{12}k/^{13}k$	ε (‰)
$CO_2 + H_2O \rightarrow H^+ + HCO_3^-$	k_{+1}/k'_{+1}	1.013^a	13^a
$H^+ + HCO_3^- \rightarrow CO_2 + H_2O$	k_{-1}/k'_{-1}	1.022^a	22^a
$CO_2 + OH^- \rightarrow HCO_3^-$	k_{+4}/k'_{+4}	1.011^b	11^b
		1.027^c	27^c
		1.039^d	39^d
$HCO_3^- \rightarrow CO_2 + OH^-$	k_{-4}/k'_{-4}	1.020^b	20^b

In the presence of
carbonic anhydrase (CA):

$CO_2 + H_2O \xrightarrow{CA} H^+ + HCO_3^-$	k_{+1}/k'_{+1}	1.001^e	1^e
$H^+ + HCO_3^- \xrightarrow{CA} CO_2 + H_2O$	k_{-1}/k'_{-1}	1.010^e	10^e

Air-sea gas exchange

Process	α	$^{13}C/^{12}C$	ε (‰)
$CO_2(g) \rightarrow CO_2(aq.)$	α_{as}	0.998^f	-2.0^f
		0.998^g	-2.1^g
$\Sigma CO_2 \rightarrow CO_2(g)$	α_{sa}	0.990^h	-10.3^h

[a] O'Leary et al. (1992), $T = 24°C$.
[b] O'Leary (pers. comm. 1998), $T = 24°C$.
[c] Siegenthaler and Münnich (1981), $T = 20°C$.
[d] Usdowski et al. (1982), $T = 18°C$.
[e] Paneth and O'Leary (1985), $T = 25°C$.
[f] Mook (1986), based on Inoue and Sugimura (1985) and Wanninkhof (1985).
[g] Zhang (1995).
[h] Mook (1986), $T = 20°C$. Note that α_{sa} strongly depends on temperature.

Exercise 3.11 (*)

Figure 3.2.12 shows the carbon isotope fractionation between CO_2, CO_3^{2-}, and HCO_3^- (in sequence from isotopically light to heavy). Considering a general rule, in which compound is the carbon atom bound most strongly?

Exercise 3.12 (***)
Determine the overall carbon isotope fractionation between a reference phase, say $CO_2(g)$, and ΣCO_2 at $pH \sim 0$, ~ 7, and ~ 14 and $T = 25°C$, i.e. assume the $\delta^{13}C$ of $CO_2(g)$ to be constant and calculate $\varepsilon_{(CO_2(g)-\Sigma CO_2)}$ (use the fractionation factors by Zhang et al. (1995)). Sketch the graph of $\varepsilon_{(CO_2(g)-\Sigma CO_2)}$ over the full pH range from 0 to 14.

There are many more interesting examples of the use of carbon isotopes in e.g. paleoceanography as well as many more interesting features of carbon isotope fractionation within the carbonate system. However, the elaboration of those topics is beyond the scope of this book. We proceed to the discussion of stable oxygen isotopes.

3.3 Oxygen

Stable oxygen isotopes contained in gases, water, minerals, and organic matter are used in studies of the ocean, atmosphere, cryosphere (ice), and the continents. Oxygen isotopes became very famous in the 1950s because the ratio of the stable isotopes $^{18}O/^{16}O$ preserved in the carbonate skeletons of fossil organisms provide a 'thermometer', allowing us to estimate the temperature of the earth in the geological past (Section 3.3.3). Biological aspects such as respiration and photosynthesis and their influence on, for example, the oxygen isotopic composition of atmospheric O_2 (the so-called Dole effect, Section 3.3.1) and on atmospheric CO_2 (e.g. Farquhar et al., 1993) are studied using stable oxygen isotopes. An understanding of many of these aspects demands comprehension of the oxygen isotope fractionation between different phases of the system under consideration, one example being the fractionation between water and the dissolved carbonate species.

Whereas the principles of carbon isotope partitioning among the dissolved carbonate species are rather simple (cf. Section 3.2.5), the principles of oxygen isotopes partitioning among the carbonate species and a comprehension of their isotopic composition with respect to water is more complicated. This subject will be discussed in detail in Section 3.3.5. The application of those principles leads to interesting consequences for the interpretation of oxygen isotope fractionation in foraminifera which has implications for paleothermometry (Section 3.3.7).

3.3.1 Natural variations

The natural variations of the oxygen isotopic composition of some terrestrial samples is summarized in Figure 3.3.15. The primary source of the water of the hydrological cycle is the ocean, which has an isotopic composition of about 0‰ relative to the V-SMOW scale (Vienna-Standard Mean Ocean Water, see Section 3.1.4). In general, the water vapor of the atmosphere is depleted in ^{18}O since $H_2^{16}O$ has a higher vapor pressure than $H_2^{18}O$. Snow accumulating on the ice sheets of Greenland and Antarctica - its origin is water vapor of the atmosphere - is heavily depleted in ^{18}O. To first order, this feature can be explained by a Rayleigh process (Section 3.1.6) in which the water of a cloud traveling from low to high latitudes is successively depleted in ^{18}O because the rain forming from that cloud is enriched in ^{18}O relative to the remaining isotopic composition of the water vapor in the cloud.

The analogous depletion of deuterium ($D = {^2H}$) in meteoric water leads to a linear relationship between $\delta^{18}O$ and δD because H_2O has a higher vapor pressure than HDO. The hydrogen isotopes are therefore fractionated in proportion to the oxygen isotopes. This relationship which is observed in the δD and $\delta^{18}O$ values of the global precipitation is called the meteoric water line (Figure 3.3.16). It may be written as (Craig, 1961):

$$\delta D = 8\ \delta^{18}O + 10\ .$$

Atmospheric oxygen has a fairly constant $\delta^{18}O$ value of $\sim 23.5‰$ with respect to the V-SMOW scale. The difference between the $\delta^{18}O$ of atmospheric O_2 and the $\delta^{18}O$ of contemporaneous seawater is called the Dole effect (see Section 3.3.2). Organic matter has an oxygen isotope composition that ranges between $\sim 15‰$ and $\sim 35‰$ (e.g. Epstein et al., 1977; Galimov, 1985). Marine carbonates mainly reflect the isotopic composition of calcium carbonate precipitated in isotopic equilibrium with seawater at temperatures between approximately 0° and 30°C (see Section 3.3.4).

Atmospheric CO_2

Atmospheric CO_2 has a $\delta^{18}O$ value very close to the value expected in thermodynamic equilibrium with water (41.2‰ at 25°C, Friedman and O'Neil, 1977). Deviations from equilibrium with the ocean are particularly observed at high latitudes in the Northern Hemisphere where atmospheric CO_2 is lighter by some 4‰ than the value expected in equilibrium with the surface ocean (Francey and Tans, 1987). This feature has been explained by equilibration of atmospheric CO_2 with leaf water in terrestrial plants (Farquhar et al., 1993). Briefly, precipitation and soil water is most depleted

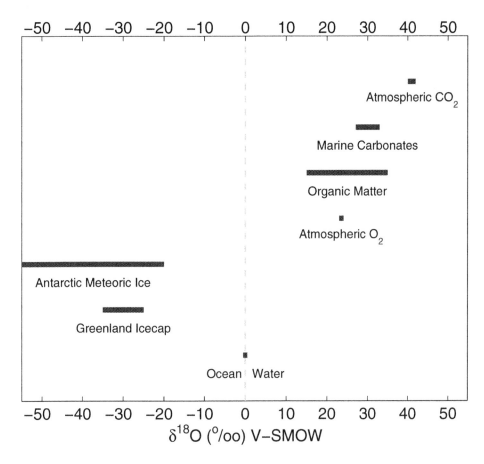

Figure 3.3.15: Natural variations in oxygen isotopic composition of some terrestrial compounds.

in ^{18}O at high latitudes which leads to ^{18}O depletion of atmospheric CO_2 after exchange with the leaf water. This can be explained as follows.

Roughly only half of the CO_2 entering the leaf cells is fixed into organic matter, whereas the other half is leaking back out to the atmosphere after exchanging oxygen atoms with the leaf water. The equilibration between CO_2 and the leaf water is catalyzed by the enzyme carbonic anhydrase which dramatically speeds up the hydrolysis reaction. This mechanism suggests an exchange time of oxygen atoms of the atmospheric CO_2 pool of only 2 years, whereas the exchange time with the ocean is in the order of 10 years. The 'light CO_2' (light with respect to the expected $\delta^{18}O$ value of CO_2 in equilibrium with the ocean) at high northern latitudes can therefore be

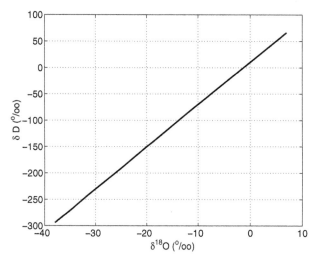

Figure 3.3.16: Meteoric water line: $\delta D = 8\ \delta^{18}O+10$ (Craig, 1961), reflecting the relationship between δD and $\delta^{18}O$ of the global precipitation.

explained by exchange of atmospheric CO_2 with the terrestrial biosphere.

The asymmetry in oxygen isotope composition of CO_2 between the northern and southern hemisphere is a result of the distribution of ocean and continents. At high southern latitudes the $\delta^{18}O$ of atmospheric CO_2 is dominated by the ocean. The details of the processes involved are rather complicated, including fractionation effects such as ^{18}O enrichment of leaf water relative to soil water by evapotranspiration because $H_2^{18}O$ evaporates less readily than $H_2^{16}O$ (cf. also Ciais et al., 1997).

3.3.2   The Dole effect

The difference between the $\delta^{18}O$ of atmospheric O_2 and the $\delta^{18}O$ of contemporaneous seawater is defined as the Dole effect (Figure 3.3.17). Atmospheric O_2 has a constant oxygen isotopic composition of $\sim 23.5‰$ with respect to the V-SMOW scale (Dole et al., 1954; Kroopnick and Craig, 1972). The major processes affecting this value on a time scale of 1,200 years (the residence time of atmospheric O_2) are photosynthesis and respiration of the terrestrial and marine biosphere (Bender et al., 1994). During photosynthesis CO_2 and H_2O are consumed whereas molecular oxygen is produced, originating from the water. The isotopic composition of the O_2 is therefore very close to that of the water at the site of carboxylation. Respiration, on the other hand, utilizes O_2, whereas CO_2 is released. It has been shown that the O_2 consumed during respiration is depleted in the heavier isotope ^{18}O by some 20‰ (Guy et al., 1993; Kiddon et al., 1993).

Bender et al. (1994) reconsidered the Dole effect, using estimates of isotope fractionation during photosynthesis and respiration based on much better data than was available to e.g. Lane and Dole (1956). Balancing today's photosynthesis and respiration on a global basis and taking into account the isotopic fractionation associated with those fluxes should give an estimate of the $\delta^{18}O$ of atmospheric O_2. The values for production and uptake of O_2 and associated fractionations as used by Bender et al. (1994) are summarized in Figure 3.3.17. A mass-balance calculation, estimating the Dole effect is given in the box on page 191.

The calculated $\delta^{18}O$ value of atmospheric O_2 is 20.8‰, whereas the observed value is 23.5‰. As pointed out by Bender et al. (1994), the largest uncertainty probably lies in the $\delta^{18}O$ of leaf water, which determines the $\delta^{18}O$ of O_2 produced during terrestrial photosynthesis. Fractionation effects of processes such as evapotranspiration (see preceding section) are difficult to estimate on a global scale. Assuming a $\delta^{18}O$ of leaf water of 8.5‰ instead of 4.4‰ (Eq. 3.3.35), one can explain the observed value of the Dole effect.

Using the stable oxygen isotopes ^{18}O, ^{17}O, and ^{16}O, and their fractionation associated with the Dole effect, Luz et al. (1999) recently proposed a method to reconstruct the productivity of the biosphere in the past. Their approach utilizes differences between mass-dependent and mass-independent fractionation effects. For most chemical reactions the fractionation effect for ^{18}O is twice as large as the corresponding effect for ^{17}O - this is called mass-dependent. On the other hand, in some photochemical reactions the fractionation of ^{18}O and ^{17}O is nearly the same which is called mass-independent. Interestingly, the fractionations of oxygen atoms associated with photosynthesis and respiration are mass-dependent, whereas the isotope fractionation due to stratospheric ^{18}O exchange between O_2 and CO_2 is mass-independent. Changes in the relative contributions of these processes therefore lead to changes in the isotopic composition of atmospheric O_2 with respect to the mass numbers 16, 17, and 18. Based on this mechanism Luz et al. (1999) estimated that global biosphere productivity varied between a minimum value of 89% and a maximum value of 97% during the time interval from the last interglacial to the last glacial period, relative to today's value of 100%.

Exercise 3.13 (*)

Why is Antarctic meteoric ice partly more depleted in ^{18}O than ice on the Greenland icecap?

Exercise 3.14 (*)

Which processes control the $\delta^{18}O$ of atmospheric O_2 on a time-scale of 10^3 years?

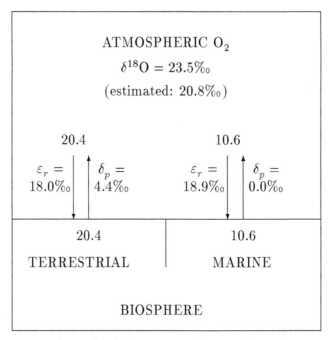

Mass-Balance (including stratospheric exchange):

$$0 = [\delta_p - (\delta_A - \varepsilon_p)] - 0.4\%_0$$

Inserting terrestrial and marine contributions:

$$0 = [4.4\%_0 - (\delta_A - 18.0\%_0)] \times 20.4$$
$$+ [0.0\%_0 - (\delta_A - 18.9\%_0)] \times 10.6 - 0.4\%_0$$

Figure 3.3.17: Illustration of the Dole effect. Numbers refer to gross production (photosynthesis) and uptake (respiration) of O_2 in 10^{15} mol yr^{-1}. ε_r is the fractionation during respiration, δ_p is the $\delta^{18}O$ value of O_2 produced during photosynthesis (V-SMOW scale), and δ_A is the $\delta^{18}O$ of atmospheric O_2. Solving the mass-balance (cf. box on page 191), yields $\delta_A = 20.8\%_0$. The term $-0.4\%_0$ in the mass-balance arises from stratospheric oxygen exchange between O_2 and CO_2.

Estimating the Dole effect. The concentration of atmospheric O_2 is constant over approximately 1,200 yr (= residence time of O_2, Bender et al., 1994). Considering smaller time scales, one may write:

$$\frac{d[O]}{dt} = F_p - F_r = 0$$

where [O] is the total number of oxygen atoms in atmospheric O_2 and F_p and F_r are the fluxes of oxygen through photosynthesis and respiration, respectively. Since $[^{18}O]$ may always be approximated by $[^{18}O] = R[O]$, where $R = [^{18}O]/[^{16}O]$, a similar equation holds for ^{18}O:

$$\frac{d[^{18}O]}{dt} = R_p F_p - R_r F_r = 0$$

where R_p and R_r refer to the isotopic ratios of the fluxes of oxygen through photosynthesis and respiration, respectively. Since long-term fluxes of photosynthesis and respiration must be equal, it follows that the isotopic ratios of those fluxes must be equal also:

$$R_p = R_r \qquad \text{or} \qquad \delta_p = \delta_r \ .$$

The δ value of consumed oxygen during respiration is about $18 - 19\%_0$ less than that of the source O_2, which is the atmospheric $\delta^{18}O$ (δ_A) here:

$$\delta_r \approx \delta_A - \varepsilon_r$$

where $\varepsilon_r \approx 18 - 19\%_0$. Finally, the isotopic composition of atmospheric O_2 is given by:

$$\delta_A = \delta_p + \varepsilon_r \ .$$

The difference between the $\delta^{18}O$ of atmospheric O_2 and the $\delta^{18}O$ of contemporaneous seawater is defined as the Dole effect. On the time scale considered, the Dole effect is therefore almost completely controlled by biology. Global values for δ_p and ε_r are given by the individual values of the terrestrial and marine contribution multiplied by their relative fluxes (cf. Figure 3.3.17):

$$\delta_p = (4.4\%_0 \times 20.4 + 0.0\%_0 \times 10.6)/31.0 = 2.9\%_0 \tag{3.3.35}$$
$$\varepsilon_r = (18.0\%_0 \times 20.4 + 18.9\%_0 \times 10.6)/31.0 = 18.3\%_0 \ . \tag{3.3.36}$$

Using a value of $-0.4\%_0$ for stratospheric exchange between O_2 and CO_2, one obtains:

$$\delta_A = 20.8\%_0$$

which is only approximately equal to the observed value of $23.5\%_0$. The difference could be explained by an underestimation of the $\delta^{18}O$ of leaf water (terrestrial O_2 production) which might by higher than $4.4\%_0$.

3.3.3 Paleotemperature scale

Since the pioneering work of Urey (1947), McCrea (1950), and Epstein et al. (1953), the use of stable oxygen isotope ratios ($^{18}O/^{16}O$) as a paleo-

temperature indicator has become a standard tool in paleoceanography. The
isotopic composition of sedimentary marine carbonates, particularly that of
foraminiferal calcite ($CaCO_3$) is widely used to reconstruct the temperature
of ancient oceans (e.g. Emiliani, 1955; Shackleton and Opdyke, 1973, Miller
et al., 1987; Zachos et al., 1994). Based on the theoretical prediction by Urey
(1947) that the fractionation of oxygen isotopes between calcium carbonate
and water should vary with the temperature of the water, McCrea (1950)
determined the first paleotemperature equation experimentally. Briefly, he
measured the oxygen isotopic composition of slowly formed calcium carbonate in Florida and Cape Cod water at temperatures between $-1.2°$ and
$79.8°C$. He demonstrated that the $\delta^{18}O$ of the precipitated $CaCO_3$ varied
systematically by about 14‰ over the considered temperature range. The
equation given by McCrea may be written as (see Bemis et al. (1998) for
summary):

$$T(°C) = 16.0 - 5.17 \, (\delta_c - \delta_w) + 0.092 \, (\delta_c - \delta_w)^2$$

where δ_c represents the $\delta^{18}O$ of CO_2 released from the calcite after reaction with phosphoric acid, and δ_w represents the $\delta^{18}O$ of CO_2 equilibrated
with the water at $25°C$. (Figure 3.3.18). Concluding his paper, McCrea
states: "The isotopic composition of calcium carbonate slowly formed from
aqueous solution has been noted to be usually the same as that produced by
organisms at the same temperature. As a result a relation giving the growth
temperature in terms of isotopic composition has been obtained, and with
certain restrictions it may be used to determine paleotemperatures". The
carbonate paleotemperature scale was invented.

McCrea (1950) also invented the technique to determine the isotopic
composition of calcium carbonate which is still a standard method in today's laboratories. Calcium carbonate is reacted with 100% phosphoric
acid (H_3PO_4) and the liberated CO_2 is analyzed on a mass spectrometer
(cf. Section 3.1.4). Wachter and Hayes (1985) demonstrated that the phosphoric acid must be carefully prepared in order to minimize oxygen isotope
exchange between CO_2 and phosphoric acid. They concluded that phosphoric acids with nominal concentrations of H_3PO_4 approaching 105% are
preferable for the phosphorolyses of carbonate minerals.

Epstein et al. (1953) determined the $\delta^{18}O$ of calcium carbonate in marine
shells and found that the relationship between oxygen isotopic composition
of the shell and growth temperature agreed essentially with the relationship determined by McCrea (1950). They also showed that the relative ^{18}O

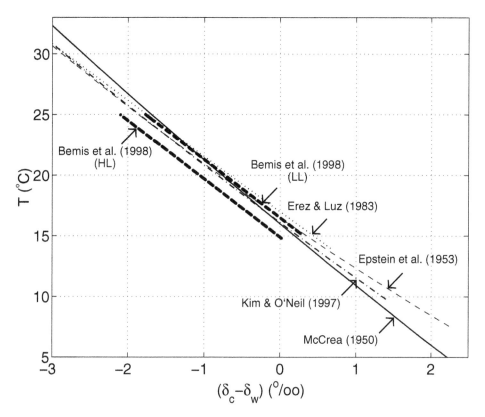

Figure 3.3.18: Paleotemperature equations as given by various authors (after Bemis et al. (1998)). Curves are shown in the temperature range where the respective experimental values had been obtained. Inorganic: McCrea (1950) and Kim and O'Neil (1997). Mollusca: Epstein et al. (1953). Foraminifera: Erez and Luz (1983) and Bemis et al. (1998) (HL: high light conditions, LL: low light conditions).

contents of marine waters increase with salinity which is mainly a result of evaporation/precipitation of the water of ocean and atmosphere, continental runoff and addition of polar meltwater. (The details are much more complicated, including processes such as water mass transport, mixing, and sea ice formation, e.g. Epstein and Mayeda (1953); Craig and Gordon (1965); Broecker (1974)). One important aspect of this finding is that the reliability of paleotemperature determinations from isotopic compositions of marine calcium carbonate decreases. Since the *difference* in the isotopic composition of the water and $CaCO_3$ enters the paleotemperature equation one should know the isotopic composition of the water in which the calcium carbonate was formed. Unfortunately, the isotopic composition of ancient ocean water can only be estimated.

For instance, during glacial periods a large amount of ocean water is deposited as ice on the continents, resulting in an increase of salinity and $\delta^{18}O$ of the ocean, so-called ice volume effect. The enrichment of ^{18}O in the ocean during glaciations is due to the fact that water vapor of the atmosphere - which is ultimately deposited as ice on the ice sheets - is depleted in ^{18}O because $H_2^{16}O$ has a higher vapor pressure than $H_2^{18}O$. Thus, the accumulation of ^{18}O depleted H_2O during expansion of continental ice sheets results in a relative enrichment of the ^{18}O of the ocean. It has been estimated that the $\delta^{18}O$ of the ocean varied by $0.4 - 1.1\%_0$ during glacial-interglacial times in the Pleistocene (Emiliani and Shackleton, 1974). At the last glacial maximum (LGM) at about 18,000 yr B.P. sea level is thought to have been ca. 120 m lower than today (Figure 3.3.19) and $\delta^{18}O$ of the ocean was about $1.25\%_0$ higher than today (e.g. Fairbanks, 1989; Rohling et al., 1998). Note that based on pore fluid measurements, Schrag et al. (1996) suggested a smaller $\delta^{18}O$ value of $1.0\%_0$ for the global ocean during the LGM.

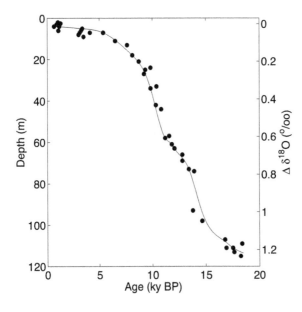

Figure 3.3.19: Sea level curve and seawater $\delta^{18}O$ as derived from Barbados corals for the time period from 18,000 yr B.P. until today (after Fairbanks, 1989).

Emiliani (1955) analyzed the stable oxygen isotopic composition of planktonic foraminifera from Atlantic, Caribbean, and Pacific deep-sea cores and concluded that the temperature of the equatorial Atlantic and Caribbean had changed periodically by about 6°C during the last 280,000 years. Since then isotopic paleotemperatures derived from foraminiferal calcite have become a standard tool in paleoceanography. As already said, it is not possible to distinguish between temperature and salinity effects on the $\delta^{18}O$ of foraminiferal calcite from the sedimentary record. Hence, changes of $\delta^{18}O$ in the past may be interpreted as temperature variations and/or variations of the ice volume of the earth and are therefore equivocal. Nevertheless, these changes are indicators of climate variability in the past, irrespective of whether temperature or ice volume changes caused the variations of the oxygen isotopic composition in the fossil record.

Emiliani (1955) had applied the paleotemperature scale as given by Epstein et al. (1953) to planktonic foraminifera, although this equation had been derived from the isotopic composition of mollusca. It was not until 1983 that a paleotemperature equation was determined for a planktonic foraminifer (*Globigerinoides sacculifer*) in the laboratory by Erez and Luz (1983) (see Figure 3.3.18). Laboratory studies and field calibrations with the planktonic foraminifer *Orbulina universa* were made by Bouvier-Soumagnac and Duplessy (1985), whereas the most recent paleotemperature equations for the planktonic foraminifera *Orbulina universa* and *Globigerina bulloides*, respectively, determined in the laboratory was given by Bemis et al. (1998). The latter paper also contains a nice summary of published paleotemperature equations.

As mentioned before, paleotemperature equations have been determined by various authors - see Figure 3.3.18 which displays some of those equations in the temperature range where the respective experimental values have been obtained. McCrea (1950) and Kim and O'Neil (1997) analyzed inorganic calcite, whereas Epstein et al. (1953) analyzed mollusca. The equation of Erez and Luz (1983) and Bemis et al. (1998) refer to the planktonic foraminifera *G. sacculifer* and *O. universa*, respectively. The curves of Bemis et al. (1998) for high light conditions (HL) and low light conditions (LL) are quite different. Obviously, the oxygen isotopic composition of the calcite is affected by life processes of the organism that change when the foraminiferal calcite is produced under high light or low light conditions. Those phenomena are called 'vital effects'. In the present case the depletion of ^{18}O in the shell under high light conditions might be attributed to a local increase of pH due to symbiont photosynthesis which is consistent with the effect of higher pH values of the ambient medium (Spero et al., 1997) which can be understood on a theoretical basis (Zeebe (1999), cf. Section 3.3.7).

Drawbacks of ^{18}O paleothermometry

Vital effects significantly influence stable oxygen and carbon isotope ratios in foraminifera. For example, the enrichment of ^{13}C in the microenvironment of the foraminifer during photosynthesis of the symbiotic algae of *O. universa* has been shown to enrich the shell calcite in ^{13}C by about 1.5‰ in high light compared to dark conditions (Spero and Williams, 1988). Moreover, laboratory experiments demonstrated that the interaction between the life processes of the organism and the seawater carbonate chemistry may lead to fractionation effects up to 1.5‰ and 4‰ in oxygen and carbon, respectively (Spero et al., 1997). The underlying mechanisms of fractionation effects of carbon and oxygen isotopes associated with the seawater chemistry and vital effects in foraminifera are discussed in detail in Wolf-Gladrow et al. (1999a), Zeebe et al. (1999a), and Zeebe (1999).

The interpretation of stable oxygen isotope values as a temperature proxy is also complicated by diagenesis: the isotopic alteration of primary oxygen in the sedimentary carbonates over time. Calcite or aragonite shells may dissolve and recrystallize during diagenesis and their oxygen isotopic composition may be altered (Killingley, 1983). Considering paleoreconstructions based on foraminifera it is of great importance to use unaltered shells (e.g. Shackleton, 1984; Zachos et al., 1994). Another aspect that influences oxygen isotope fractionation between carbonates and water is the solution chemistry which is described in detail in Section 3.3.5. This effect which has widely been overlooked was thoroughly investigated by Usdowski et al. (1991) and Usdowski and Hoefs (1993). Zeebe (1999) demonstrated that oxygen isotope effects associated with the solution chemistry in both foraminiferal calcite and in synthetic carbonates can be explained on the theoretical basis provided by Usdowski and co-workers (Section 3.3.7 and Section 3.3.8).

In spite of the differences between the $CaCO_3 - H_2O$ systems for which the paleotemperature scales shown in Figure 3.3.18 were derived, temperatures based on $\delta^{18}O$ of inorganic $CaCO_3$ in fresh water (McCrea (1950); Kim and O'Neil (1997)) and biogenic $CaCO_3$ in seawater (Epstein et al. (1953); Erez and Luz (1983); Bemis et al. (1998) (low light)) show good agreement around 20°C. *A priori* this is not to be expected due to (1) differences in the carbonate system of fresh water and seawater (see Section 3.3.5), (2) organic and inorganic precipitation of $CaCO_3$ (cf. discussion of 'vital effects'), (3) different animals and species (mollusca, foraminifera), and (4) differences in mineralogy: whereas planktonic foraminifera produce calcite, most mollusca produce aragonite; inorganic studies refer to calcite. Tarutani et al. (1969) reported an enrichment of ^{18}O in inorganic aragonite relative to calcite of 0.6‰ at 25°C. The good agreement of published paleotemperature equations at about 20°C, which is nearly independent of mineralogy and the organism considered, supports the validity of this paleoproxy in this temperature range.

The isotopic paleothermometer has its advantages and its drawbacks. It is a powerful tool which has provided an irreplaceable contribution to our current knowledge of climate variability in the past. On the other hand, it suffers from several problems which make necessary additional assumptions about physiological parameters of the considered organism and about the environment in which the organism lived.

Remark: Other temperature proxies, the CLIMAP data

Maps of paleotemperature reconstructions for the last glacial maximum produced by the CLIMAP project (Climate/Long-Range Investigation, Mapping and Prediction) (CLIMAP, 1976; 1981) have widely been used as the basis for paleotemperature estimates for the past 20 years. The charts of sea surface temperatures (SSTs) for the LGM were compiled using biological transfer functions and oxygen isotope stratigraphy. The transfer function method is based on biotic assemblages of foraminifera, coccolithophorids, and radiolaria which reflect the distribution of surface water masses (Imbrie and Kipp, 1971). A numerical description of the planktonic biota is then translated into estimates of past SSTs. The stable oxygen isotope composition of calcareous shells was used for stratigraphy and was checked against ^{14}C dates (CLIMAP, 1976).

The CLIMAP data are in part controversial (e.g. Guilderson et al., 1994, Hostetler and Mix, 1999; Bard, 1999). For instance, CLIMAP data indicate that large areas of the subtropical Pacific Ocean were warmer during the last glacial period than today. On the contrary, Lee and Slowey (1999) recently analyzed oxygen isotopic compositions and species assemblages of planktonic foraminifera in a shallow-water core near Hawaii, indicating that annual average SST was ca. 2 K cooler during the last glaciation than today. Similarly, estimates of SST utilizing the more recent temperature proxy $U_{37}^{K'}$ show clear deviations from the estimates of CLIMAP for the last glacial maximum (e.g. Rosell-Melé et al., 1998). The $U_{37}^{K'}$ proxy is based on the relative abundance of double bonds in alkenones (ketones with 37 carbon atoms) which are produced by marine algae such as the coccolithophorid *Emiliania huxleyi*. The relative abundance of the double bonds is related to growth temperature - at warm SSTs the $U_{37}^{K'}$ value is close to unity, whereas at low temperatures this ratio is close to zero (the number of biosynthesized molecules with three double bonds is increasing at low temperatures).

The CLIMAP data are currently revised by the EPILOG Project (Environmental Processes of the Ice Age: Land, Ocean, and Glaciers). Revised charts produced by EPILOG will include (among other proxy estimates) sea surface temperature estimates derived from planktonic foraminifera census counts via the modern analog technique and from diatom and radiolarian counts using transfer functions (for comparison of those tools, see e.g. Ortiz and Mix (1997)). The results indicate that tropical SSTs during the LGM as given by CLIMAP are probably too high by up to $3 - 4$ K in some regions (Sarnthein and the Working Group GLAMAP-2000, 1999). Geochemical paleoproxies such as Sr/Ca ratios in corals (e.g. Guilderson et al., 1994; McCulloch et al., 1999) or Mg/Ca ratios in foraminifera (Nürnberg et al., 1996; Hastings et al., 1998; Mashiotta et al., 1999) are also promising candidates for paleotemperature reconstructions.

3.3.4   Temperature dependence of ^{18}O fractionation between H_2O, CO_2, and $CaCO_3$

As discussed in Section 3.3.3, the oxygen isotope fractionation between H_2O and $CaCO_3$ depends on the temperature at which the carbonate is formed. This behavior is a general feature of isotopic fractionation which can be understood on the basis of thermodynamics (Section 3.5). Figure 3.3.20 shows the temperature dependence of the ^{18}O fractionation between H_2O, CO_2, and $CaCO_3$ - the equations used to calculate the values are given in Table 3.3.6. Water vapor is depleted in ^{18}O relative to liquid water because $H_2^{16}O$ has a higher vapor pressure than $H_2^{18}O$. Gaseous carbon dioxide and $CaCO_3$ are both enriched in ^{18}O relative to liquid water, where $CO_2(g)$ is the isotopically heavier species.

The fractionation factors between H_2O and $CO_2(g)$ (Brenninkmeijer et al., 1983) and H_2O and calcite (Kim and O'Neil, 1997) were determined using deionized water. The fractionation factors obtained for deionized water and for seawater are obviously not very different. The $\delta^{18}O$ of atmospheric CO_2 in equilibrium with the seawater of the oceans is close to the laboratory value for deionized water (Francey and Tans, 1987). An analogous statement apparently holds for the oxygen fractionation between deionized/sea-water and $CaCO_3$ (cf. Figure 3.3.18).

Table 3.3.6: Equations used to calculate oxygen isotope fractionation factors.

Compounds	Equation (T in Kelvin)	$\alpha(25°C)$
$H_2O(l)$-$H_2O(v)$	$\ln \alpha = \frac{1.137 \times 10^3}{T^2} - \frac{0.4156}{T} - \frac{2.0667}{10^3}$ [a]	1.0094
$CO_2(g)$-$H_2O(l)$	$\alpha = \frac{17.604}{T} + 0.98211$ [b]	1.0412
$CaCO_3$ (calcite) - $H_2O(l)$	$\ln \alpha = \frac{18.03}{T} - 0.03242$ [c]	1.0285

[a] Majoube, 1971.

[b] Brenninkmeijer et al., 1983.

[c] Kim and O'Neil, 1997.

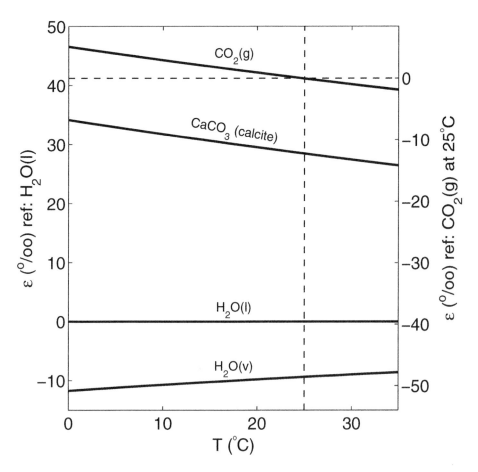

Figure 3.3.20: Oxygen isotope fractionation between H_2O, $CO_2(g)$, and $CaCO_3$ as a function of temperature. Values are calculated using the equations given in Table 3.3.6 (note that $\varepsilon = (\alpha - 1) \times 10^3$). $H_2O(l)$ and $H_2O(v)$ refer to liquid water and water vapor, respectively.

Exercise 3.15 (*)

Considering the paleotemperature equation given by McCrea (1950), how large is the uncertainty in the estimated temperature if the uncertainty in δ_w is 1‰?

Exercise 3.16 (**)

It has been demonstrated that the $\delta^{18}O$ of foraminiferal calcite is decreasing with increasing seawater pH (ca. -1.42‰ per pH unit, Zeebe (1999; 2001b)). Assume that the pH of the ocean was 0.5 units lower than today at some time in the past. What follows for the temperature estimates based on foraminiferal oxygen isotopes for this time period?

3.3.5   Equilibrium ^{18}O fractionation in the carbonate system

In comparison to the carbon isotope partitioning in the system $CO_2 - H_2O$ (cf. Figure 3.2.14, page 181), the comprehension of the oxygen isotope partitioning in the system requires a little more thought. Firstly, in natural waters, the vast majority of oxygen atoms of the system are contained in the water - and not in the carbon compounds H_2CO_3, CO_2, HCO_3^-, and CO_3^{2-} (Figure 3.3.21). Considering oxygen isotope exchange between water and the carbonate species, the isotopic composition of the water can be assumed constant which in turn means that independent of pH the carbonate species equilibrate with the water without changing its isotopic composition. For instance, as the dominant species in solution changes from CO_2 to HCO_3^- as pH increases, the water is marginally enriched in ^{18}O because CO_2 is isotopically heavier than HCO_3^-. The enrichment is, however, virtually equal to zero: ca. 0.0009‰ for typical seawater conditions. Secondly, there are up to three oxygen atoms in a single compound, whereas there is only one carbon atom in a single compound. This feature makes possible a variety of combinations of the stable oxygen isotopes ^{18}O and ^{16}O in a single molecule. Considering carbon dioxide, for example, the combinations $C^{18}O^{18}O$, $C^{18}O^{16}O$, and $C^{16}O^{16}O$ are possible (including ^{13}C and ^{17}O, there are even more, see Table 3.3.7). Thirdly, since oxygen isotope exchange between the reservoir (H_2O) and the carbon compounds exclusively occurs via CO_2 (see below), also carbonic acid has to be taken into account. Although the concentration of H_2CO_3 is very small and can be neglected in the mass balance, it is important in this context because it is involved in the oxygen isotope equilibrium via hydration of CO_2.

One goal of the discussion of isotope partitioning in a chemical system is to obtain values for the fractionation between the different chemical species of the system. In the case of oxygen isotopes one might be interested in the isotopic fractionation between water (reference phase) and the dissolved carbonate species (H_2CO_3, CO_2, HCO_3^-, CO_3^{2-}), that is, the ratio of the total number of oxygen-18 and oxygen-16 atoms in a certain carbon compound, divided by the corresponding ratio in water - which reads for CO_2:

$$\alpha_{(CO_2-H_2O)} = \frac{2[C^{18}O^{18}O] + [C^{18}O^{16}O]}{2[C^{16}O^{16}O] + [C^{18}O^{16}O]} \bigg/ \frac{[H_2^{18}O]}{[H_2^{16}O]} \ .$$

This expression follows from the definition of the fractionation factor. For example, the number of ^{18}O atoms in $C^{18}O^{18}O$ is two, whereas the number of ^{18}O atoms in $C^{18}O^{16}O$ is one. In order to count the total number of ^{18}O atoms present in CO_2, the concentration of $C^{18}O^{18}O$ and $C^{18}O^{16}O$ has to be multiplied by a factor of two and one, respectively.

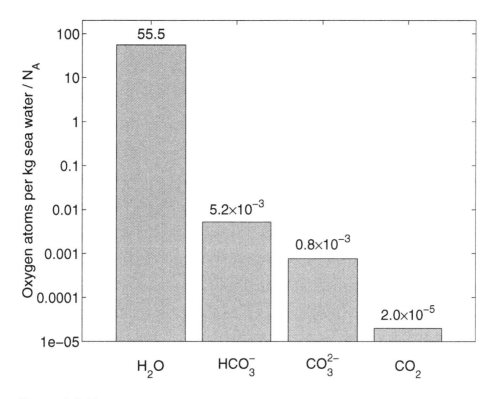

Figure 3.3.21: Number of oxygen atoms contained in the compounds of the carbonate system per kg of seawater divided by N_A (note the logarithmic vertical scale). Shown are typical values at $T = 25°C$, $S = 35$, $\Sigma CO_2 = 2$ mmol kg^{-1}, and $pH = 8.2$. The number of oxygen atoms contained in the water is about four orders of magnitude greater than in e.g. HCO_3^-.

In order to determine the overall oxygen isotope fractionation between water and the carbonate species, it appears that one has to know the individual concentrations of all oxygen bearing compounds in solution (e.g. $[HC^{18}O^{18}O^{18}O^-]$, $[HC^{18}O^{18}O^{16}O^-]$, $[HC^{18}O^{16}O^{16}O^-]$, $[HC^{16}O^{16}O^{16}O^-]$, ...) at any pH. However, these fractions are not known. (Note that the concentrations may be estimated analogous to the results for CO_2 shown in Table 3.3.7. Unfortunately, these estimates are obtained by neglecting possible fractionations between isotopically substituted species, and they may cause differences of the order of per mil.) Fortunately, Usdowski et al. (1991) found a different approach to the problem. They derived a relationship for the oxygen isotopic equilibrium between water and the quantity $S = [H_2CO_3] + [HCO_3^-] + [CO_3^{2-}]$ and $\ln(S/[CO_2])$ which is a simple function of pH (see Section C.9). Their approach enabled them to calculate

Table 3.3.7: Mass distribution of stable carbon and oxygen isotopes in natural CO_2.[a]

Mass	Molecule	Abundance[b]	Abundance of given mass
44	$^{12}C^{16}O^{16}O$	0.9842	0.9842
45	$^{13}C^{16}O^{16}O$	1.105×10^{-2}	0.01179
	$^{12}C^{16}O^{17}O$	7.399×10^{-4}	
46	$^{12}C^{16}O^{18}O$	3.936×10^{-3}	0.00394
	$^{13}C^{16}O^{17}O$	8.305×10^{-6}	
	$^{12}C^{17}O^{17}O$	1.391×10^{-7}	
47	$^{13}C^{16}O^{18}O$	4.418×10^{-5}	4.57×10^{-5}
	$^{12}C^{17}O^{18}O$	1.480×10^{-6}	
	$^{13}C^{17}O^{17}O$	1.561×10^{-9}	
48	$^{12}C^{18}O^{18}O$	3.936×10^{-6}	3.95×10^{-6}
	$^{13}C^{17}O^{18}O$	1.661×10^{-8}	
49	$^{13}C^{18}O^{18}O$	4.418×10^{-8}	4.42×10^{-8}
All	All	1.0000	1.0000

[a]After McCrea (1950), assuming natural abundances of ^{13}C: 1.11%, ^{17}O: 0.0375%, and ^{18}O: 0.1995%.
[b]Calculated values are based on natural abundances only, i.e., there is no isotopic fractionation among the different molecules (cf. also Exercise 3.19).

the individual fractionation factors between water and H_2CO_3, HCO_3^-, and CO_3^{2-}.

Some remarks on the quantity $\mathcal{S} = [H_2CO_3] + [HCO_3^-] + [CO_3^{2-}]$ are in order. This quantity, which represents the sum of the species containing the carbonate group, is very useful for the determination of the ^{18}O fractionation between water and the carbonate species. This is because (i) the overall isotope exchange can be written in a simple form (Eq. C.9.36), and (ii) the major component equilibrium and the overall isotopic equilibrium are linearly related (Eq. C.9.40). The definition of $\mathcal{S}$ is very similar to the definition of $\Sigma CO_2 = [CO_2] + [HCO_3^-] + [CO_3^{2-}]$. Note, however, that this similarity is in part accidental and that $\mathcal{S}$ does not have to be equal to ΣCO_2 - any quantity that allows the calculation of the fractionation factors could be used. In the following, the term 'sum of the dissolved carbonate species' will be used for $\mathcal{S}$. The calligraphic symbol $\mathcal{S}$ and $^{18}\mathcal{S}$ are used here rather than S and ^{18}S, respectively (cf. Usdowski et al. (1991)) in order to avoid any confusion with sulfur or sulfur isotopes.

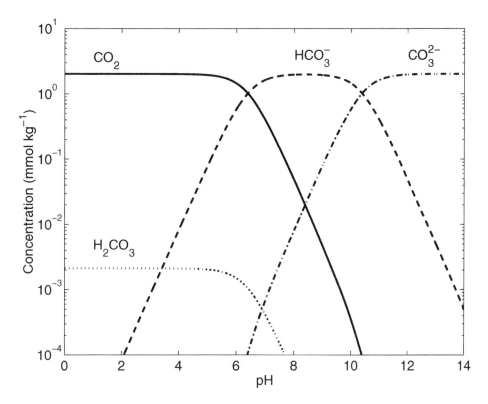

Figure 3.3.22: The concentration of the dissolved carbonate species as a function of pH: CO_2 (solid line), H_2CO_3 (dotted line), HCO_3^- (dashed line), and CO_3^{2-} (dot-dashed line). The values shown correspond to fresh water conditions ($T = 19°C$, $\Sigma CO_2 \approx 2$ mmol kg^{-1}).

The essence of the results of Usdowski et al. (1991) and Usdowski and Hoefs (1993) is briefly summarized here, whereas the details of the oxygen isotope partitioning in the system CO_2-H_2O are examined in Appendix C.9. Usdowski et al. (1991) studied a fresh water system. To date, similar studies have not been made for the seawater carbonate system. Thus, in contrast to the seawater system at $T = 25°C$ and $S = 35$ which is adopted as a standard system throughout the book, a fresh water system at 19°C is studied in the following. Since there are differences in ion activities and equilibrium constants between fresh water and seawater it might be difficult to compare absolute values of isotope fractionation directly. However, the basic mechanisms of oxygen isotope partitioning in the carbonate system do apply to fresh water systems as well as to seawater systems (cf. Zeebe (1999) and Sections 3.3.3 and 3.3.7).

The concentrations of the dissolved carbonate species as a function of pH

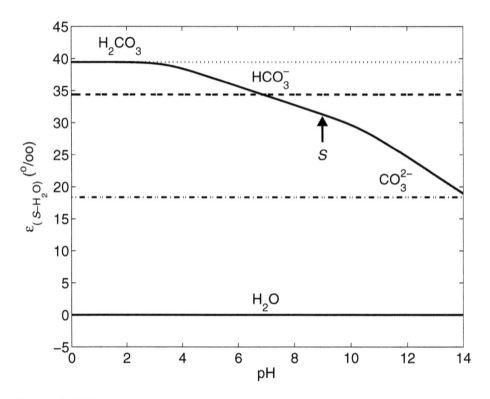

Figure 3.3.23: Oxygen isotope partitioning between water and $\mathcal{S} = [\mathrm{H_2CO_3}] + [\mathrm{HCO_3^-}] + [\mathrm{CO_3^{2-}}]$ as a function of pH (after Zeebe (1999)). As pH increases the concentration of the carbonate species in solution changes from mainly $\mathrm{H_2CO_3}$ to $\mathrm{HCO_3^-}$ and $\mathrm{CO_3^{2-}}$ resulting in a decrease of $\varepsilon_{(\mathcal{S}-\mathrm{H_2O})}$ with increasing pH. At pH 0, 6.9, and 14, $\varepsilon_{(\mathcal{S}-\mathrm{H_2O})}$ equals the individual fractionation factors between $\mathrm{H_2O}$ and $\mathrm{H_2CO_3}$, $\mathrm{HCO_3^-}$, and $\mathrm{CO_3^{2-}}$.

for fresh water ($T = 19°\mathrm{C}$) are shown in Figure 3.3.22. At low pH, aqueous $\mathrm{CO_2}$ is the dominant species, whereas $\mathrm{HCO_3^-}$ is most abundant at intermediate pH, and $\mathrm{CO_3^{2-}}$ is the dominant carbonate species in solution at high pH. The concentration of carbonic acid ($\mathrm{H_2CO_3}$) is small compared to $[\mathrm{CO_2}]$ ($\sim 0.1\%$) The fractionation between water and the sum of the dissolved carbonate species $\varepsilon_{(\mathcal{S}-\mathrm{H_2O})} = (\alpha_{(\mathcal{S}-\mathrm{H_2O})} - 1) \times 10^3$, as derived from precipitation experiments (Usdowski et al., 1991) is displayed in Figure 3.3.23 as a function of pH (solid line). At a limiting pH $= 0$ the fractionation factor $\alpha_{(\mathcal{S}-\mathrm{H_2O})}$ is equal to the fractionation between $\mathrm{H_2CO_3}$ and $\mathrm{H_2O}$, (all dissolved carbonate is essentially carbonic acid), whereas at a limiting pH $= 14$, $\alpha_{(\mathcal{S}-\mathrm{H_2O})}$ equals $\alpha_{(\mathrm{CO_3^{2-}}-\mathrm{H_2O})}$ (all dissolved carbonate is essentially carbonate ion). At intermediate pH of 6.90, $\alpha_{(\mathcal{S}-\mathrm{H_2O})}$ largely reflects the fractionation between $\mathrm{HCO_3^-}$ and $\mathrm{H_2O}$.

Thus, individual fractionation factors between water and H_2CO_3, HCO_3^-, and CO_3^{2-} can be deduced from $\alpha_{(S-H_2O)}$ at $pH = 0.00, 6.90$, and 14.00. The ratios $[H_2CO_3]/S$, $[HCO_3^-]/S$, and $[CO_3^{2-}]/S$ are at maximum at these pH values $(0.9996, 0.9994$ and $0.9998)$, implying that other carbonate species exhibit negligible contributions to S here. The individual fractionation factors obtained in this way are summarized in Table 3.3.8 and in Figure 3.3.23 (horizontal lines).

Table 3.3.8: Oxygen isotope fractionation factors for the carbonate species with respect to water $(19°C)$.

Species	$\varepsilon_{(x-H_2O)}$ (‰)
$CO_2(aq)$	57.9^a
$CO_2(g)$	41.2^b $(25°C)$
H_2CO_3	39.5^a
HCO_3^-	34.3^c
CO_3^{2-}	18.4^a

[a] Usdowski et al. (1991).
[b] Kim and O'Neil (1997).
[c] Recalculated after Usdowski et al. (1991) (Usdowski, pers. comm.).

The bottom line of the discussion so far is as follows. The individual oxygen isotope fractionation factors between water and the dissolved carbonate species can be determined by the approach used by Usdowski and co-workers (Table 3.3.8). The isotope fractionation between water and the sum of the dissolved carbonate species (S) is decreasing with pH (Figure 3.3.23).

Before getting into the practical aspects of the ^{18}O partitioning in the carbonate system, i.e. its application in paleoceanography, Section 3.3.7, oxygen isotope equilibration times are discussed. This completes the theoretical part of this section.

3.3.6 Time required for oxygen isotope equilibration

The time required for oxygen isotope equilibration in the carbonate system is very different from the time required for carbon isotope equilibration (cf. Section 2.5). This is because oxygen isotopic equilibrium has to be established between the solvent, H_2O, and the dissolved carbonate species, whereas carbon isotope equilibrium only has to be established among the dissolved carbonate species. Considering the reactions involved in oxygen equilibration in the carbonate system (C.9.34), it is obvious that CO_2 plays a key role in this process because oxygen isotope equilibration between H_2O and the dissolved carbonate species occurs only via hydration or hydroxylation of CO_2:

$$CO_2 + H_2O \quad \underset{k_{-2}}{\overset{k_{+2}}{\rightleftharpoons}} \quad H_2CO_3 \tag{3.3.37}$$

$$CO_2 + OH^- \quad \underset{k_{-4}}{\overset{k_{+4}}{\rightleftharpoons}} \quad HCO_3^- . \tag{3.3.38}$$

Note that equilibration between OH^- and H_2O is very rapid; for numbering of the $k's$, see Section 2.3.1. Usdowski et al. (1991) gave an equation from which the equilibration time can be calculated (see also Mills and Urey, 1940):

$$\ln \left(\frac{R_S - R_{S,eq.}}{R_{S,0} - R_{S,eq.}} \right) = -t/\tau \tag{3.3.39}$$

where $R_{S,0}$, R_S, and $R_{S,eq}$ refer to the oxygen isotope ratio of S at time t, $t = 0$, and at equilibrium $(t \to \infty)$, respectively. The time constant τ is given by:

$$\tau^{-1} = 0.5 \times k \times \left\{ 1 + \frac{[CO_2]}{S} - \left[1 + \frac{2}{3}\frac{[CO_2]}{S} + \left(\frac{[CO_2]}{S} \right)^2 \right]^{1/2} \right\}$$

where

$$k = k_{+2} + k_{+4}[OH^-]$$

and

$$S = [H_2CO_3] + [HCO_3^-] + [CO_3^{2-}] . \tag{3.3.40}$$

For example, if the time for 99% equilibration is to be calculated, Eq. (3.3.39) may be rearranged to yield:

$$t_{99\%} = -\ln(0.01) \times \tau$$

Table 3.3.9 and Figure 3.3.24 summarize the oxygen isotope equilibration time in the carbonate system (in fresh and in seawater) as a function of pH. Since the equilibration time $(t_{99\%})$ depends on the rate of oxygen isotope exchange via aqueous carbon dioxide, $t_{99\%}$ increases dramatically with increasing pH as $[CO_2]$ decreases.

Exercise 3.17 (*)

Why does the oxygen isotope fractionation between H_2O and $S = [H_2CO_3] + [HCO_3^-] + [CO_3^{2-}]$ decrease with pH?

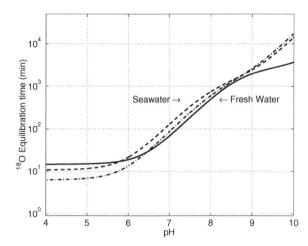

Figure 3.3.24: Oxygen isotope equilibration time (99%) in the carbonate system in fresh water ($T = 19°C$, solid line) and seawater ($T = 19°C$, $S = 35$, dashed line; $T = 25°C$, $S = 35$, dot-dashed line) as a function of pH.

Table 3.3.9: Time for oxygen isotope equilibration in the carbonate system.

pH	$t_{99\%}$					
	$(T = 19°C, S = 0)^a$		$(T = 19°C, S = 35)^b$		$(T = 25°C, S = 35)^b$	
	(min)	(h)	(min)	(h)	(min)	(h)
4	15	0.25	11	0.18	6	0.10
5	15	0.25	12	0.19	7	0.11
6	19	0.32	21	0.35	14	0.23
7	68	1.14	128	2.13	87	1.44
8	498	8.30	745	12.4	601	10.0
9	1,968	32.8	2,455	40.9	2,630	43.8
10	3,726	62.1	14,400	240	17,684	295

[a] Calculated using fresh water constants (k, K) (Usdowski et al., 1991).
[b] Calculated using seawater constants (k, K^*) (Johnson, 1982; DOE, 1994).

Exercise 3.18 (*)

Why is the oxygen equilibration time at a high pH longer than at low pH?

Exercise 3.19 (***)

Calculate the mass distribution of stable carbon and oxygen isotopes in natural CO_2 as given in Table 3.3.7. The problem is analogous to the following: Calculate the probability of picking three particular balls (e.g. with specific colors) from different pools.

3.3.7   The effect of *p*H on foraminiferal oxygen isotopes

The mechanism described in Section 3.3.5 provides an explanation of an effect observed by Spero et al. (1997): higher *p*H values or increasing CO_3^{2-} concentrations of the culture medium resulted in isotopically lighter shells in the planktonic foraminifera *Orbulina universa* (Figure 3.3.25) and *Globigerina bulloides*. In another series of experiments this effect was also observed in *Globigerinoides sacculifer* and *Globigerinoides ruber* (Spero et al., 1999). In paleoreconstructions this effect can have important consequences. For example, higher *p*H values during the last glacial maximum as indicated by boron isotope analysis in foraminifera (Sanyal et al., 1995), would result in ^{18}O depleted shells. Lower $\delta^{18}O$ values are interpreted as higher temperatures and therefore result in sea surface temperature (SST) estimates that might be up to 1°C too high. Correcting for this effect would bring tropical SST closer to estimates based on other marine proxies, e.g., Sr/Ca ratios in corals and terrestrial indicators such as snowline and ice-core $\delta^{18}O$. More dramatic and opposite effects on SST estimates are to be expected for periods of high atmospheric pCO_2 in the geological past such as the Cretaceous (Zeebe, 2001b).

The basic mechanism to explain the effect observed by Spero et al. (1997) can be described as follows: the total dissolved inorganic carbon (ΣCO_2 = $[CO_2]+[H_2CO_3]+[HCO_3^-]+[CO_3^{2-}]$) present in solution is mainly in the form of HCO_3^- at intermediate *p*H and mainly in the form of CO_3^{2-} at high *p*H (Figure 3.3.22, page 203). Since HCO_3^- is isotopically heavier than CO_3^{2-} (Figure 3.3.23, page 204), the oxygen isotopic composition of the total dissolved carbonate species decreases with increasing *p*H. Provided that the calcite is formed from a mixture of the carbonate species in proportion to their relative contribution to $\mathcal{S} = [H_2CO_3]+[HCO_3^-]+[CO_3^{2-}]$, the oxygen isotopic composition of the calcite also decreases with increasing *p*H (Zeebe, 1999).

If calcium carbonate is quantitatively precipitated from a bicarbonate-carbonate solution, i.e., none of the carbonate in solution escapes precipitation, the oxygen isotopic composition of the precipitate is the average of the oxygen present in the bicarbonate and carbonate at the time of precipitation (McCrea, 1950). Thus, the calcium carbonate simply reflects the isotopic composition of $\mathcal{S}$ and is also decreasing with *p*H. The key hypothesis of this discussion is that also in a non-quantitative precipitation of biogenic carbonate (foraminiferal calcite) and synthetic carbonate (calcite, witherite, see Section 3.3.8) the calcium carbonate is built from a mixture of bicarbonate and carbonate proportional to their respective contributions to $\mathcal{S}$.

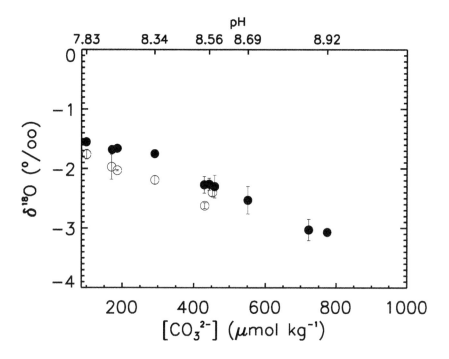

Figure 3.3.25: Oxygen isotopic composition of foraminiferal calcite (*O. universa*) as a function of $[CO_3^{2-}]$ ($\Sigma CO_2=$ const. $\sim$ 2,032 μmol kg^{-1}). The shells are depleted in ^{18}O as $[CO_3^{2-}]$ or pH increases. The mean slope given by Spero et al. (1997) of shell $\delta^{18}O$ vs. $[CO_3^{2-}]$ is $-0.0022 \pm 0.0004\%_0$ (μmol kg^{-1})$^{-1}$ in the dark (filled circles) and $-0.0015 \pm 0.0008\%_0$ (μmol kg^{-1})$^{-1}$ in the light (open circles).

The validity of this hypothesis is confirmed by the coincidence of the observed and calculated slope of $\delta^{18}O$ vs. $[CO_3^{2-}]$ in foraminiferal calcite and $\mathcal{S}$, respectively, and the agreement between oxygen isotope effects observed in synthetic carbonates (Kim and O'Neil, 1997) and theoretical predictions (Section 3.3.8).

Comparison with culture data of *O. universa*

The results of the oxygen isotope partitioning (Section C.9) of inorganic precipitation can be compared to shell $\delta^{18}O$ vs. $[CO_3^{2-}]$ observed in culture experiments with living foraminifera (Spero et al., 1997). For comparison, the fractionation between the sum of the carbonate species ($\mathcal{S}$) and water must merely be plotted vs. $[CO_3^{2-}]$ instead of pH. In order to (a) avoid complications associated with metabolic effects such as photosynthesis of the symbiotic algae in light, and (b) ensure that ΣCO_2 of the culture medium

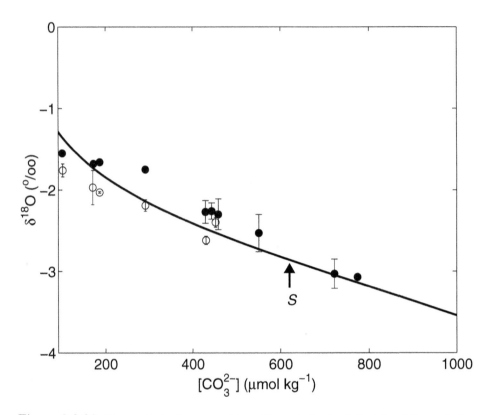

Figure 3.3.26: Oxygen isotopic composition of foraminiferal calcite (closed/open circles: dark/high light) and of $\mathcal{S} = [\mathrm{H_2CO_3}] + [\mathrm{HCO_3^-}] + [\mathrm{CO_3^{2-}}]$ vs. $[\mathrm{CO_3^{2-}}]$ as predicted from theory (solid line). The mean slope given by Spero et al. (1997) of shell $\delta^{18}\mathrm{O}$ vs. $[\mathrm{CO_3^{2-}}]$ for their constant $\Sigma\mathrm{CO_2}$ experiments is $-0.0022 \pm (\mu\mathrm{mol\ kg^{-1}})^{-1}$ in the dark. The mean slope predicted from theory is $-0.0024 \pm (\mu\mathrm{mol\ kg^{-1}})^{-1}$.

remained constant, the dark experiments of Spero et al. (1997) ($\Sigma\mathrm{CO_2}=$ const. $\sim 2{,}032\ \mu\mathrm{mol\ kg^{-1}}$) are chosen for comparison. The fractionation factors reported by Usdowski et al. (1991) refer to a fresh water system at 19°C; the dependence on temperature was not examined in their experiments. On the other hand, foraminifera were cultured in seawater at 22°C. Since both systems have different temperatures and salinities, absolute values ($\delta^{18}\mathrm{O}$) of the carbonate species in solution and of the calcite are not compared. However, the calculated slope of the fractionation between the major carbonate component in solution ($\mathcal{S}$) and the water can be compared to the slope of foraminiferal $\delta^{18}\mathrm{O}$ as a function of $[\mathrm{CO_3^{2-}}]$ (Figure 3.3.26). The intercept of the calculated curve of $\mathcal{S}$ (solid line) with the vertical axis is chosen arbitrarily whereas the slope is predicted by theory.

The observed slope of shell $\delta^{18}O$ vs. $[CO_3^{2-}]$ in *O. universa* is $-0.0022 \pm 0.0004\permil$ $(\mu mol\ kg^{-1})^{-1}$ in dark experiments, whereas the mean slope of $\mathcal{S}$ vs. $[CO_3^{2-}]$ within the interval from $100 - 1000\ \mu mol\ kg^{-1}$ calculated from theory is $-0.0024\permil$ $(\mu mol\ kg^{-1})^{-1}$. From this coincidence the following conclusion is drawn: the depletion of ^{18}O in *O. universa* with increasing carbonate ion concentration can be explained by the uptake of bicarbonate and carbonate for calcification proportional to their respective contribution to $\mathcal{S}$. (Note that this description is simplified - the calculation of the isotope partitioning is not a simple mass-balance calculation, see Section C.9). As the species in solution change from mainly HCO_3^- (isotopically heavy) at intermediate pH to mainly CO_3^{2-} (isotopically light) at high pH the isotopic composition of $\mathcal{S}$ decreases and so does the isotopic composition of the calcite. The presented results suggest that the calcite is obviously formed from a mixture of bicarbonate and carbonate such that the oxygen isotope composition of the calcite (solid) and of $\mathcal{S}$ (dissolved) are equally decreasing with increasing $[CO_3^{2-}]$.

3.3.8 The effect of pH on synthetic carbonates

Similar to the comparison described in the previous section, the results of the inorganic oxygen isotope partitioning (quantitative precipitation) are compared to oxygen isotope effects in synthetic carbonates that have been precipitated slowly from solution (non-quantitative precipitation) as reported by Kim and O'Neil (1997). In this case, however, the analysis of the experimental data is not as straightforward as the analysis of the culture data of *O. universa*.

Kim and O'Neil (1997) thoroughly examined equilibrium oxygen isotope effects in synthetic carbonates and determined new values for the oxygen isotope fractionation between water and several carbonates at low temperatures. Surprisingly, they also found that the isotopic fractionation between water and carbonates such as calcite ($CaCO_3$) and witherite ($BaCO_3$) increased with increasing initial concentrations of Ca^{2+}/Ba^{2+} and HCO_3^- at a given temperature (Figure 3.3.27). They judged these effects to be nonequilibrium fractionations since there should be only one equilibrium fractionation at any temperature. However, it is suggested that the observed effects are equilibrium fractionations expressed at different pH of the solution.

In the following discussion calcite is considered, however, the results presented should hold for carbonates in general as will be demonstrated for witherite. Kim and O'Neil (1997) prepared their solutions by dissolving

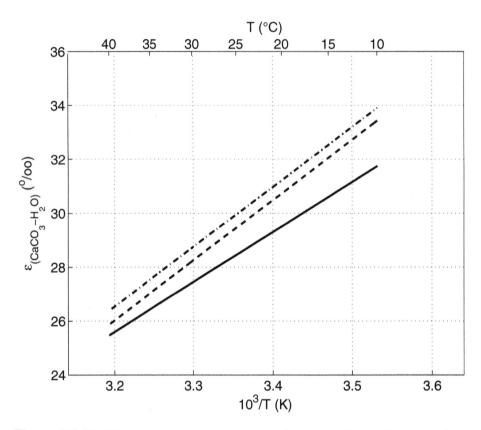

Figure 3.3.27: The oxygen isotope fractionation between calcite and water as determined by Kim and O'Neil (1997) as a function of temperature and different initial concentrations of Ca^{2+} and HCO_3^-: 5 mmol kg^{-1} (solid line), 15 mmol kg^{-1} (dashed line), and 25 mmol kg^{-1} (dot-dashed line).

equimolar concentrations of $CaCl_2$ and $NaHCO_3$ in deionized water, with concentrations ranging from 5 to 25 mmol kg^{-1}. Slow precipitation was promoted by bubbling nitrogen gas through the solution to remove CO_2, therefore increasing the supersaturation of the solution. Removal of CO_2 is accompanied by increase of pH and $[CO_3^{2-}]$ until a critical state of supersaturation ($[CO_3^{2-}]_c$) is reached and calcium carbonate starts to precipitate (no seeds were used). Since the corresponding critical pH is a function of the total dissolved inorganic carbon, the critical state of supersaturation depends on the initial concentration of HCO_3^-. This effect can be illustrated utilizing the concentrations of the dissolved carbonate species as a function of pH shown in Figure 3.3.22 on page 203. As pH increases, HCO_3^- is converted to CO_3^{2-} and the carbonate ion concentration increases until all dissolved carbonate is essentially carbonate ion. For the composition of the solution displayed in Figure 3.3.22 ($\Sigma CO_2 = 2$ mmol kg^{-1}) pH must increase to about 8 to reach a critical state of supersaturation of e.g. $[CO_3^{2-}]_c = 10^{-2}$ mmol kg^{-1}. However, if ΣCO_2 is smaller (corresponding to a lower initial concentration of HCO$_3^-$) pH must increase further to reach the same critical state of supersaturation.

As described in detail in Zeebe (1999) the critical pH values of the solutions studied by Kim and O'Neil (1997) can be estimated as 7.8, 6.9, and 6.6 for initial concentrations of Ca^{2+} and HCO$_3^-$ of 5, 15, and 25 mmol kg^{-1}, respectively. These values were calculated for calcite precipitation, assuming a critical supersaturation of $s_c = 2$. The critical supersaturation may be written as (Stumm and Morgan, 1996):

$$ s_c = \left(\frac{\gamma_{Ca^{2+}}[Ca^{2+}] \, \gamma_{CO_3^{2-}}[CO_3^{2-}]_c}{K'_{sp}} \right)^{\frac{1}{2}} \tag{3.3.41} $$

where γ's are activity coefficients, $[CO_3^{2-}]_c$ is the critical carbonate ion concentration, and $\log K_{sp} = -8.45$ is the solubility product of calcite in fresh water at 19°C (Plummer and Busenberg, 1982).

The essential result of the consideration presented so far is as follows. Provided that the critical pH represents a mean pH at which calcite was precipitated and that at every pH oxygen isotope equilibrium was established between the carbonate species and water, it follows from the theory presented in Section 3.3.5 (cf. Figure 3.3.23, page 204) that the isotopic composition of $\mathcal{S}$ and thus of the calcite decreases with critical pH (or lower initial concentration).

The values of the oxygen isotope fractionation between calcite/witherite and water as given by Kim and O'Neil (1997) for different initial concentrations at 19°C (diamonds) and the fractionation between $\mathcal{S}$ and water as calculated from the critical pH values at 19°C and a critical supersaturation of $s_c = 2$ (solid line) are shown in Figure 3.3.28. In order to plot the values for calcite and witherite in the same graph, the difference of the equilibrium fractionation between calcite and water, and witherite and water ($\sim 2.59\%_0$) at 19°C and an initial concentration of 5 mmol kg^{-1} was added to the values for witherite. It is emphasized that the intercept of the calculated curve of $\mathcal{S}$ with the vertical axis is chosen arbitrarily. Since the actual critical pH values of the experiments are not known, absolute values cannot be compared. The influence of different critical pH values was examined by varying the supersaturation s_c between 1 and 8 which had only a marginal effect on the shape of the curve. Keeping in mind the assumptions necessary to

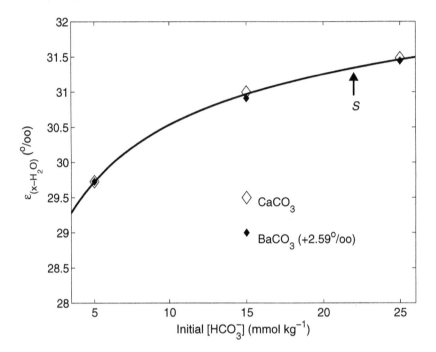

Figure 3.3.28: The oxygen isotope fractionation between water and calcite (open diamonds), and water and witherite (filled diamonds) as determined by Kim and O'Neil (1997) and of $\mathcal{S}$ (solid line) as a function of the initial concentration of HCO_3^- at 19°C. The intercept of the calculated curve was chosen arbitrarily; the supersaturation was set at $s_c = 2$ (see text). A value of 2.59‰ was added to the values for witherite. In this representation, the values for calcite and witherite for an initial concentration of 5 mmol kg^{-1} are identical.

obtain the theoretically derived fractionation the calculated curve and the experimental data show excellent agreement.

Even though the absolute fractionation between water and $CaCO_3$, and water and $BaCO_3$ are quite different, the increase of the fractionation with increasing initial concentration of HCO_3^- in solution is almost identical for these minerals. This result suggests that the described effect is caused by the solution chemistry rather than by different nonequilibrium (kinetic) fractionation effects associated with crystal growth. In summary, the nonequilibrium fractionation as described by Kim and O'Neil (1997) can be explained by multiple equilibrium fractionations at a single temperature but at different pH.

Only a few examples of the use of stable oxygen isotopes in the marine science could be presented in this section. Although there are many other applications, they certainly cannot all be discussed in this book. Nevertheless, the discussion so far has hopefully demonstrated the power of this manifold tool. We will now turn to the discussion of stable boron isotopes in the marine environment.

3.4 Boron

Boron isotopes have received much attention in chemical oceanography during the last few years, partly due to their use as a means of reconstructing paleo-pH. Spivack et al. (1993) utilized the boron isotopic composition of planktonic foraminifera to reconstruct the pH of the surface ocean during the past 20 million years. Using a similar approach, Palmer et al. (1998) reconstructed pH-depth profiles of the past ocean during the same time interval. It is noted that paleo-pH reconstructions based on boron isotopes are limited by the residence time of boron in seawater which is about 10 to 20 million years (cf. box on page 218). On time scales longer than this, the boron isotopic composition of the ocean might have been different from today and cannot be estimated by the use of a paleoproxy - for a modelling approach, see Lemarchand et al. (2000). Sanyal et al. (1995) utilized stable boron isotope analysis to investigate climate changes during a more recent period of earth's history. They found evidence for a higher pH in the glacial surface and deep ocean during the last glacial maximum (LGM, $\sim 18,000$ years B.P.). Recently, seasonal variations in the boron isotopic composition of corals have been used to shed light on coral calcification processes and coral $\delta^{13}C$ records (Hemming et al., 1998b).

In this section the natural variations of boron isotopes in some terrestrial compounds are briefly summarized (for review see e.g. Palmer and Swihart (1996)). Then the principles of boron isotope partitioning among the dissolved species of boron and the incorporation of boron into biogenic (e.g. foraminiferal) and inorganic calcium carbonate will be discussed.

3.4.1 Natural variations

Figure 3.4.29 shows natural variations in boron isotopic compositions of some terrestrial compounds. The $\delta^{11}B$ value of the total dissolved boron (B_T) in the ocean relative to the NBS SRM 951 boric acid standard is 39.5‰

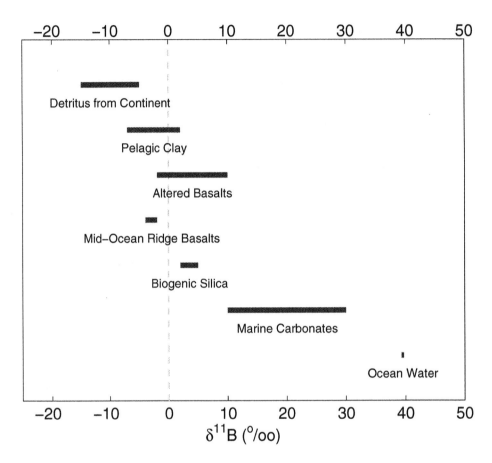

Figure 3.4.29: Boron isotopic composition of some terrestrial compounds (relative to NBS SRM 951 boric acid standard).

and is fairly constant throughout the world ocean (Spivack and Edmond, 1987; Hemming and Hanson, 1992) due to the long residence time of boron in the ocean (cf. box on page 218). The 'heavy' boron isotope composition of boron in the ocean (heavy with respect to the average continental crust) constitutes a seeming enigma because the main boron input into the ocean is derived from detritus from continents which has $\delta^{11}B$ values of approximately $-15\%_0$ to $-8\%_0$ (Ishikawa and Nakamura, 1993). Provided that the geochemical cycle of boron is in steady state, an amount of boron (equal to the boron input per time interval) has to be removed from the ocean with an isotopic composition equal to that of the input.

Thus, mechanisms must be active which preferentially remove ^{10}B, thereby enriching the boron isotope composition of dissolved seawater boron in the ocean relative to the input. Three major sinks of boron in the ocean have been recognized which also alter the $\delta^{11}B$ of dissolved seawater boron: (1) adsorption of boron onto clays, (2) exchange with the oceanic crust, and (3) coprecipitation with calcium carbonate (see e.g. Schwarcz et al. (1969); Spivack and Edmond (1987); Palmer et al. (1987); Vengosh et al. (1991); Ishikawa and Nakamura (1993)).

It was experimentally demonstrated by Schwarcz et al. (1969) and Palmer et al. (1987) that boron adsorbed on marine clay is depleted in ^{11}B with respect to the dissolved boron in the seawater - the magnitude of the fractionation being pH dependent (see Section 3.4.2). Ishikawa and Nakamura (1993) measured boron concentrations and $\delta^{11}B$ values of clays in deep-sea sediments and reported high concentrations ($96 - 132$ ppm) and $\delta^{11}B$ values between $-6\%_0$ and $+2.8\%_0$ (Figure 3.4.29). The adsorption of boron on clays is therefore an important process removing isotopically light boron from the ocean.

Unaltered (fresh) mid-ocean ridge basalts (MORB) have $\delta^{11}B$ values between $-2\%_0$ and $-4\%_0$, whereas altered basalts have $\delta^{11}B$ values between $0\%_0$ and $+10\%_0$ (Spivack and Edmond, 1987). The enrichment of ^{11}B in altered basalts presumably results from reaction of the basalts with seawater. Spivack and Edmond (1987) estimated that more than 95% of the boron found in a number of their samples of altered basalt may have been derived from seawater. The boron isotopic composition of the secondary boron in the basalt which is derived from seawater is about $30\%_0$ lighter than the boron in seawater. As a result, when a parcel of seawater reacts with fresh basalt the boron concentration of the water will decrease and its $\delta^{11}B$ will increase. This process removes large portions of boron from the ocean and increases its boron isotopic composition.

The boron isotopic composition of modern marine biogenic carbonates (e.g. foraminifera, pteropods, corals) and of modern deep-sea carbonate sediments range approximately from $+10\%_0$ to $+30\%_0$ (Vengosh et al., 1991; Hemming and Hanson, 1992; Ishikawa and Nakamura, 1993). The depletion of ^{11}B in marine carbonates with respect to the isotopic composition of B_T may be explained by the preferential uptake of $B(OH)_4^-$ into the carbonates which is isotopically lighter than B_T at seawater pH (see Section 3.4.3). Coprecipitation of boron in marine carbonates is believed to contribute $\sim 20\%$ to the total sinks of boron in the ocean (Vengosh et al., 1991).

Residence time of Boron in the ocean. At steady state, the residence time, τ, of an element in the ocean can be estimated from the ratio of the total inventory of this element in seawater divided by the input/output per year, e.g.:

$$\tau_B = \frac{B_T \times M_O}{B_{input}} ,$$

where $B_T = 4.52$ ppm $(= 4.52 \times 10^{-6}$ g B per g seawater) (Hemming and Hanson, 1992), $M_O \approx 1.4 \times 10^{24}$ g is the mass of the ocean, and B_{input} is the input of boron into the ocean in gram per year. Using a boron input of $30 - 50 \times 10^{10}$ g yr^{-1} (Vengosh et al., 1991), the residence time of boron in the ocean is $13 - 21$ million years (Ma). The most recent estimate by Lemarchand et al. (2000) is 14 Ma.

In summary, the enrichment of ^{11}B in dissolved boron in the ocean with respect to the isotopic composition of the input may be explained by preferential removal of ^{10}B associated with processes such as adsorption of boron onto clays, exchange with the oceanic crust, and coprecipitation with calcium carbonate.

The boron isotopic composition of biogenic silica (radiolaria and diatoms) has also been reported. It ranges between 2‰ and 5‰ which is consistent with boron isotopic equilibrium between seawater and lattice-bonded boron in silica (Ishikawa and Nakamura, 1993).

Exercise 3.20 (*)

The boron isotope composition of a sample A of fresh basalt is -2‰. Calculate the fractionation, $\varepsilon_{(A-B)}$, between sample A and seawater (sample B, $\delta_B = 39.5$‰). Also calculate $\varepsilon_{(B-A)}$ and compare the results to the difference of the δ values.

3.4.2   Boron isotope partitioning

Boron has two stable isotopes (cf. Section 3.1.4), ^{10}B and ^{11}B, which make up 19.82% and 80.18% of the total boron (IUPAC, 1998). The major dissolved species of boron in seawater are $B(OH)_3$ and $B(OH)_4^-$, which are connected by the chemical equilibrium:

$$B(OH)_3 + H_2O \rightleftharpoons B(OH)_4^- + H^+ \tag{3.4.42}$$

Typical seawater concentrations of these species as a function of pH are displayed in Figure 3.4.30a. At low pH all dissolved boron is essentially boric acid, $B(OH)_3$, whereas at high pH all dissolved boron is essentially borate ion, $B(OH)_4^-$.[9] The pK^* value of boric acid (pK_B^*) where $[B(OH)_3] =$

---

[9]Polynuclear boron species such as $B_3O_3(OH)_4^-$, $B_4O_5(OH)_4^{2-}$, and $B_5O_6(OH)_4^-$ are negligible at concentrations smaller than 25 mmol kg^{-1} (Su and Suarez, 1995) and can be safely ignored at typical seawater concentrations of ~ 0.4 mmol kg^{-1}.

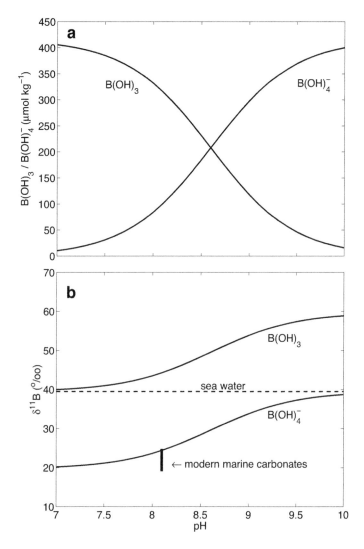

Figure 3.4.30: (a) The concentration of dissolved boron species as a function of pH ($T = 25°C$, $S = 35$). The dissociation constant of boric acid pK_B^* is 8.60; the total boron concentration is 416 μmol kg^{-1} (DOE, 1994). (b) Boron isotopic composition of B(OH)$_3$ and B(OH)$_4^-$ as a function of pH (cf. Hemming and Hanson (1992)).

[B(OH)$_4^-$] is 8.60 in seawater at $T = 25°C$, $S = 35$ (DOE, 1994). A value of 8.830 for pK_B^* at 25°C for the dissociation constant of boric acid in NaCl solutions (Hershey et al., 1986) with ionic strength $I = 0.68$ has also been used for seawater in the literature (ionic strength of seawater is 0.697). Hershey et al. (1986) also determined pK_B^* for Na-Ca-Cl and Na-Mg-Cl solutions at $T = 25°C$ ($I = 0.68$) to be 8.675 and 8.725 which is closer to the seawater value of 8.60 (DOE, 1994), demonstrating the influence of Ca^{2+} and Mg^{2+} ions on the pK^*'s (~ 0.1 units). Concerning calculations which involve the concentrations of the dissolved boron species in the ocean, it is therefore probably more accurate to use a pK_B^* that has been determined in seawater.

Utilizing the total boron concentration of seawater

$$B_T = [B(OH)_3] + [B(OH)_4^-] \qquad (3.4.43)$$

where $B_T = 416.0 \times (S/35)$ μmol kg^{-1} (DOE, 1994), the concentration of the dissolved species can be calculated. Using the dissociation constant of boric acid:

$$K_B^{'*} = \frac{[B(OH)_4^-][H^+]}{[B(OH)_3]} , \qquad (3.4.44)$$

the concentrations of $B(OH)_3$ and $B(OH)_4^-$ are given at any pH by:

$$[B(OH)_3] = \frac{B_T}{1 + K_B^*/[H^+]} \qquad (3.4.45)$$

$$[B(OH)_4^-] = \frac{B_T}{1 + [H^+]/K_B^*} \qquad (3.4.46)$$

Because the stable isotope ^{11}B is enriched in $B(OH)_3$ compared to $B(OH)_4^-$, the isotopic composition of the dissolved species change with pH (Figure 3.4.30b). At low pH the isotopic composition of $B(OH)_3$ is equal to the isotopic composition of the total dissolved boron, $\sim 39.5\%_0$. In contrast, at high pH the isotopic composition of $B(OH)_4^-$ is equal to the isotopic composition of the total dissolved boron. In between, the $\delta^{11}B$ of both species change, with $B(OH)_3$ being enriched at equilibrium by ca. $20\%_0$ (Kakihana et al., 1977) with respect to $B(OH)_4^-$ at any pH. The boron isotope partitioning between $B(OH)_3$ and $B(OH)_4^-$ as a function of pH is shown in Figure 3.4.30b.

The isotope partitioning as depicted in Figure 3.4.30b can be calculated as follows. The $\delta^{11}B$ of the total dissolved boron is given by a mass-balance relation[10]:

$$\delta^{11}B_{(B_T)} \times B_T = \delta^{11}B_{(B(OH)_3)} \times [B(OH)_3]$$
$$+ \delta^{11}B_{(B(OH)_4^-)} \times [B(OH)_4^-] \qquad (3.4.47)$$

---

[10]The error introduced in the δ values of the dissolved boron species by using δ values instead of fractional abundances ($r = R/(1 + R)$) for the mass-balance equation (cf. Section 3.1.5) is smaller than $0.08\%_0$.

The fractionation factor ε_B between $B(OH)_3$ and $B(OH)_4^-$ is given by

$$\varepsilon_B = \left(\frac{R_{(B(OH)_3)}}{R_{(B(OH)_4^-)}} - 1 \right) \times 10^3$$

$$= (\alpha_B - 1) \times 10^3 \tag{3.4.48}$$

where $\alpha_B = 1.0194$ at $25°C$ (Kakihana et al., 1977). Using the relation between isotopic ratios and δ values of two substances 1 and 2:

$$\frac{R_1}{R_2} = \frac{\delta_1 + 10^3}{\delta_2 + 10^3}$$

we have:

$$\delta^{11}B_{(B(OH)_3)} = \delta^{11}B_{(B(OH)_4^-)} \, \alpha_B + \varepsilon_B \tag{3.4.49}$$

Inserting Eq. (3.4.49) into Eq. (3.4.47) and solving for $\delta^{11}B_{(B(OH)_4^-)}$ one obtains:

$$\delta^{11}B_{(B(OH)_4^-)} = \frac{\delta^{11}B_{B_T} B_T - \varepsilon_B [B(OH)_3]}{[B(OH)_4^-] + \alpha_B [B(OH)_3]} \,. \tag{3.4.50}$$

Since α_B is very close to unity, this may be approximated by:

$$\delta^{11}B_{(B(OH)_4^-)} = \delta^{11}B_{B_T} - \varepsilon_B [B(OH)_3]/B_T \,,$$

and $\delta^{11}B_{(B(OH)_3)}$ may be written as:

$$\delta^{11}B_{(B(OH)_3)} = \delta^{11}B_{(B(OH)_4^-)} + \varepsilon_B \,.$$

Thus, given the total boron concentration B_T, the isotopic composition of B_T, and the fractionation factor ε_B, the isotopic composition of $B(OH)_3$ and $B(OH)_4^-$ can be calculated at any pH.

The fractionation factor $\alpha_B = 1.0194$ (at $25°C$) for the isotope partitioning between the dissolved boron species given by Kakihana et al. (1977) shows a weak dependence on temperature (Table 3.4.10) (Kakihana and Kotaka, 1977; Kotaka and Kakihana, 1977). It was calculated from the thermodynamic properties of isotopic substances (Urey, 1947) which is discussed in detail in Section 3.5. To the best of our knowledge, no experimental value has yet been determined. Hence the theoretically derived value is widely used in the literature.

It is emphasized that a value of 1.032 for α_B which has also been employed (taken from a paper by Palmer et al. (1987) on boron adsorption on marine clay) should not be used for the boron isotope fractionation between the dissolved species $B(OH)_3$ and $B(OH)_4^-$. Palmer et al. (1987) determined boron distribution coefficients and isotope fractionation between dissolved boron in seawater and boron adsorbed on marine clay at pH values ranging from 6.5 to 8.5. The fractionation factors given by Palmer et al. (1987), 0.969 and 0.992 (equivalent to 1.032 and 1.008)[11], refer to the individual isotope fractionation factors between the dissolved boron species in solution and the respective adsorbed boron species on the clay. That is, the fractionation between $B(OH)_3$ in solution and the respective adsorbed species is $-32\%_0$, whereas the fractionation between $B(OH)_4^-$ in solution and the respective adsorbed species is $-8\%_0$. Hence the value of 1.0320 is not the fractionation factor α_B between dissolved $B(OH)_3$ and $B(OH)_4^-$. One might, however, estimate α_B from the data of Palmer et al. (cf. remark at the end of this section).

The example discussed above points out the usefulness of the theoretical approach for determining fractionation factors from first principles (for review see e.g. Richet et al., 1977) if experimental values are not available. Crucial for the calculation of fractionation factors are the vibrational frequencies of the molecules under consideration.

Table 3.4.10: Calculated fractionation factor between $B(OH)_3$ and $B(OH)_4^-$ for boron isotope exchange (Kakihana et al., 1977).

T_c (°C)	$\alpha_{(B(OH)_3 - B(OH)_4^-)}$
0	1.0206
5	1.0204
15	1.0199
25	1.0194
40	1.0187
50	1.0181
60	1.0177

In order to simplify the calculations for boron isotope partitioning between the dissolved boron species, Kakihana and Kotaka (1977) approxi-

---

[11]Fractionation factors greater than one, i.e. $1.0320 = 0.969^{-1}$ are used here to compare the values of Palmer et al. (1987) (they considered $R(\text{adsorbed})/R(\text{solution})$ which is smaller than 1) to calculated values by Kakihana et al. (1977) (they considered $R(B(OH)_3)/R(B(OH)_4^-)$ which is greater than 1, cf. Eq. (3.4.48)).

mated the trigonal planar $B(OH)_3$ molecule which consists of 7 atoms (Figure 3.4.34, page 231) by a molecule consisting of only 4 atoms, that is, they approximated each OH group by a single mass point. A similar approach was chosen for the tetrahedral molecule $B(OH)_4^-$. It is unclear if and to what degree these approximations influence the calculated boron isotope fractionation factor. Since the fractionation factor between the dissolved boron species enters almost every calculation associated with boron isotope partitioning in aqueous solutions, future work should either determine the fractionation factor experimentally or should refine the calculation by Kakihana and Kotaka (1977).

Remark:
Based on several assumptions one may derive a value for α_B from the findings of Palmer et al. (1987) as follows. It is assumed that $B(OH)_4^-$ is the sole species adsorbed on the clay which is supported by the increase of adsorbed boron by a factor of 2 as the fraction of $B(OH)_4^-$ in solution increases from 0 to 40%. Assuming further that there is a constant fractionation between dissolved $B(OH)_4^-$ and the adsorbed boron, the isotopic composition of adsorbed boron would follow the curve of the $\delta^{11}B$ of $B(OH)_4^-$ vs. pH (cf. Fig 3.4.30) with a constant offset of $-8\%_0$ (the fractionation between $B(OH)_4^-$ and adsorbed boron as given by Palmer et al. (1987) is $-8\%_0$). At very high pH the adsorbed boron is therefore about $8\%_0$ lighter than $\delta^{11}B$ of the total dissolved boron (all dissolved boron is essentially $B(OH)_4^-$), whereas at very low pH the adsorbed boron is lighter than $\delta^{11}B_{B_T}$ by $8\%_0$ plus an amount which is equal to the fractionation between $B(OH)_3$ and $B(OH)_4^-$ (α_B). As the fractionation between $B(OH)_3$ ($= B_T$ at very low pH) and adsorbed boron as given by Palmer et al. (1987) is $-32\%_0$, α_B is $\sim 24\%_0$. In summary, α_B obtained in this way is equal to the *difference* of the fractionation factors given by Palmer et al., i.e. about $24\%_0$, which is similar (considering the error due to interpolation by Palmer et al.) to the fractionation of $\sim 20\%_0$ calculated by Kakihana et al. (1977).

Exercise 3.21 (*)
In nature, the 'light' stable isotope ^{12}C and ^{16}O is more abundant than the 'heavy' isotope ^{13}C and ^{17}O, respectively, whereas for boron the opposite is true (^{11}B is more abundant than ^{10}B). Considering the number of protons and neutrons in the nuclei of those stable isotopes (cf. Section 3.1), what is different for boron?

Exercise 3.22 (**)
Consider Figure 3.4.30a. Assume that the pK^* value of boric acid is 9.0. How would the curves of $B(OH)_4^-$ and $B(OH)_3$ look like? Would the $\delta^{11}B$ value of $B(OH)_4^-$ at pH 8.2 be heavier or lighter than the value calculated using $pK_B^* = 8.6$ (Figure 3.4.30b).

Exercise 3.23 (***)
Show that the error introduced in the δ values of the dissolved boron species by using δ

values instead of fractional abundances ($r = R/(1 + R)$) for the mass-balance equation
(3.4.47) is smaller than 0.08‰ (derive equations for $r_{B(OH)_4^-}$ and $r_{B(OH)_3}$, calculate
δ values from these equations and compare the results to the δ values calculated using
Eqs. (3.4.49) and (3.4.50)).

3.4.3   Boron in calcium carbonate

During precipitation of $CaCO_3$ from aqueous solution, boron is incorporated into the calcite or aragonite lattice. The concentration of boron in
the crystal is small (in the order of parts per million) compared to the concentration of carbon in the crystal. Boron concentrations in modern marine
carbonates as measured by Hemming and Hanson (1992) varied over a considerable range from 10.9 to 75.1 ppm.[12] In contrast, the variation in the
boron isotopic composition of these carbonates is small (Figure 3.4.30b).

Provided that only the charged species is adsorbed on the calcite it
follows that the boron isotopic composition of calcite increases with pH,
similar to the increase of $\delta^{11}B$ of $B(OH)_4^-$ with pH. In recent precipitation
experiments with inorganic calcite at different pH this hypothesis has been
corroborated (Sanyal et al., 2000) (Figure 3.4.31). Inorganic calcite was
precipitated from artificial seawater with a dissolved boron concentration
~ 17 times higher than in seawater. As seeds were used for precipitation of
which the overgrowth had to be analyzed the high boron concentration was
necessary to yield enough boron for mass spectrometry. Sanyal et al. (2000)
studied boron isotope fractionation at three different pH values, 7.9, 8.3, and
8.6, and observed a clear increase of $\delta^{11}B$ with increasing pH - the slope of
$\delta^{11}B$ of the inorganic calcite vs. pH is compatible with the calculated slope
of $\delta^{11}B$ of the dissolved species $B(OH)_4^-$ vs. pH (Figure 3.4.31)[13]. It is noted,
however, that if the theoretically calculated fractionation between $B(OH)_4^-$
and $B(OH)_3$ (Kakihana et al., 1977) is correct, then the boron isotopic
composition of the inorganic calcite is about 5‰ lighter than $B(OH)_4^-$.

---

[12]Here the unit ppm means parts per million by weight. For instance, the total boron
concentration in modern seawater is 4.52 ppm, which may be expressed as 4.52×10^{-6} kg B
per kg seawater = 4.52 mg kg^{-1}. As the atomic weight of boron is 10.811, this translates
into 418 μmol kg^{-1}.

[13]The $\delta^{11}B$ values obtained by Sanyal et al. (2000) who used artificial seawater with
a $\delta^{11}B$ of total dissolved boron (B_T) of $+1$‰ were converted to the natural seawater
scale ($\delta^{11}B_{B_T} = 39.5$‰) by $\delta^{11}B_{NS} = 1.03846 \times \delta^{11}B_{AS} + 38.46$, where 1.03846 is a
fractionation factor expressing the difference between the artificial and natural seawater:
$\alpha_{NS-AS} = (\delta_{NS} + 10^3)/(\delta_{AS} + 10^3) = (39.5 + 10^3)/(1 + 10^3) = 1.03846$. Finally, the $\delta^{11}B$
values of -19.2, -16.8, and -13.8‰ (artificial seawater) at pH 7.9, 8.3, and 8.6 give 18.5,
21.0, and 24.1‰ (natural seawater, cf. Figure 3.4.31)

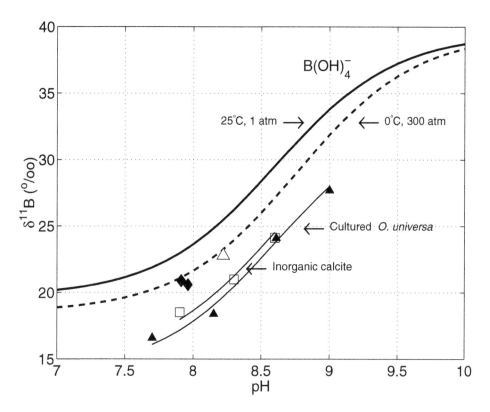

Figure 3.4.31: Boron isotopic composition of $B(OH)_4^-$ at 25°C, $P = 1$ atm (solid line, $\alpha_B = 1.0194$, $pK_B^* = 8.60$) and at 0°C, $P = 300$ atm (dashed line, $\alpha_B = 1.0206$, $pK_B^* = 8.75$). Also shown are $\delta^{11}B$ values of inorganic calcite (squares), cultured plank-tonic foraminifera (*Orbulina universa*, filled triangles), Holocene planktonic foraminifera (*Globigerinoides sacculifer*, open triangle), and Holocene mixed benthic foraminifera (di-amonds). Data are from Sanyal et al. (1995; 1996; 2000).

A similar behavior has been observed in biogenic calcite of the plank-tonic foraminifer *Orbulina universa* (Figure 3.4.31) which was cultured at different pH (Sanyal et al., 1996). In this case, the $\delta^{11}B$ also follows the slope of the calculated line but the calcite is about 6‰ lighter than $B(OH)_4^-$. It appears that the boron isotopic composition of *Orbulina universa* is lighter than that of the planktonic foraminifer *Globigerinoides sacculifer* (open tri-angle in Figure 3.4.31; data from Holocene sediment core samples from the Atlantic and Pacific Ocean, Sanyal et al., 1995). This feature is unlikely to be an artifact of the culture experiments since living *O. universa* from the Southern California Bight and shells from sediment coretop samples have $\delta^{11}B$ values very similar to those $\delta^{11}B$ values of cultured *O. universa* at the same pH (~ 8.15). Most importantly, recent results of culture experiments with *G. sacculifer* showed that the response of δ^{11}B in *G. sacculifer* to changes in the ambient seawater *p*H parallels the response in *O. universa* - the δ^{11}B values in *G. sacculifer* indeed being consistently heavier than those in *O. universa* (A. Sanyal, J. Bijma, pers. comm.). The most simple explanation of the difference between *O. universa* and *G. sacculifer* is a vital effect, possibly adding a constant (negative) offset at any *p*H to the isotopic signal of $B(OH)_4^-$ which is thought to be adsorbed on the calcite (Sanyal et al., 1996).

In summary, the δ^{11}B of biological (and of inorganic) $CaCO_3$ shows a clear trend with *p*H of the solution from which it is formed - this trend is compatible with the theoretically predicted slope of δ^{11}B of $B(OH)_4^-$ vs. *p*H (Figure 3.4.31). Consequently, the boron isotopic composition of foraminiferal calcite may be used as a paleo-*p*H recorder (e.g. Sanyal et al., 1995; Palmer et al., 1998). This statement has recently been corroborated by demonstrating that kinetic effects associated with boron incorporation into biogenic $CaCO_3$ are unlikely (Zeebe et al., 2001). It is important to note, however, that in order to employ the potential of stable boron isotopes, other crucial parameters such as the residence time of boron in the ocean (Lemarchand et al., 2000) and vital effects have to be carefully evaluated as well.

The seawater *p*H at which the carbonate formed may be calculated from the boron isotopic composition of the carbonate as follows. Based on the assumption that the isotopic composition of the carbonate, $\delta^{11}B_c$, is equal to the isotopic composition of the charged species[14], $\delta^{11}B_{(B(OH)_4^-)}$, Eqs. (3.4.44) and (3.4.45) can be used to solve Eq. (3.4.50) for $[H^+]$:

$$pH = -\log([H^+])$$

$$= pK_B^* - \log\left(-\frac{\delta^{11}B_{B_T} - \delta^{11}B_c}{\delta^{11}B_{B_T} - \alpha_B\delta^{11}B_c - \varepsilon_B}\right) \tag{3.4.51}$$

where pK_B^* is the pK^* value of boric acid ($pK_B^* = 8.60$ at $T = 25°C$, $S = 35$ (DOE, 1994)), $\delta^{11}B_{B_T} = 39.5‰$ is the boron isotopic composition of modern seawater, $\alpha_B = 1.0194$ is the fractionation factor between $B(OH)_4^-$ and $B(OH)_3$ at $25°C$ (Kakihana et al., 1977), and $\varepsilon_B = (\alpha_B - 1) \times 10^3$. For example, if the measured boron isotopic composition of the carbonate, $\delta^{11}B_c$, is $24‰$, we have:

$$pH = 8.60 - \log\left(-\frac{39.5 - 24}{39.5 - 1.0194 \times 24 - 19.4}\right)$$

$$= 8.05 . \tag{3.4.52}$$

---

[14]Otherwise a constant offset may be added to the isotopic composition of the carbonate.

The calculated pH value depends on pK_B^*, for which different values have been used in the literature (cf. Section 3.4.2). One way to avoid such complications is to investigate changes in seawater pH but not absolute values (e.g. Sanyal et al., 1995). The pH estimates are also influenced by the fractionation factor α_B (see discussion in Section 3.4.2). For instance, a change of α_B by 0.001 (ε_B changes by 1‰) results in a change of the estimated pH in Eq. (3.4.52) of approximately 0.1 pH units.

Exercise 3.24 (**)
The boron isotope composition of *Globigerinoides sacculifer* from sediment cores (last glacial) have been analyzed to be ~ 25‰, whereas modern *Globigerinoides sacculifer* have δ^{11}B values of about 23‰. What would you conclude about the surface ocean pH in the glacial?

Exercise 3.25 (*)
Provided that the surface ocean pH did change according to the boron isotopes, what happened to the carbonate ion concentration if ΣCO_2 was approximately the same as today? Is this consistent with lower CO_2 concentrations in the glacial atmosphere as observed in ice cores?

3.4.4   Boron abundances in calcium carbonate

The isotopic composition of the marine carbonates as observed by Hemming and Hanson (1992) appears to be correlated to the boron abundances (Figure 3.4.32). Values of δ^{11}B increase with increasing abundances of boron in the carbonates. A similar behavior was observed in the shells of foraminifera (Sanyal et al., 1996) and in inorganic calcite (Sanyal et al., 2000) with a strong correlation between boron concentration and δ^{11}B (Figure 3.4.32). This behavior is consistent with a model in which $B(OH)_4^-$ is adsorbed on the carbonate surface. As pH increases the concentration of $B(OH)_4^-$ and its isotopic composition increase (Figure 3.4.30). The charged species $B(OH)_4^-$ is more likely to be attracted to the crystal surface - increased concentrations of $B(OH)_4^-$ in solution (at higher pH) would therefore result in higher abundances of boron in the carbonates. Simultaneously, the boron isotopic composition of $B(OH)_4^-$ and consequently that of the carbonate increases. This model is compatible with the relationship between boron abundances and boron isotopic composition observed in various carbonates (Figure 3.4.32). The relationship is, however, different for example for calcite and aragonite - this subject is discussed below.

Hemming and Hanson (1992) suggested the following mechanism for boron substitution in the CO_3^{2-} site in carbonate:

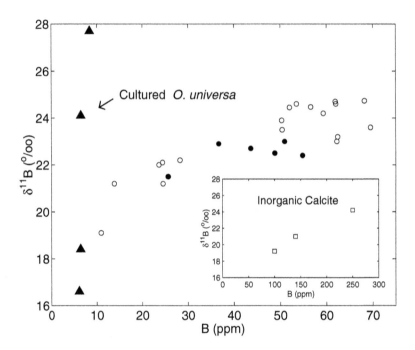

Figure 3.4.32: Boron isotopic compositions and boron abundances in carbonates. Filled circles: marine high-Mg calcite and calcite, open circles: marine aragonite (Hemming and Hanson, 1992). Squares: inorganic calcite (Sanyal et al., 2000). Triangles: foraminiferal calcite (Sanyal et al., 1996). As pH (and $[B(OH)_4^-]$) increase, $\delta^{11}B$ and the boron abundances of the carbonates increase. This behavior is consistent with $B(OH)_4^-$ being adsorbed on the carbonate surface.

$$CaCO_3 + B(OH)_4^- \rightarrow Ca(HBO_3) + HCO_3^- + H_2O \qquad (3.4.53)$$

An exchange distribution coefficient, K_D, for boron substitution in carbonate may then be expressed as:

$$K_D = \frac{[HBO_3^{2-}/CO_3^{2-}]_{\text{solid}}}{[B(OH)_4^-/HCO_3^-]_{\text{fluid}}} , \qquad (3.4.54)$$

describing the ratio of the boron-to-carbon ratio in the solid vs. the borate-to-bicarbonate ratio in the fluid. It is important here that the ratio in the fluid refers to the borate ion concentration and not to the total dissolved boron. The boron distribution coefficient can be calculated from, e.g., the data of Hemming and Hanson (1992) (Table 3.4.11) as follows:

$$K_D = \frac{[B(ppm)] \times 10^{-6} \times M_{CaCO_3}/M_B}{[B(OH)_4^-/HCO_3^-]}$$

where $M_{CaCO_3} = 100$ g and $M_B = 10.8$ g are the molar masses of $CaCO_3$ and B, respectively. Using the boron isotopic composition and Eq. (3.4.51), the pH and the concentration of $B(OH)_4^-$ can be estimated. Hemming and Hanson (1992) found that their data can be best approximated by assuming a HCO_3^- concentration of 142 ppm (~ 2330 μmol kg^{-1}), which gives an average value for K_D of 0.012.

Table 3.4.11: Examples of calculated boron distribution coefficient K_D (data from Hemming and Hanson (1992)).

$\delta^{11}B_{CaCO_3}$ (‰)	B (ppm)	$B(OH)_4^-$ [a] (μmol kg^{-1})	HCO_3^- (μmol kg^{-1})	K_D
22.10	24.4	51	2330	0.010
22.90	36.7	68	2330	0.012
24.60	53.5	104	2330	0.011

[a] $B(OH)_4^-$ was estimated from $\delta^{11}B_{CaCO_3}$ using Eqs. (3.4.46) and (3.4.51).

Table 3.4.11 lists three examples of the data given by Hemming and Hanson (1992), whereas the complete data set is shown in Figure 3.4.33. Also shown are results from inorganic calcite and foraminiferal calcite for which the boron distribution coefficient K_D was determined from the slope of the linear fit to the data.[15] The inorganic and foraminiferal calcite ($K_D = 0.001$ and 0.0001, respectively) obviously have a much smaller K_D than the aragonite and the high-Mg calcite ($K_D = 0.012$, cf. also Hemming et al., 1995). This behavior might be explained by differences in the coordination of boron in aragonite and calcite (see Section 3.4.5). It is noted that the linear fit to the data of the planktonic foraminifer *Orbulina universa* does not intersect the vertical axis at zero, suggesting that even if $B(OH)_4^-$ in solution goes to zero, the concentration of boron in the calcite does not go to zero. If $B(OH)_4^-$ was the sole species adsorbed on the calcite, however, it should. Considering foraminifera, the model (Eq. (3.4.53)) might therefore not work adequately at low concentrations of boron in the carbonate and $B(OH)_4^-$ in solution. Nevertheless, there is a clear correlation between the boron-to-carbon ratio in the solid and the borate-to-bicarbonate ratio in the fluid in all carbonates studied (Figure 3.4.33), suggesting that the model does a good job in general.

---

[15] The K_D of the inorganic calcite was calculated assuming $\Sigma CO_2 = 2200$ μmol kg^{-1}, $T = 25°C$, $S = 35$.

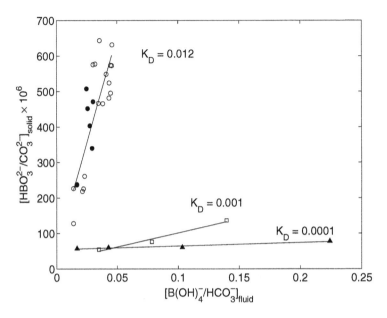

Figure 3.4.33: Boron distribution coefficient K_D in carbonates. Filled circles: marine high-Mg calcite and calcite, open circles: marine aragonite (Hemming and Hanson, 1992). Squares: inorganic calcite precipitated in artificial seawater (Sanyal et al., 2000). Triangles: foraminiferal calcite from culture experiments with *Orbulina universa* (Sanyal et al., 1996). In order to plot the data of inorganic calcite into the same graph, the boron concentration of the inorganic calcite and of the artificial seawater (fluid) was divided by 17 (this does not affect the value of K_D, cf. Eq. (3.4.54)).

3.4.5   Boron coordination

Boron is predominantly in trigonal coordination in calcite, whereas boron is in tetrahedral coordination in aragonite (Hemming et al., 1998a). As discussed in the preceding section, boron isotope analyses of calcium carbonate revealed that the charged species $(B(OH)_4^-)$ is most likely taken up during precipitation which is evident from the $\delta^{11}B$ of e.g. various marine carbonates.

The dissolved forms of boron, $B(OH)_3$ and $B(OH)_4^-$, exhibit planar trigonal and tetrahedral coordination, respectively (Figure 3.4.34). Thus, the incorporation of boron into the calcite lattice requires a transition from the tetrahedral structure of the adsorbed species $(B(OH)_4^-)$ to an (ultimately) trigonal coordination in the calcite lattice. No such transition is necessary for the incorporation into aragonite, in which boron is in tetrahedral coordination. This might explain the higher boron distribution coefficient in marine aragonite relative to foraminiferal calcite (Figure 3.4.33). The same behavior was observed by Hemming et al. (1995) in synthetic aragonite, high Mg-calcite, and calcite (listed according to their relative boron abundances).

Species   % at pH 8.2   Structure

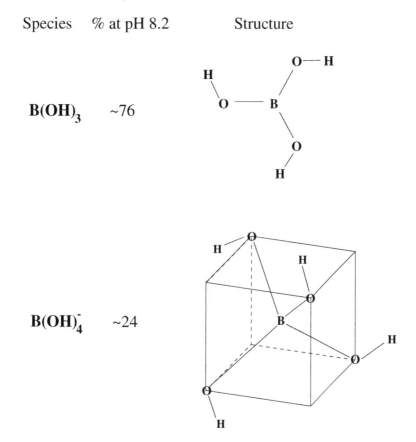

B(OH)$_3$ ~76

B(OH)$_4^-$ ~24

Figure 3.4.34: Structure and percentage (at *p*H 8.2) of the dissolved boron species B(OH)$_3$ (planar trigonal) and B(OH)$_4^-$ (tetrahedral). Bond lengths and bond angles are only schematic.

With the discussion of boron isotopes we finish our survey on selected elements of the carbonate system. The next section describes the theoretical background of isotope fractionation in chemical equilibrium. It is primarily addressed to the reader who is interested in understanding physico-chemical phenomena from first principles. In order to comprehend the details of the calculations involved (Appendix C.10), it is probably advantageous to have a basic knowledge of physics. The theory is applied to the carbon isotope equilibrium between dissolved CO_3^{2-} ion and $CO_2(g)$ (Section 3.5.3) which represents a contribution to a current problem - the results of this section should be accessible to everybody.

3.5 Thermodynamic properties of isotopic substances

This section comprises an introduction to the theoretical basis of isotopic fractionation under equilibrium conditions. It is of particular interest, since several fractionation phenomena can be understood on this theoretical basis. Moreover, since several isotopic fractionation factors have not been determined experimentally until today (e.g. between the dissolved boron species), theoretically derived values are widely used. The theory described below permits the calculation of these factors. Besides this rather practical motivation it is also fascinating to see how isotopic fractionation (which has become such a powerful tool in modern science) can be derived from first principles. The section contains mathematical calculations which may be skipped by a reader who is more interested in the practical application of the theory.

In 1947, Harold C. Urey published a paper on the thermodynamic properties of isotopic substances in which he laid the foundations for the utilization of stable isotopes in various scientific disciplines such as stable isotope geology or paleoceanography (Urey, 1947). Urey had obviously foreseen the impact of his theoretical predictions on the reconstruction of paleotemperatures. The conclusive sentence of his paper reads: "These small differences (in thermodynamic properties of isotopes) make possible the concentration and separation of the isotopes of some of the elements and may have important applications as a means of determining the temperatures at which geological formations were laid down." A professor, working on biomineralization in Israel, had once talked to Harold Urey and reported that Urey invented the stable isotope paleothermometer because he wanted to find out why the dinosaurs had disappeared.

The following introductory remarks are meant to elucidate the motivation for the discussion and calculations presented in this section.

3.5.1 Physical background

The physical background of isotope fractionation can be summarized as follows. Chemical equilibrium (or more general: thermodynamic equilibrium) is a function of the energy of the system which in turn is a function of the energy of the molecules. Consider two molecules of the same chemical species, one of them containing the heavy isotope, the other one containing the light isotope. The difference in the masses of those molecules change the energy of the molecules which therefore leads to a shift in chemical equilibrium. This phenomenon is called isotopic fractionation (in equilibrium) - the theory describing the underlying physics is quantum mechanics.

As first demonstrated by Waldmann (1943), Urey (1947) and Bigeleisen and Mayer (1947) isotopic fractionation factors can be derived from first principles of statistical thermodynamics and quantum mechanics. The chemical and isotopic equilibrium of a thermodynamic system is characterized by a minimum of the Gibbs free energy. Every single molecule contributes to this energy through its translational, rotational, and vibrational energy. It will be shown in detail how isotopic fractionation can be calculated from the energies of isotopic molecules. Strictly, the theory is valid only for ideal gases. Under certain assumptions, however, it may also be applied to solutions or solid states.

Isotopic fractionation is a quantum-mechanical effect; it is therefore necessary to repeat some basic features of the calculation of energy levels in a quantum-mechanical system. The energy spectrum of such a system is not continuous - it is quantized - which means that the total energy is an integral multiple of elementary discrete energy units, called quanta. The population of a certain energy level is a function of temperature. When the temperature is at absolute zero, only the lowest energy level is populated (zero-point energy); with increasing temperature, however, also higher energy levels will be populated. (The calculation of the distribution of quanta among energy levels requires the calculation of the partition function, see Appendix C.10).

Applying this feature to a system of two gases A and B in chemical equilibrium, the energy of the system is characterized by the population of the energy levels of the respective molecules. At very low temperature the

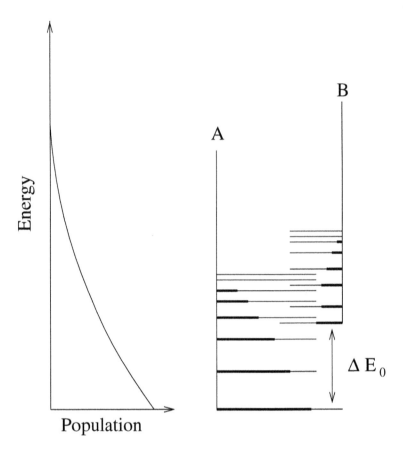

Figure 3.5.35: Schematic presentation of the energy states of two molecules A and B (after Atkins, 1998). The population (Boltzmann distribution, left graph) corresponds to the number of quanta of a certain energy. Energy states of molecules A and B are discrete and populated according to the Boltzmann distribution (thick lines, right graph). The energy difference between ground states of the molecules is ΔE_0.

zero-point energy level of the one molecule (e.g of molecule A) is populated which has the lower zero-point energy (Figure 3.5.35, cf. Atkins, 1998). As temperature rises, higher energy levels of A are populated, until the energy difference of the ground state (ΔE_0) between A and B is achieved. Subsequently, more energy levels of molecule B are populated, meaning that the concentration of B increases with temperature. Though simplified, this is a quantum mechanical interpretation of the temperature dependence of the chemical equilibrium constant (for a more detailed description also the density of states has be taken into account). The energy levels of the system are characterized by the energy levels of translation, rotation, and vibration of the molecules[16].

The translational energy of a gaseous substance results from the motion of each molecule. In classical mechanics, and in one dimension, this is expressed by the well-known formula $E_{kin} = mv^2/2$, where E_{kin} is the kinetic energy, and m and v are the mass and the velocity of the molecule. At a given temperature the distribution of velocities of the molecules is not homogeneous. There is a large number of molecules with velocities (absolute values) close to a certain mean velocity, some with higher and some with a smaller velocity than the mean. As is described in Appendix C.10, the quantum mechanical calculation of the translational energy of a molecule yields a dependence on the mass of the molecule. Thus, differences in the mass of isotopic substances lead to different translational energies at a given temperature. Since the equilibrium distribution of isotopes among different chemical species is a function of the energy of the molecules, different energies lead to isotopic fractionation.

A similar quantum mechanical calculation of the rotational energy of a molecule reveals a dependence on the moments of inertia of the molecule. For a homonuclear diatomic molecule (one may picture such a molecule as a dumb-bell) the moment of inertia is simply $mr^2/2$, where m is the mass of an atom and r is the bond length (i.e. the distance between the nuclei). As for the case of translation, different masses of isotopes lead to different rotational energies and thus to fractionation effects.

The vibrational energy is closely related to the internal structure of the molecule. The atoms of a molecule are bound such that the position of each atom relative to the others appears to be fixed at its equilibrium position. However, there is always movement of the atoms around their equilibrium position (even at a temperature of absolute zero). A diatomic molecule for example is no rigid dumb-bell; its behavior is better characterized by two masses connected by a spring, with each mass vibrating around its equilibrium position. The zero-point energy of a diatomic molecule is $h\nu/2$, where $h = 6.626 \times 10^{-34}$ J s is Planck's constant and ν is the fundamental frequency of vibration. The typical frequencies of molecular vibrations are on the order of 10^{13} s^{-1}, whereas the corresponding wavelengths are on the order of 10^{-5} m (near infrared spectrum).

Depending on the structure of the molecule a variety of complicated vi-

---

[16]The electronic contribution (transitions of electrons in the shell of atoms) can usually be neglected in this context because the excited electronic levels of atoms or molecules lie far above the ground state. Only at temperatures of several thousand degrees these levels become populated.

brations are possible which may be hard to describe mathematically. However, one knows from classical mechanics that every vibrational state of a system of N masses can be described by a superposition of the $3N - 6$ fundamental eigenmodes of the system ($3N - 5$ for a linear molecule). For example, the linear CO_2 molecule has $3 \times 3 - 5 = 4$ fundamentals of which two have the same energy (this is called a twofold degeneracy, see Figure 3.5.36 on page 240). Every vibrational state of the CO_2 molecule can therefore be described by a superposition of the 4 fundamental eigenmodes. The essence of the discussion so far is: the vibrational energy of a molecule is completely described by the fundamental vibration modes (energy states) of the molecule and the mean number of energy quanta populating a certain energy state.

The population of higher energy levels of translation, rotation and vibration when temperature rises from absolute zero to e.g. room temperature is not simultaneous. With increasing temperature higher energy levels of translation are populated first, followed by higher rotational and vibrational energy levels being populated at higher temperatures. This fact can be understood by comparing the spacing between adjacent energy levels (ΔE) with the mean thermal energy of a molecule (kT). For translation the ratio of ($\Delta E/kT$) is in the order of 10^{-20} at room temperature (see Appendix C.10, Figure C.10.5), i.e. the spacing is very small and an enormous number of higher energy levels are populated - the spectrum is virtually continuous. For rotation, the ratio is in the order of 10^{-3} at room temperature - also a significant number of higher energy levels of rotation are populated (the ratio is unity at about 1 K). In contrast, for vibration this ratio is $\sim 10^1$ at room temperature, i.e. the gap between adjacent energy levels is large compared to the mean thermal energy. The transition between these states is not continuous - the spectrum is discrete.

In summary, given the molecular masses, moments of inertia, and the fundamentals of vibration of the molecular species in a chemical system, the equilibrium distribution of the concentrations (and of isotopes) can be calculated as a function of temperature. The derivation of the theory is demonstrated explicitly in Appendix C.10.

3.5.2   Oxygen isotope equilibrium between $CO_2(g)$ and $H_2O(l)$ at 25°C

It is shown in Appendix C.10 that the isotopic fractionation in equilibrium between two gaseous species can be calculated from the reduced partition function ratio. For diatomic molecules this ratio $(Q'/Q)_r$ is given by:

$$\left(\frac{Q'}{Q}\right)_r = \frac{s}{s'}\frac{u'}{u}\frac{\exp(-u'/2)}{1-\exp(-u')}\frac{1-\exp(-u)}{\exp(-u/2)}$$

with s and s' being the symmetry number of the molecule containing the light and heavy isotope, respectively, and

$$u = hc\omega/kT$$
$$u' = hc\omega'/kT$$

where h is Planck's constant, c is the speed of light, k is Boltzmann's constant, T is the absolute temperature in Kelvin, and ω and ω' being the wavenumbers of vibration of the molecule containing the light and heavy isotope, respectively.

As an example of the application of the theory the exchange of ^{18}O and ^{16}O between $CO_2(g)$ and liquid H_2O is presented. The value of this isotope equilibrium has been determined experimentally with high precision, hence the theoretically derived value can be compared to the measured value. The oxygen fractionation factor between gaseous CO_2 and water vapor is determined first, then a correction factor is applied to account for the isotope equilibrium between liquid water and water vapor.

The chemical species of the system are $C^{16}O^{16}O$, $C^{18}O^{16}O$, $C^{18}O^{18}O$, $H_2^{16}O$, and $H_2^{18}O$, where labels such as 'g' for gaseous or 'v' for vapor have been omitted. The fractionation factor α between CO_2 and water is defined by the ratio of the concentrations of ^{18}O to ^{16}O present in CO_2 divided by the corresponding ratio in H_2O:

$$\alpha_{(CO_2-H_2O)} = \frac{2[C^{18}O^{18}O] + [C^{18}O^{16}O]}{2[C^{16}O^{16}O] + [C^{18}O^{16}O]} \bigg/ \frac{[H_2^{18}O]}{[H_2^{16}O]} \ .$$

Note that this expression follows from the definition of the fractionation factor. For example, the number of ^{18}O atoms in $C^{18}O^{18}O$ is two, whereas the number of ^{18}O atoms in $C^{18}O^{16}O$ is one. To count the total number of ^{18}O atoms present in CO_2, the concentrations of $C^{18}O^{18}O$ and $C^{18}O^{16}O$ have to be multiplied by a factor of two and one, respectively. It can be shown that α may be very well approximated by (Urey, 1947):

$$\alpha_{(CO_2-H_2O)} = \frac{[C^{18}O_2]^{\frac{1}{2}}}{[C^{16}O_2]^{\frac{1}{2}}} \bigg/ \frac{[H_2^{18}O]}{[H_2^{16}O]}$$

which corresponds to the isotopic exchange reaction

$$1/2\ C^{16}O_2 + H_2^{18}O \ \rightleftharpoons \ 1/2\ C^{18}O_2 + H_2^{16}O$$

To calculate the fractionation factor α:

$$\alpha_{(CO_2-H_2O)} = \left(\frac{Q(C^{18}O_2)}{Q(C^{16}O_2)}\right)^{\frac{1}{2}} \Big/ \left(\frac{Q(H_2^{18}O)}{Q(H_2^{16}O)}\right) .$$

the partition function ratio $Q(C^{18}O_2)/Q(C^{16}O_2)$ and $Q(H_2^{16}O)/Q(H_2^{18}O)$ must be determined.

Because confusion often arises from the relation between the fractionation factor α, the partition function ratio (cf. Appendix C.10), the reduced partition function ratio, and β factors, the equations connecting these quantities are briefly summarized here. The fractionation factor ($\alpha_{(A-B)}$) between two chemical species A and B is commonly expressed as the ratio of β factors:

$$\alpha_{(A-B)} = \frac{\beta_A}{\beta_B} .$$

On the other hand, the β factor of e.g. A is related to the reduced partition function ratio by

$$\beta_A = \left(\frac{Q'_A}{Q_A}\right)_r^{\frac{1}{n}}$$

where the prime indicates the presence of the heavy isotope and n is the number of isotopic atoms being exchanged. Since the reduced partition function ratio is given by:

$$\left(\frac{Q'_A}{Q_A}\right)_r = \left(\frac{Q'_A}{Q_A}\right) \Big/ \left(\frac{m'}{m}\right)^{\frac{3}{2}n} ,$$

where m' and m are the masses of the heavy and light isotopic atoms being exchanged, it follows:

$$\beta_A = \left(\frac{Q'_A}{Q_A}\right)^{\frac{1}{n}} \Big/ \left(\frac{m'}{m}\right)^{\frac{3}{2}} . \tag{3.5.55}$$

The fractionation factor is the ratio of β factors, thus terms such as the last of Eq. (3.5.55) cancel, yielding:

$$\alpha_{(A-B)} = \left(\frac{Q'_A}{Q_A}\right)^{\frac{1}{n}} \Big/ \left(\frac{Q'_B}{Q_B}\right)^{\frac{1}{k}}$$

where n and k are the number of isotopic atoms being exchanged in molecule A and B, respectively. For the example considered:

$$1/2\ C^{16}O_2 + H_2^{18}O \rightleftharpoons 1/2\ C^{18}O_2 + H_2^{16}O$$

we have $n = 2$ for the CO_2 molecule, whereas $k = 1$ for H_2O. Thus, the fractionation factor is given by

$$\alpha_{(CO_2-H_2O)} = \left(\frac{Q(C^{18}O_2)}{Q(C^{16}O_2)}\right)^{\frac{1}{2}} \Big/ \left(\frac{Q(H_2^{18}O)}{Q(H_2^{16}O)}\right) .$$

The β factor of $CO_2(g)$ (O-exchange)

The β factor for $CO_2(g)$ is calculated according to Eq. (C.10.54) to evaluate separately the translational, rotational, and vibrational contribution to the partition function ratio. The partition function ratio is (note that the symmetry numbers of $C^{18}O_2$ and $C^{16}O_2$ are the same):

$$\frac{Q'}{Q} = \left(\frac{M'}{M}\right)^{\frac{3}{2}} \left(\frac{I'}{I}\right) \prod_i \frac{\exp(-u_i'/2)}{1 - \exp(-u_i')} \frac{1 - \exp(-u_i)}{\exp(-u_i/2)} \qquad (3.5.56)$$

where it has been used that the ratio of principal moments of inertia of a linear molecule such as CO_2 reduces to I'/I.

The translational contribution which is independent of temperature is readily calculated:

$$\left(\frac{Q'}{Q}\right)_{\text{tr}} = \left(\frac{M_{C^{18}O_2}}{M_{C^{16}O_2}}\right)^{\frac{3}{2}} = \left(\frac{47.9983}{43.9898}\right)^{\frac{3}{2}} = 1.1398 \;.$$

The rotational contribution is:

$$\left(\frac{Q'}{Q}\right)_{\text{rot}} = \frac{I'}{I} = \frac{2m'r^2}{2mr^2}$$

where $r = 1.189$ Å is the C$-$O bond length and m' and m are the masses of ^{16}O and ^{18}O, respectively. Thus, we have:

$$\left(\frac{Q'}{Q}\right)_{\text{rot}} = \frac{m'}{m} = \frac{17.9992}{15.9949} = 1.1253 \;.$$

The vibrational contribution is calculated from the normal modes of the $CO_2(g)$ molecule. Carbon dioxide is a linear molecule and has 4 fundamentals of which two have the same energy (twofold degeneracy, see Figure 3.5.36). Vibrational mode I is an 'in phase' vibration of the oxygen atoms, whereas the carbon atom is fixed relative to the center-of-mass of the molecule. The vibrations of the atoms of mode IIa and b occur perpendicular to the axis of the molecule and have the same energy. Mode III is an 'out of phase' vibration.

Several measurements (and calculations for the isotopically substituted molecule) have been carried out to determine the frequencies of the fundamentals of $CO_2(g)$. The wavenumbers reported by Richet et al. (1977) are given in Table 3.5.12. Using these wavenumbers, the vibrational contribution to the partition function can be calculated:

$$\left(\frac{Q'}{Q}\right)_{\text{vib}} = \prod_i \frac{\exp(-u_i'/2)}{1 - \exp(-u_i')} \frac{1 - \exp(-u_i)}{\exp(-u_i/2)}$$

I:

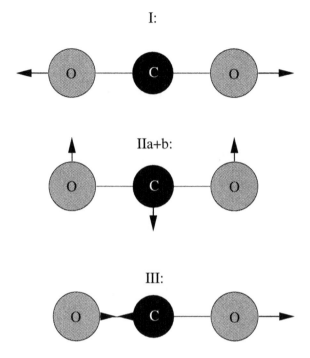

IIa+b:

III:

Figure 3.5.36: Fundamentals of the $CO_2(g)$ molecule. The vibrational mode IIb is analogous to mode IIa, but perpendicular to the plane of the paper (cf. Atkins (1998)).

where $u_i = hc\omega_i/kT$. At a temperature of 25°C, we have:

$$\left(\frac{Q'}{Q}\right)_{vib} = (1.2032)(1.0263)^2(1.0920) = 1.3838 \ .$$

Obviously, the vibrational contribution is larger than the translational and rotational contribution at room temperature.

Finally, the β factor or the reduced partition function ratio $((Q'/Q)^{1/2}$

Table 3.5.12: Wavenumbers of normal modes of the $CO_2(g)$ molecule in cm^{-1} (Richet et al., 1977).

Molecule	ω_1	ω_2	ω_3
$^{12}C^{16}O_2$	1353.637	672.625(2)[a]	2396.269
$^{12}C^{18}O_2$	1277.250	662.700(2)	2359.810

[a] The twofold degeneracy of ω_2 is indicated by '(2)'.

divided by the factor $(m'/m)^{3/2}$, which is 1.1937 in this case) is:

$$\beta_{CO_2} = \left(\frac{Q'}{Q}\right)_r^{\frac{1}{2}} = [(1.1398)(1.1253)(1.3838)]^{\frac{1}{2}}(1.1937)^{-1} = 1.1160$$

A refined calculation, including e.g. anharmonicity constants for the potential energy gives a value of 1.1174 at 25°C (Richet et al., 1977).

The β factor of $H_2O(v)$ (O-exchange)

The β factor for water vapor is calculated analogously to the calculation of the β factor for CO_2. The partition function ratio is (note that the water molecule has three principal moments of inertia):

$$\frac{Q'}{Q} = \left(\frac{M'}{M}\right)^{\frac{3}{2}} \left(\frac{I_1'I_2'I_3'}{I_1I_2I_3}\right)^{\frac{1}{2}} \prod_i \frac{\exp(-u_i'/2)}{1-\exp(-u_i')}\frac{1-\exp(-u_i)}{\exp(-u_i/2)} \qquad (3.5.57)$$

The translational contribution is:

$$\left(\frac{Q'}{Q}\right)_{tr} = \left(\frac{M_{H_2^{18}O}}{M_{H_2^{16}O}}\right)^{\frac{3}{2}} = \left(\frac{20.0148}{18.0106}\right)^{\frac{3}{2}} = 1.1715 .$$

The rotational contribution is

$$\left(\frac{Q'}{Q}\right)_{rot} = \left(\frac{I_1'I_2'I_3'}{I_1I_2I_3}\right)^{\frac{1}{2}}$$

where I's are the three principal moments of inertia. They can be calculated using the definition $I = \sum m_i r_i^2$, where r_i is the distance of the mass i from the axis of rotation. After some geometry one obtains:

$$I_1 = 2m_H R^2 \sin^2\theta$$
$$I_2 = 2m_H R^2 \cos^2\theta \,(1 - 2m_H/M)$$
$$I_3 = 2m_H R^2 \,(1 - 2m_H \cos^2\theta/M)$$

where m_H is the mass of the hydrogen atom, $R = 0.957$ Å is the O−H bond length, $2\theta = 104.5°$ is the bond angle (HOH), and M is the mass of the molecule. Thus, the rotational contribution reduces to:

$$\left(\frac{Q'}{Q}\right)_{rot} = \left[\frac{(1 - 2m_H/M')}{(1 - 2m_H/M)}\frac{(1 - 2m_H \cos^2\theta/M')}{(1 - 2m_H \cos^2\theta/M)}\right]^{\frac{1}{2}}$$

$$= \left(\frac{0.8993}{0.8881}\frac{0.9623}{0.9581}\right)^{\frac{1}{2}}$$

$$= 1.0085 .$$

I:

II:

III:

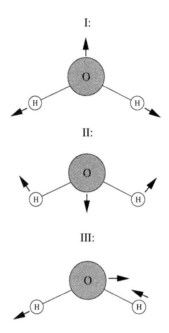

Figure 3.5.37: Fundamentals of the H_2O molecule (cf. Atkins (1998)).

the rotational contribution is small in this case (oxygen exchange) since the position of the oxygen atom virtually coincides with the center-of-mass of the molecule about which it rotates.

The vibrational contribution is calculated from the 3 normal modes of the H_2O molecule (Figure 3.5.37). The wavenumbers as reported by Richet et al. (1977) are given in Table 3.5.13. Using these wavenumbers, the vibrational contribution to the partition function can be calculated:

$$\left(\frac{Q'}{Q}\right)_{\text{vib}} = \prod_i \frac{\exp(-u_i'/2)}{1 - \exp(-u_i')} \frac{1 - \exp(-u_i)}{\exp(-u_i/2)}$$

At a temperature of 25°C, we have:

$$\left(\frac{Q'}{Q}\right)_{\text{vib}} = (1.0189)(1.0170)(1.0395) = 1.0772 \ .$$

Finally, the β factor or the reduced partition function ratio ((Q'/Q) divided by the factor $(m'/m)^{3/2}$ which is 1.1937 in this case) is:

$$\beta_{H_2O(v)} = \left(\frac{Q'}{Q}\right)_r = (1.1715)(1.0085)(1.0772)(1.1937)^{-1} = 1.0661 \ .$$

A refined calculation gives a value of 1.0635 at 25°C (Richet et al., 1977).

Oxygen isotope equilibrium between $CO_2(g)$ and $H_2O(l)$ at 25°C

The calculated β factors for CO_2 and H_2O refer to gaseous CO_2 and water vapor. Since we are interested in the fractionation between $CO_2(g)$ and liquid water, the oxygen isotope fractionation between water vapor and liquid water has to be taken into account. At 25°C, the liquid water is enriched in ^{18}O by ca. 9‰ (the α value measured by Majoube (1971) is 1.00937). Thus, the fractionation factor between $CO_2(g)$ and liquid water at 25°C is:

$$\alpha_{(CO_2(g)-H_2O(l))} = \frac{\beta_{CO_2(g)}}{\beta_{H_2O(v)}} \frac{1}{\alpha_{(H_2O(l)-H_2O(v))}}$$

$$= \frac{1.1174}{1.0635} \frac{1}{1.00937}$$

$$= 1.0409 \ .$$

The experimentally determined value by O'Neil et al. (1975) is 1.0412, demonstrating good agreement between theory and experiment (better than 0.4‰ in this case).

3.5.3 Carbon isotope equilibrium between dissolved CO_3^{2-} ion and $CO_2(g)$

The second example considered is the isotopic exchange of ^{13}C and ^{12}C between CO_3^{2-} ion and $CO_2(g)$ in dilute solutions (for the discussion of seawater, see Section 3.2.3). The oxygen isotope equilibrium between CO_2 and H_2O as discussed in Section 3.5.2 is very well known. However, the carbon isotope exchange between carbon dioxide and carbonate ion is still under debate. Thus, the following discussion is a contribution to a current problem rather than a description of a well known problem.

It is emphasized that the theory is strictly valid only for ideal gases. Thus, errors may arise from the fact that CO_3^{2-} is an ion dissolved in water

Table 3.5.13: Wavenumbers of normal modes of the H_2O molecule in cm^{-1} (Richet et al., 1977).

Molecule	ω_1	ω_2	ω_3
$H_2^{16}O$	3835.37	1647.59	3938.74
$H_2^{18}O$	3827.59	1640.62	3922.69

(this problem is discussed below). The overall equilibrium may be written as:

$$^{13}CO_2(g) + {}^{12}CO_3^{2-} \quad \rightleftharpoons \quad {}^{12}CO_2(g) + {}^{13}CO_3^{2-} \ .$$

The equilibrium constant is given by

$$K = \frac{[^{12}CO_2(g)][^{13}CO_3^{2-}]}{[^{13}CO_2(g)][^{12}CO_3^{2-}]}$$

which is equal to the fractionation factor α in this case (note that this is not true in general, e.g. if more than one atom is exchanged). Thus, the equilibrium constant K can be written as:

$$K \ = \ \alpha \ = \ \frac{(Q'_{CO_3^{2-}}/Q_{CO_3^{2-}})_r}{(Q'_{CO_2}/Q_{CO_2})_r} \ =: \ \frac{\beta_{CO_3^{2-}}}{\beta_{CO_2}}$$

where ratios of Q_i's refer to reduced partition function ratios and the primes indicate the presence of the heavy isotope ^{13}C.

The β factor of $CO_2(g)$ (C-exchange)

The β factor for $CO_2(g)$ is calculated according to Eq. (C.10.56) from the frequencies of the molecules only. Note that the translational and rotational contribution are accounted for by Eq. (C.10.56); the terms of the translational and rotational contributions have merely been expressed in terms of vibrational frequencies using the Redlich-Teller product rule (Appendix C.10). For carbon exchange, the rotational contribution is negligible since the position of the carbon atom coincides with the center-of-mass of the molecule about which it rotates. The reduced partition function ratio therefore is (note that the symmetry numbers of $^{13}CO_2$ and $^{12}CO_2$ are the same):

$$\left(\frac{Q'}{Q}\right)_r = \prod_i \frac{u'_i}{u_i} \frac{\exp(-u'_i/2)}{1 - \exp(-u'_i)} \frac{1 - \exp(-u_i)}{\exp(-u_i/2)} \tag{3.5.58}$$

where $u = hc\omega/kT$ and ω and ω' are the wavenumbers of the normal modes of the $^{13}CO_2(g)$ and $^{12}CO_2(g)$, respectively (see Figure 3.5.36). The wavenumbers reported by Richet et al. (1977) are given in Table 3.5.14. Using these wavenumbers, we have $\prod_i u'_i/u_i = 0.9179$ and the reduced partition function ratio at a temperature of 0°C is (note the degeneracy of ω_2):

$$\left(\frac{Q'}{Q}\right)_r = (0.9179)(1.0000)(1.0542)^2(1.1969) = 1.2207 \ .$$

Table 3.5.14: Wavenumbers of normal modes of the $CO_2(g)$ molecule in cm^{-1} (Richet et al., 1977).

Molecule	ω_1	ω_2	ω_3
$^{12}C^{16}O_2$	1353.637	672.625(2)^a	2396.269
$^{13}C^{16}O_2$	1353.680	653.771(2)	2328.037

^{a} The twofold degeneracy of ω_2 is indicated by '(2)'.

A refined calculation, including e.g. anharmonicity constants for the potential energy gives a value of 1.2171 at 0°C (Richet et al., 1977). Table 3.5.15 summarizes the values of the reduced partition function of CO_2 for carbon exchange as given by Richet et al. (1977) in the temperature range 0 to 75°C.

Table 3.5.15: β factors for carbon exchange ($^{13}C/^{12}C$) of $CO_2(g)$ (after Richet et al., 1977).

T_c (°C)	β_{CO_2} (C-exchange)
0	1.2171
10	1.2060
20	1.1958
25	1.1910^a
30	1.1864
40	1.1777
50	1.1697
75	1.1519

^{a} Value was calculated using a fit to the data of Richet et al. (1977).

The β factor of CO_3^{2-} ion (C-exchange)

Analogously to the calculation for $CO_2(g)$ the reduced partition function of the CO_3^{2-} ion is calculated (for chemical structure of CO_3^{2-}, see Figure 3.2.11 on page 171). The vibrational contribution is determined using measured frequencies of the ^{12}C carbonate ion in solution as given by Davis and Oliver (1972). The frequencies of the $^{13}CO_3^{2-}$ ion are calculated by applying the isotopic shift for $^{12}CO_3^{2-}/^{13}CO_3^{2-}$ frequencies as given by Bottinga (1968) and refined by Chacko et al. (1991). The wavenumbers obtained in this way are given in Table 3.5.16. Chacko et al. (1991) used the Redlich-Teller product rule to refine the frequencies given by Bottinga (1968). The logarithmic form of the product rule reads in this case:

$$\frac{3}{2} \ln \left(\frac{M'}{M}\right) + \frac{3}{2} \ln \left(\frac{m}{m'}\right) = \sum_i g_i \ln \frac{\omega_i'}{\omega_i} \qquad (3.5.59)$$

where M's and m's are the masses of the molecules and exchanged isotopes, respectively. Since the same isotopic shift as given by Chacko et al. (1991) was applied to the frequencies used here, they also obey the product rule:

$$\frac{3}{2} \ln \left(\frac{M'}{M}\right) + \frac{3}{2} \ln \left(\frac{m}{m'}\right) = -0.095568$$

$$\sum_i g_i \ln \frac{\omega_i'}{\omega_i} = -0.095574 \ .$$

This result underlines the reliability of the wavenumber shift of the isotopically substituted molecule as given by Chacko et al. (1991).

Table 3.5.16: Fundamentals of the CO_3^{2-} in cm^{-1}, obtained by using data of Davis and Oliver (1972) for the carbonate ion ($^{12}CO_3^{2-}$) in solution and applying the frequency shift of Chacko et al. (1991) for $^{13}CO_3^{2-}$ to these frequencies.

Molecule	ω_1	ω_2	ω_3	ω_4
$^{12}CO_3^{2-}$	1064.00	880.00	1436.00(2)[a]	684.00(2)
$^{13}CO_3^{2-}$	1064.00	852.41	1395.89(2)	681.59(2)

[a] The twofold degeneracy of ω_3 and ω_4 is indicated by '(2)'.

With these wavenumbers, the reduced partition function is calculated as:

$$\left(\frac{Q'}{Q}\right)_r = (0.9089)(1.0000)(1.0770)(1.1116)^2(1.0067)^2 = 1.2258 \ .$$

The β factor for the temperature range from 0 to 75°C is summarized in Table 3.5.17.

As mentioned before, the calculated values can be in error, since a CO_3^{2-} ion in solution is not an ideal gas. Firstly, the applicability of the theory presented requires that there are no electrostatic interactions between the particles of the system considered. In this case, the energy of a single particle can be determined as if it were alone in the box. However, as an ion is charged there are interactions between particles - as a result the ion is e.g. surrounded by a number of water molecules of hydration in solution. Fortunately, the effect of hydration concerns both the molecule containing the heavy isotope and the molecule containing the light isotope. Secondly,

Table 3.5.17: Calculated β factor of CO_3^{2-} ion for C-exchange ($^{13}C/^{12}C$) after Eq. (C.10.54).

T_c (C)	$\beta_{CO_3^{2-}}$ (C-exchange)
0	1.2258
10	1.2134
20	1.2020
25	1.1967
30	1.1916
40	1.1819
50	1.1730
75	1.1533

the volume occupied by the molecules may not be negligible compared to the total volume of the system as it is required for an ideal gas.

The hydration of the ion affects the translational and rotational but also the vibrational energy of the molecule. For example, the wavenumbers of the carbonate ion in solution as given by Davis and Oliver (1972) are 0 to 4% (maximum) lower than those given by Chacko et al. (1991) for the CO_3^{2-} ion in the calcite lattice. This effect may be understood by a simple consideration. The frequency ν of a harmonic oscillator is proportional to $\sqrt{k/m}$, where k is the force constant and m is the mass. Everything else being equal, an increase of the effective mass (e.g. through water molecules of hydration) results in a lower frequency of vibration. Again, this effect holds for both the molecule containing the heavy and the light isotope.

The major contribution to the reduced partition function ratio is the difference in the vibrational frequencies of the $^{13}CO_3^{2-}/^{12}CO_3^{2-}$ molecules. Since this ratio should be independent of hydration effects to a large degree, we are confident that the calculated β factors are reasonable. However, we cannot rule out the possibility that these values are in error because the CO_3^{2-} ion in solution is not an ideal gas.

Carbon isotope equilibrium between CO_3^{2-} and $CO_2(g)$

The fractionation factor α between CO_3^{2-} and $CO_2(g)$ is given by the ratio of the β factors:

$$\alpha = \frac{\beta_{CO_3^{2-}}}{\beta_{CO_2}}$$

At 0°C, we have:

$$\alpha = \frac{1.2258}{1.2171} = 1.0071$$

In words, the theoretical calculation of the equilibrium fractionation between carbon dioxide and carbonate ion gives an enrichment of ^{13}C in the carbonate ion of 7.1‰ at 0°C.

Table 3.5.18: Calculated fractionation factor between CO_3^{2-} and $CO_2(g)$ ion for C-exchange ($^{13}C/^{12}C$).

T_c (°C)	$\alpha_{(CO_3^{2-}-CO_2)}$
0	1.0071
10	1.0061
20	1.0052
25	1.0048
30	1.0044
40	1.0036
50	1.0028
75	1.0012

The calculated values of the fractionation factor in the temperature range from 0 to 75°C are given in Table 3.5.18 and shown in Figure 3.5.38 (solid line). Also shown are results of theoretical work (Thode et al., 1965) and experimental work (Leśniak and Sakai, 1989; Turner, 1982; Zhang et al., 1995; Halas et al., 1997). Whereas the results of Thode et al. (1965) and Leśniak and Sakai (1989) favor a larger ^{13}C fractionation between CO_3^{2-} and $CO_2(g)$ ($\sim$8‰ at 25°C), the results of Zhang et al. (1995), Halas et al. (1997), and the calculation presented here indicate a smaller fractionation of about $5-6$‰ at 25°C.

It is difficult to judge which of the data or calculations are most reliable. However, there are some points to be discussed which might help to answer this question. According to the data of Leśniak and Sakai

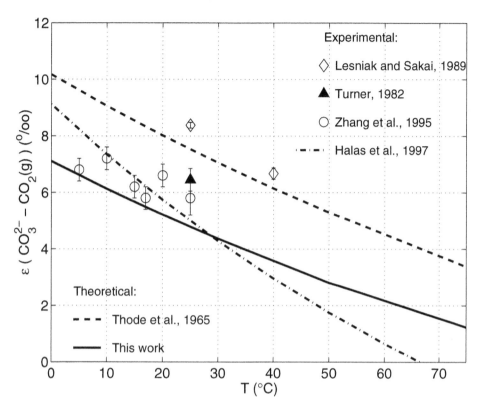

Figure 3.5.38: Carbon isotope equilibrium fractionation between CO_3^{2-} and $CO_2(g)$ as a function of temperature.

(1989), HCO_3^- is lighter in ^{13}C ($\varepsilon_{(HCO_3^- - CO_2(g))} = 8.21 \pm 0.1$) than CO_3^{2-} ($\varepsilon_{(CO_3^{2-} - CO_2(g))} = 8.38 \pm 0.1$) at 25°C. This is in contradiction to the results of all other studies mentioned above. The frequencies used by Thode et al. (1965) for the calculation of the β factors for $CO_2(g)$ and CO_3^{2-} are from 1959 and 1938, respectively. It is likely, though not certain, that the frequencies used by Richet et al. (1977) for $CO_2(g)$ and those used here for CO_3^{2-} (Davis and Oliver, 1972) are more accurate since they were determined using more recent technology. The points discussed so far therefore seem to argue against a large fractionation of $\sim 8\permil$ at 25°C.

On the other hand, provided that the small fractionation is correct ($5 - 6\permil$ at 25°C), it follows that the carbon isotope equilibrium fractionation between $CO_2(g)$ and HCO_3^- is about 1.3 to 1.6 times greater than the equilibrium fractionation between $CO_2(g)$ and CO_3^{2-}. Considering the large difference in the molecular structure between $CO_2(g)$ and HCO_3^- (see

Figure 3.2.11) vs. the small difference between HCO_3^- and CO_3^{2-} (just a proton-transfer) the much larger fractionation between $CO_2(g)$ and HCO_3^- is not to be expected. In order to clarify this, it would be very helpful to calculate β factors also for HCO_3^- - which is not a simple task - using a set of force constants consistent with the one used for CO_3^{2-}. At present, the agreement between the theoretical derived values of this work and the experimental results of Zhang et al. (1995) and Halas et al. (1997) appear to vote for the smaller fractionation between CO_3^{2-} and $CO_2(g)$ ($5-6‰$ at 25°C).

One aspect that should also be clarified in future work is the influence of ion complexes such as $MgCO_3^0$ on the fractionation between CO_3^{2-} and $CO_2(g)$ in seawater. As already pointed out in Section 3.2.3, the experimental values as reported above have been measured in distilled water and should be taken with caution when applied to seawater. The presence of Mg^{2+} ions leads to larger fractionation between $CO_2(g)$ and the dissolved carbon species (Thode et al., 1965) which is presumably due to the enrichment of ^{13}C in carbonate complexes (cf. also Zhang et al., 1995).

Appendix A

Equilibrium constants

This appendix is meant mainly for reference. After some remarks, expressions for various stoichiometric equilibrium constants are listed. These equilibrium constants, which are functions of temperature, pressure, and salinity, have been derived by fitting experimental data. The use of a consistent set of equilibrium constants is absolutely necessary. The constants used should be derived from measurements in the same medium, i.e. natural or artificial seawater, converted to the same concentration unit and the same pH scale. The equilibrium constants summarized in DOE (1994) (artificial seawater, concentrations in mol $(\text{kg-soln})^{-1}$; total pH-scale = $p\text{H}_\text{T}$) are recommended (see discussion below). A MATLAB file (equic.m) containing the equilibrium constants is available on our web-page: 'http://www.awi-bremerhaven.de/Carbon/co2book.html'.

Before the various equilibrium constants are listed, a few remarks are in order. The empirical formulas for the equilibrium constants usually apply to surface pressure (1 atm) and are often restricted to certain ranges of temperature and salinity (see original papers for detailed information); hence do not extrapolate to $S = 0$ (freshwater)! Equations to calculate the effect of pressure on the equilibrium constants as summarized by Millero (1995) are given in Section A.11.

The recommended concentration unit is mol $(\text{kg-solution})^{-1}$ (short mol $(\text{kg-soln})^{-1}$ or even shorter mol kg^{-1}) as in DOE (1994). The use of gravimetric units such as mol $(\text{kg-soln})^{-1}$ yields temperature and pressure independent values of DIC and TA. If not otherwise stated this unit is used. Other concentration units (not recommended) are mol l^{-1} (molarity) and mol $(\text{kg-H}_2\text{O})^{-1}(\text{molality}^1)$. The conversion between mol l^{-1} (molarity), mol $(\text{kg-H}_2\text{O})^{-1}$(molality), and mol $(\text{kg-soln})^{-1}$ (Millero, 1995) is:

$$\ln K_i^*[\text{mol } (\text{kg-soln})^{-1}]$$

$$= \ln K_i^*[\text{mol } (\text{kg-H}_2\text{O})^{-1}] + \ln(1 - 0.001005\ S) \tag{A.0.1}$$

$$K_i^*[\text{mol } \text{l}^{-1}] = K_i^*[\text{mol } (\text{kg-soln})^{-1}] \cdot \frac{\rho_{\text{SW}}(T,S)[\text{kg m}^{-3}]}{1000} \tag{A.0.2}$$

where $\rho_{\text{SW}}(T,S)$ is the density of seawater (see, for example, UNESCO (1983) or Gill (1982)).

We shall also briefly discuss the dimensions (or units) of the quantities used in mathematical equations. In general, two different types of equations must be recognized: dimensionally correct and dimensionally incorrect equations. Equations are preferentially formulated in a dimensionally correct form, i.e. the form of an equation does not depend on the units used, provided that the same units apply to all terms of the equation. For example, Newton's famous equation relating the acceleration, $\vec{a}$, of a body with mass m to the sum of applied forces, $\sum_i \vec{F}_i$:

$$m \cdot \vec{a} = \sum_i \vec{F}_i \tag{A.0.3}$$

is dimensionally correct and therefore it does not matter whether length is measured in m, cm, foot, or yard. In dimensionally correct expressions the arguments of exponential or logarithmic functions are dimensionless. This is the case in the Arrhenius equation (Eq. 2.2.7):

$$k = A \exp(-E_a/RT) \tag{A.0.4}$$

where E_a/RT is dimensionless and E_a and RT may be given in units of J mol^{-1} or cal mol^{-1}. In an equation with several additive terms such as in Eq. (A.0.3) each summand must be of the same dimension. This property can be used to find inconsistencies in given equations. However, despite of this advantageous feature of dimensionally correct equations, the second general type of equations, namely dimensionally incorrect equations are widely used (by the way, this book is no exception in this regard). This type of equations gives correct values only when certain units are applied. These units have to be explicitly provided together with the equation or have to be well known to the community (generally accepted convention).

The expressions for the equilibrium constants, K_i^*, given in this appendix are in dimensionally incorrect form. For example, they yield correct values

---

[1]Molality instead of molarity is often used in laboratory experiments.

only when K_i^* is in mol (kg-soln)$^{-1}$, K_W^* is in mol^2 (kg-soln)$^{-2}$, T is in K (a generally accepted convention) and so forth. On the contrary, they yield incorrect values when K_i^* is in e.g. μmol l^{-1} and T is in °C. Another example of a dimensionally incorrect expression is the definition of pH:

$$p\text{H} := -\log[\text{H}^+] \ .$$

This expression only makes sense when certain units for [H$^+$] are used (mol (kg-soln)$^{-1}$ by convention). The pH expression can be made dimensionally correct as follows

$$p\text{H} := -\log\left([\text{H}^+]/k^\circ\right)$$

where $k^\circ = 1$ mol (kg-soln)$^{-1}$. This approach is, for instance, used in DOE (1994) for the empirical expressions of the equilibrium constants. We will, however, explicitly write down the units of the equilibrium constants given in this appendix.

A.1   CO₂: Acidity constants K_1^* and K_2^*

The first and second acidity constants of carbonic acid

$$K_1^* := [\text{H}^+][\text{HCO}_3^-]/[\text{CO}_2]$$

and

$$K_2^* := [\text{H}^+][\text{CO}_3^{2-}]/[\text{HCO}_3^-]$$

were determined by Mehrbach et al. (1973), Hansson (1973b) (the original data were refitted by Dickson and Millero (1987); these refits will be referred to as 'Mehrbach' and 'Hansson' constants), Goyet and Poisson (1989), and Roy et al. (1993a). Mehrbach et al. (1973) used natural seawater whereas all other workers used artificial seawater. Different sets of constants appear to do a better job for different parameters (cf. discussion in Section 1.1.6). For example, Wanninkhof et al. (1999), Johnson et al. (1999), Lee et al. (2000), and Lueker et al. (2000) demonstrated that Mehrbach constants yield better consistency when determining the partial pressure (more precisely the fugacity) of carbon dioxide, $f\text{CO}_2$, from DIC and TA (Figure A.1.1). The combination of these three parameter is of major importance because it is used in global carbon cycle models. These models contain prognostic equations for DIC and TA from which $f\text{CO}_2$ is calculated. This in turn determines the gas exchange with the atmosphere for the exchange depends

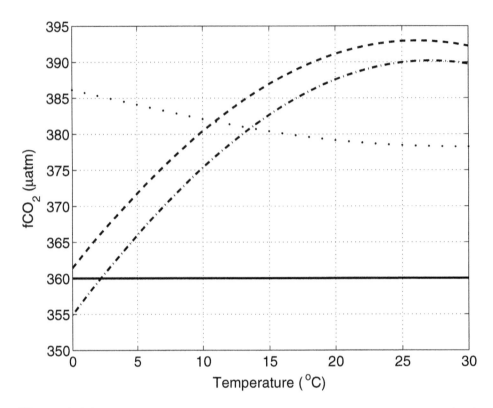

Figure A.1.1: Calculation of $f\mathrm{CO}_2$ from DIC and TA using acidity constants (pK_1^* and pK_2^*) as given by different authors. DIC (T) was calculated from TA = 2300 μmol kg^{-1} and $f\mathrm{CO}_2 = 360$ μatm using 'Mehrbach' constants. Then $f\mathrm{CO}_2$ was calculated from the same values of TA and DIC (T) using 'Hansson' constants (dotted line), Goyet and Poisson (1989; dashed-dotted line), and DOE (1994; dashed line). The differences amount up to 30 μatm and depend on temperature. Compare Wanninkhof et al. (1999), Johnson et al. (1999) and Lee et al. (2000) for further discussion.

on the partial pressure difference between water and air. Therefore the Mehrbach constants, instead of the Roy constants as summarized in DOE (1994), are recommended for this application. It is noted that Mehrbach constants are likely to do a good job when measurements in the ocean, i.e. in natural seawater are considered because these constants have been determined in natural seawater.

Currently there is no consistent set of all relevant equilibrium constants (K_1^*, K_2^*, K_W^*, K_B^*, K_F^*, K_S^*,...) derived from measurements in natural seawater. Because we had to calculate various combinations of carbonate system parameters in this book, we have used the consistent set of constants provided by DOE (1994) which is based on measurements in artificial seawater. This set of constants will be referred to as 'recommended' in this appendix.

Recommended: DOE (1994)

Based on Roy et al. (1993a), artificial seawater, pH_T, mol (kg-soln)$^{-1}$:

$$
\begin{aligned}
\ln K_1^* ={}& 2.83655 - 2307.1266/T - 1.5529413 \ln T \\
& -(0.207608410 + 4.0484/T)\sqrt{S} \\
& +0.0846834\,S - 0.00654208\,S^{3/2} + \ln(1 - 0.001005\,S)
\end{aligned}
$$

Check value: $\ln K_1^* = -13.4847$, $pK_1^* = 5.8563$ at $S = 35$, $T_c = 25°C$.

$$
\begin{aligned}
\ln K_2^* ={}& -9.226508 - 3351.6106/T - 0.2005743 \ln T \\
& -(0.106901773 + 23.9722/T)\sqrt{S} \\
& +0.1130822\,S - 0.00846934\,S^{3/2} + \ln(1 - 0.001005\,S)
\end{aligned}
$$

Check value: $\ln K_2^* = -20.5504$, $pK_2^* = 8.9249$ at $S = 35$, $T_c = 25°C$.

'Mehrbach'

'Mehrbach' = Mehrbach et al. (1973) data (natural seawater) as refitted by Dickson and Millero (1987); pH_{SWS}, mol (kg-soln)$^{-1}$:

$$pK_1^* = 3670.7/T - 62.008 + 9.7944 \ln T - 0.0118\,S + 0.000116\,S^2$$

$$pK_2^* = 1394.7/T + 4.777 - 0.0184\,S + 0.000118\,S^2.$$

Check values: $pK_1^* = 5.8372$ and $pK_2^* = 8.9554$ at $S = 35$, $T_c = 25°C$.

Mehrbach's constants have also been given on the total scale by Lueker et al. (2000); pH_T, mol (kg-soln)$^{-1}$:

$$pK_1^* = 3633.86/T - 61.2172 + 9.6777 \ln T - 0.011555\,S + 0.0001152\,S^2$$

$$pK_2^* = 471.78/T + 25.9290 - 3.16967 \ln T - 0.01781\,S + 0.0001122\,S^2$$

Check values: $pK_1^* = 5.8472$ and $pK_2^* = 8.9660$ at $S = 35$, $T_c = 25°C$.

'Hansson'

'Hansson' = Hansson (1973b) data (artificial seawater) as refitted by Dickson and Millero (1987); pH_{SWS}, mol $(kg\text{-soln})^{-1}$:

$$pK_1^* = 851.4/T + 3.237 - 0.0106\,S + 0.000105\,S^2$$

$$pK_2^* = -3885.4/T + 125.844 - 18.141\,\ln T - 0.0192\,S + 0.000132\,S^2.$$

Goyet and Poisson (1989)

Artificial seawater (including fluoride), pH_{SWS}, mol $(kg\text{-soln})^{-1}$, $10 < S < 50$, $-1°C < T_c < 40°C$:

$$pK_1^* = 807.18/T + 3.374 - 0.00175\,S \ln T + 0.000095\,S^2,$$

$$pK_2^* = 1486.6/T + 4.491 - 0.00412\,S \ln T + 0.000215\,S^2$$

(please note that these are the equations given in Table 4 of Goyet and Poisson (1989) whereas the equations in the abstract are derived from a combined data set including the results of Hansson (1973b)).

A.2   Acidity constant of true carbonic acid

Stumm and Morgan (1996, p.150/1):

$$K_{H_2CO_3}^{'*} := \frac{[H^+][HCO_3^-]}{[H_2CO_3]} \tag{A.2.5}$$

$$pK_{H_2CO_3}^{'*} = 3.8 \text{ to } 3.4$$

Example: $[HCO_3^-] = 2$ mmol kg^{-1}, $[H^+] = 10^{-8}$ mol $(kg\text{-soln})^{-1}$, $K_{H_2CO_3}^{*} = 10^{-3.8}$ mol kg^{-1} (recommended value) $\Longrightarrow [H_2CO_3] = 1.3 \cdot 10^{-7}$ mol $(kg\text{-soln})^{-1}$ or less than 1% of the CO_2 concentration.

A.3 CO_2 solubility in water (Henry's law)

In equilibrium the CO_2 concentration is proportional to the fugacity (Henry's law)

$$[CO_2] = K_0 \cdot fCO_2 \tag{A.3.6}$$

where K_0 is Henry's constant. Units: $[CO_2]$ in mol $(kg\text{-soln})^{-1}$, fCO_2 in atm, K_0 in mol $(kg\text{-soln})^{-1}$ atm^{-1}.

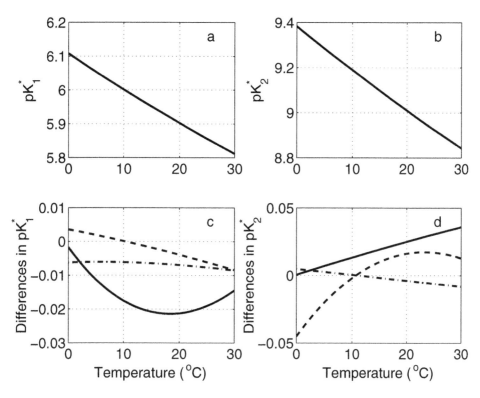

Figure A.1.2: The first and second acidity constants of carbonic acid as a function of temperature and salinity $S = 35$ as determined by Roy et al. (1993) and summarized in DOE (1994): (a) pK_1^*, (b) pK_2^*. (c) Differences between various pK_1^* with respect to DOE (1994): 'Mehrbach' (solid line), 'Hansson' (dashed line), and Goyet and Poisson (1989) (dashed-dotted line). (d) Differences between various pK_2^* with respect to DOE (1994): 'Mehrbach' (solid line), 'Hansson' (dashed line), and Goyet and Poisson (1989) (dashed-dotted line).

Recommended: Weiss (1974)

$$
\begin{aligned}
\ln K_0 &= 9345.17/T - 60.2409 + 23.3585 \ln (T/100) \\
&\quad + S \left[0.023517 - 0.00023656\, T + 0.0047036\, (T/100)^2 \right]
\end{aligned}
$$

Check value: $\ln K_0 = -3.5617$, $pK_0 = 1.5468$ at $S = 35$, $T_c = 25°C$.

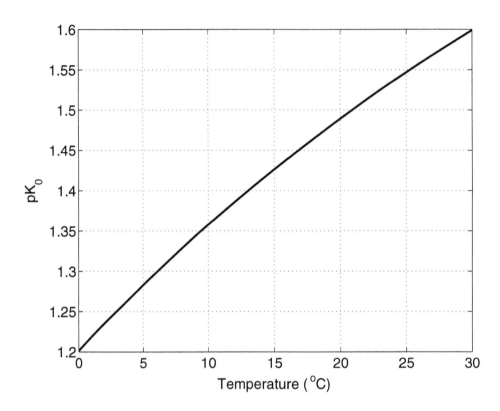

Figure A.3.3: Henry's constant pK_0 at $S = 35$.

A.4   Ion product of water: K_W^*

The equilibrium constant, K_W^*, for the reaction $H_2O \rightleftharpoons H^+ + OH^-$

$$K_W^* := [H^+][OH^-] \tag{A.4.7}$$

is called the ion product of water.

Recommended: DOE (1994)

$$
\begin{aligned}
\ln K_W^* = {} & 148.96502 - 13847.26/T - 23.6521 \ln T \\
& + [118.67/T - 5.977 + 1.0495 \ln T]S^{1/2} - 0.01615\, S
\end{aligned}
$$

K_W^* in $\text{mol}^2\ (\text{kg-soln})^{-2}$; pH-scale: pH_T.

Check value: $\ln K_W^* = -30.434$, $pK_W^* = 13.2173$ at $S = 35$, $T_c = 25°C$.

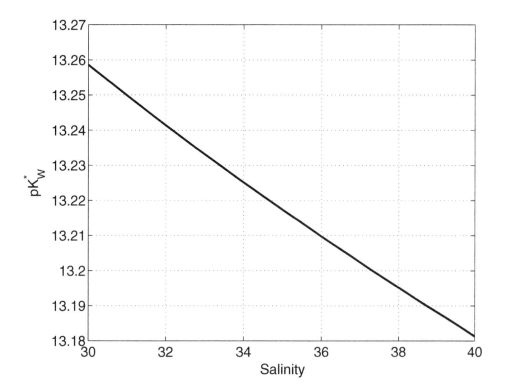

Figure A.4.4: The ion product of water, pK_W^*, as a function of salinity at $T_c = 25°C$

Millero (1995)

pH_{SWS}, mol^2 $(kg\text{-soln})^{-2}$

$$\ln K_W^* = 148.9802 - 13847.26/T - 23.6521 \ln T$$
$$+(118.67/T - 5.977 + 1.0495 \ln T)\,S^{1/2} - 0.01615\,S$$

The constant offset $(148.96502 - 148.9802 = -0.01518)$ between DOE (1994) and Millero (1995) is a result of the difference between the pH scales.

A.5   Bisulfate ion

The constant for the equilibrium $HSO_4^- \rightleftharpoons H^+ + SO_4^{2-}$, i.e.

$$K_S^* = \frac{[H^+]_F\,[SO_4^{2-}]}{[HSO_4^-]} \tag{A.5.8}$$

is given by DOE (1994)

$$
\begin{aligned}
\ln K_S^* \;=\; & -4276.1/T + 141.328 - 23.093 \ln T \\
& + (-13856/T + 324.57 - 47.986 \ln T)\, I^{1/2} \\
& + (35474/T - 771.54 + 114.723 \ln T)\, I \\
& - \frac{2698}{T} I^{3/2} + \frac{1776}{T} I^2 + \ln\left(1 - 0.001005\, S\right)
\end{aligned}
$$

where

$$
I = \frac{19.924\, S}{1000 - 1.005\, S}
$$

is the ionic strength. K_S^* in mol (kg-soln)$^{-1}$; pH-scale: 'free' scale.
Check value: $\ln K_S^* = -2.2996$, $pK_S^* = 0.9987$ at $S = 35$, $T_c = 25°C$.
The total sulfate concentration, S_T, is a function of salinity:

$$
S_T \;(\text{mol (kg-soln)}^{-1}) := [HSO_4^-] + [SO_4^{2-}] = 0.02824\,\frac{S}{35} \tag{A.5.9}
$$

DOE (1994, chapter 5, p.11).

A.6 Hydrogen fluoride

The constant for the equilibrium $HF \rightleftharpoons H^+ + F^-$, i.e.

$$
K_F^* := \frac{[H^+][F^-]}{[HF]} \tag{A.6.10}
$$

is given by Dickson and Riley (1979) (reprinted in DOE (1994)):

$$
\begin{aligned}
\ln K_F^* \;=\; & \frac{1590.2}{T} - 12.641 + 1.525\, I^{1/2} \\
& + \ln\left(1 - 0.001005\, S\right) + \ln\left(1 + S_T/K_S^*\right)
\end{aligned}
$$

where

$$
I = \frac{19.924\, S}{1000 - 1.005\, S}
$$

is the ionic strength.

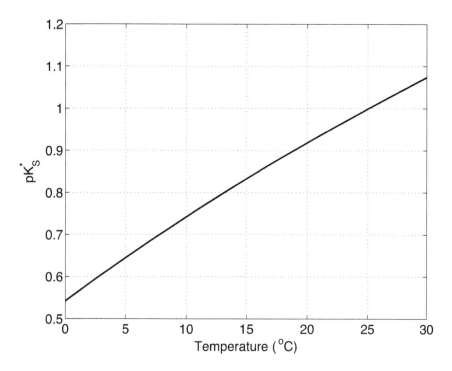

Figure A.5.5: Stability constant of hydrogen sulfate, pK_S^*, at $S = 35$ as a function of temperature.

The term $\ln(1 + S_T/K_S^*)$ converts from the 'free' hydrogen ion concentration scale to the 'total' (Hansson) scale.

Check value: $\ln K_F^* = -5.7986$, $pK_F^* = 2.5183$ at $S = 35$, $T_c = 25°C$.

Total fluoride concentration, F_T, in seawater is related to salinity by

$$F_T \text{ (mol (kg-soln)}^{-1}) \quad := \quad [HF] + [F^-] = 7 \cdot 10^{-5} \frac{S}{35} \tag{A.6.11}$$

DOE (1994, chapter 5, p.11).

A.7 Boric acid

The constant for the equilibrium

$$B(OH)_3 + H_2O \; \rightleftharpoons \; H^+ + B(OH)_4^-, \tag{A.7.12}$$

i.e.

$$K_B^* := \frac{[H^+][B(OH)_4^-]}{[B(OH)_3]}. \tag{A.7.13}$$

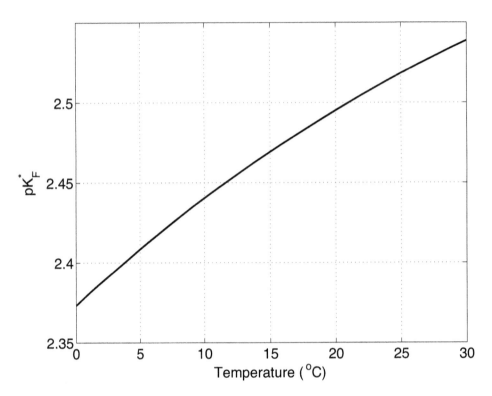

Figure A.6.6: Stability constant of hydrogen fluoride, pK_F^*, at $S = 35$ as a function of temperature.

is given by

Recommended: DOE (1994)

(based on Dickson, 1990)

$$
\begin{aligned}
\ln K_B^* = {}& \left(-8966.90 - 2890.53\, S^{1/2} - 77.942\, S + 1.728\, S^{3/2}\right. \\
& \left. -0.0996\, S^2\right)/T \\
& +148.0248 + 137.1942\, S^{1/2} + 1.62142\, S \\
& -(24.4344 + 25.085\, S^{1/2} + 0.2474\, S)\ln T \\
& +0.053105\, S^{1/2}\, T
\end{aligned}
$$

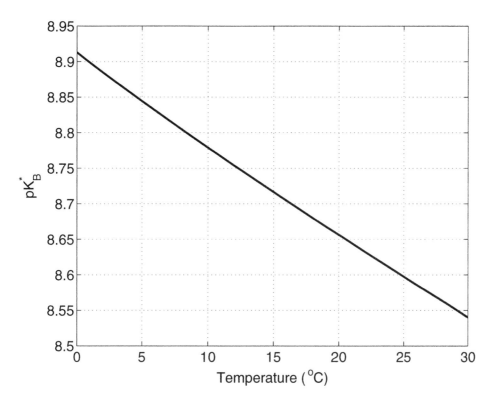

Figure A.7.7: The equilibrium constant of boric acid, pK_B^*, as a function of temperature at $S = 35$.

pH-scale: pH$_T$; units: K_B^* in mol (kg-soln)$^{-1}$; check value: $\ln K_B^* = -19.7964$, $pK_B^* = 8.5975$ at $S = 35$ and $T_c = 25°C$.

Remark: This equation is in excellent agreement with the measurements made by Roy et al. (1993b).

The total boron concentration, B_T, in seawater is related to salinity by

$$B_T(\text{mol (kg-soln)}^{-1}) \quad := \quad [B(OH)_3] + [B(OH)_4^-] \tag{A.7.14}$$

$$= \quad 4.16 \cdot 10^{-4} \frac{S}{35} \tag{A.7.15}$$

DOE (1994, chapter 5, p.11).

A.8   Phosphoric acid

The constant for the equilibrium

$$H_3PO_4 \rightleftharpoons H^+ + H_2PO_4^-, \tag{A.8.16}$$

i.e.

$$K_{1P}^{'*} := \frac{[\text{H}^+][\text{H}_2\text{PO}_4^-]}{[\text{H}_3\text{PO}_4]}. \qquad (A.8.17)$$

is given by

Recommended: DOE (1994)

$$
\begin{aligned}
\ln K_{1P}^* &= \frac{-4576.752}{T} + 115.525 - 18.453 \ln T \\
&+ \left(\frac{-106.736}{T} + 0.69171\right) S^{1/2} + \left(\frac{-0.65643}{T} - 0.01844\right) S
\end{aligned}
$$

pH-scale: pH$_T$; units: K_{1P}^* in mol (kg-soln)$^{-1}$; check value: $\ln K_{1P}^* = -3.71$, $pK_{1P}^* = 1.61$ at $S = 35$ and $T_c = 25°$C.

The constant for the equilibrium

$$\text{H}_2\text{PO}_4^- \rightleftharpoons \text{H}^+ + \text{HPO}_4^{2-}, \qquad (A.8.18)$$

i.e.

$$K_{2P}^{'*} := \frac{[\text{H}^+][\text{HPO}_4^{2-}]}{[\text{H}_2\text{PO}_4^-]}. \qquad (A.8.19)$$

is given by

Recommended: DOE (1994)

$$
\begin{aligned}
\ln K_{2P}^* &= \frac{-8814.715}{T} + 172.0883 - 27.927 \ln T \\
&+ \left(\frac{-160.34}{T} + 1.3566\right) S^{1/2} + \left(\frac{0.37335}{T} - 0.05778\right) S
\end{aligned}
$$

pH-scale: pH$_T$; units: K_{2P}^* in mol (kg-soln)$^{-1}$; check value: $\ln K_{2P}^* = -13.727$, $pK_{2P}^* = 5.96$ at $S = 35$ and $T_c = 25°$C.

The constant for the equilibrium

$$\text{HPO}_4^{2-} \rightleftharpoons \text{H}^+ + \text{PO}_4^{3-}, \qquad (A.8.20)$$

i.e.

$$K_{3P}^* := \frac{[\mathrm{H}^+][\mathrm{PO}_4^{3-}]}{[\mathrm{HPO}_4^{2-}]}. \tag{A.8.21}$$

is given by

Recommended: DOE (1994)

$$
\begin{aligned}
\ln K_{3P}^* \;=\; & \frac{-3070.75}{T} - 18.141 + \left(\frac{17.27039}{T} + 2.81197\right) S^{1/2} \\
& + \left(\frac{-44.99486}{T} - 0.09984\right) S
\end{aligned}
$$

pH-scale: pH_T; units: K_{3P}^* in mol (kg-soln)$^{-1}$; check value: $\ln K_{3P}^* = -20.24$, $pK_{3P}^* = 8.79$ at $S = 35$ and $T_c = 25°C$.

A.9   Silicic acid

The constant for the equilibrium

$$\mathrm{Si(OH)_4} \rightleftharpoons \mathrm{H}^+ + \mathrm{H_3SiO_4^-}, \tag{A.9.22}$$

i.e.

$$K_{Si}^* := \frac{[\mathrm{H}^+][\mathrm{H_3SiO_4^-}]}{[\mathrm{Si(OH)_4}]}. \tag{A.9.23}$$

is given by

Recommended: DOE (1994)

$$
\begin{aligned}
\ln K_{Si}^* \;=\; & \frac{-8904.2}{T} + 117.385 - 19.334 \ln T \\
& + \left(3.5913 - \frac{458.79}{T}\right) I^{1/2} + \left(\frac{188.74}{T} - 1.5998\right) I \\
& + \left(0.07871 - \frac{12.1652}{T}\right) I^2 + \ln\left(1 - 0.001005\, S\right)
\end{aligned}
$$

pH-scale: pH_T; units: K_{Si}^* in mol $(kg\text{-soln})^{-1}$; check value: $\ln K_{Si}^* = -21.61$, $pK_{Si}^* = 9.38$ at $S = 35$ and $T_c = 25°C$.

A.10   Solubility product of calcite and aragonite

The stoichiometric solubility product is defined as:

$$K_{sp}^* = [Ca^{2+}] \times [CO_3^{2-}], \qquad (A.10.24)$$

where e.g. $[CO_3^{2-}]$ refers to the equilibrium total (free + complexed) carbonate ion concentration.

Calcite: Mucci (1983)

$$
\begin{aligned}
\log K_{sp}^*(\text{cal}) = \ & -171.9065 - 0.077993\,T + 2839.319/T \\
& +71.595\ \log T \\
& +(-0.77712 + 0.0028426\,T + 178.34/T)\ S^{1/2} \\
& -0.07711\,S + 0.0041249\,S^{1.5}
\end{aligned}
$$

Units: $K_{sp}^*(\text{cal})$ in mol^2 $(kg\text{-soln})^{-2}$; check value: $-\log K_{sp}^*(\text{cal}) = 6.3693$ at $S = 35$ and $T_c = 25°C$.

Aragonite: Mucci (1983)

$$
\begin{aligned}
\log K_{sp}^*(\text{arg}) = \ & -171.945 - 0.077993\,T + 2903.293/T \\
& +71.595\ \log T \\
& +(-0.068393 + 0.0017276\,T + 88.135/T)\ S^{1/2} \\
& -0.10018\,S + 0.0059415\,S^{1.5}
\end{aligned}
$$

Units: $K_{sp}^*(\text{arg})$ in mol^2 $(kg\text{-soln})^{-2}$; check value: $-\log K_{sp}^*(\text{arg}) = 6.1883$ at $S = 35$ and $T_c = 25°C$.

A.11 Effect of pressure on equilibrium constants

The effect of pressure on equilibrium constants can be calculated from the equation (see Millero (1995)):

$$\ln\left(\frac{K_{i,P}^*}{K_{i,0}^*}\right) = -\frac{\Delta V_i}{\hat{R}T}\,P + 0.5\,\frac{\Delta\kappa_i}{\hat{R}T}\,P^2$$

where $K_{i,P}^*$ and $K_{i,0}^*$ are the equilibrium constants at pressure P and at reference pressure $P = 0$ bars (1 atm), respectively. The constant $\hat{R}$ is given by $\hat{R} = 83.131$ cm^3 bar mol^{-1} K^{-1}, whereas the molal volume and compressibility change are denoted by ΔV_i and $\Delta\kappa_i$ which can be fit to equations of the form ($S = 35$):

$$\Delta V_i = a_0 + a_1\,T_c + a_2\,T_c^2$$
$$\Delta\kappa_i = b_0 + b_1\,T_c + b_2\,T_c^2$$

where T_c is the temperature in °C. Values for the coefficients a_i and b_i for seawater for various equilibrium constants are given in Table A.11.1 (Millero, 1979; 1995). Check values for the effect of pressure on equilibrium constants at $P = 0$ and $P = 300$ bars (i.e. depth$\sim$ 3 km) are given in Table A.11.2.

Table A.11.1: Coefficients for the effect of pressure on various equilibrium constants.

Constant	$-a_0$	a_1	$10^3 a_2$	$-10^3 b_0$	$10^3 b_1$
K_1^*	25.50	0.1271	0.0	3.08	0.0877
K_2^*	15.82	−0.0219	0.0	−1.13	−0.1475
K_B^*	29.48	0.1622	2.6080	2.84	0.0
K_W^*	25.60	0.2324	−3.6246	5.13	0.0794
K_S^*	18.03	0.0466	0.3160	4.53	0.0900
K_F^*	9.780	−0.0090	−0.942	3.91	0.054
K_{sp}^*(calcite)	48.76	0.5304	0.0	11.76	0.3692
K_{sp}^*(aragonite)	46.00	0.5304	0.0	11.76	0.3692

Table A.11.2: Check values for the effect of pressure on equilibrium constants ($S = 35$, $T_c = 25°C$).

pK	$P = 0$ bars	$P = 300$ bars
pK_1^*	5.8563	5.7397
pK_2^*	8.9249	8.8409
pK_B^*	8.5975	8.4746
pK_W^*	13.2173	13.1039
pK_S^*	0.9987	0.9129
pK_F^*	2.5183	2.4646
pK_{sp}^*(calcite)	6.3693	6.1847
pK_{sp}^*(aragonite)	6.1883	6.0182

A.12 Chemical composition of seawater

Table A.12.3: Standard mean chemical composition of seawater at $S = 35$ (DOE, 1994).

Ion	g kg^{-1}	mol kg^{-1}
Cl$^-$	19.3524	0.54586
Na$^+$	10.7837	0.46906
Mg^{2+}	1.2837	0.05282
SO$_4^{2-}$	2.7123	0.02824
Ca^{2+}	0.4121	0.01028
K$^+$	0.3991	0.01021
CO$_2$	0.0004	0.00001
HCO$_3^-$	0.1080	0.00177
CO$_3^{2-}$	0.0156	0.00026
B(OH)$_3$	0.0198	0.00032
B(OH)$_4^-$	0.0079	0.00010
Br$^-$	0.0673	0.00084
Sr^{2+}	0.0079	0.00009
F$^-$	0.0013	0.00007
OH$^-$	0.0002	0.00001
sum of column	35.1717	1.11994
ionic strength		0.69734

A.13 The equation of state of seawater

In the following, the density of seawater, ρ (in kg m^{-3}), is given as a function of temperature, T_c (in °C), pressure, P (in bars), and practical salinity, S. The calculation of the density of seawater requires a number of steps (Millero and Poisson, 1981; Gill, 1982). A MATLAB routine for this procedure is included in the file 'equic.m' which can be found on our web-page: 'http://www.awi-bremerhaven.de/Carbon/co2book.html'.

The density of pure water ($S = 0$) is given by:

$$\rho_{\mathrm{pw}} = \quad 999.842594 + 6.793952 \times 10^{-2}\,T_c - 9.095290 \times 10^{-3}\,T_c^{\,2}$$
$$+ \quad 1.001685 \times 10^{-4}\,T_c^{\,3} - 1.120083 \times 10^{-6}\,T_c^{\,4}$$

$$+ \quad 6.536332 \times 10^{-9} \ T_c^5 \ .$$

The density of seawater at 1 atm, i.e. at $P = 0$, can be expressed as:

$$\rho(S, T_c, 0) = \rho_{pw} + AS + BS^{3/2} + CS^2$$

where

$$
\begin{aligned}
A \ &= \ 8.24493 \times 10^{-1} - 4.0899 \times 10^{-3} \ T_c + 7.6438 \times 10^{-5} \ T_c^2 \\
&\quad -8.2467 \times 10^{-7} \ T_c^3 + 5.3875 \times 10^{-9} \ T_c^4 \\
B \ &= \ -5.72466 \times 10^{-3} + 1.0227 \times 10^{-4} \ T_c - 1.6546 \times 10^{-6} \ T_c^2 \\
C \ &= \ 4.8314 \times 10^{-4} \ .
\end{aligned}
$$

The density of seawater at S, T_c, and P is finally given by:

$$\rho(S, T_c, P) = \rho(S, T_c, 0)/(1 - P/K(S, T_c, P)) \ ,$$

where K is the secant bulk modulus.

For pure water, K_{pw}, is given by:

$$
\begin{aligned}
K_{pw} \ &= \ 19652.21 + 148.4206 \ T_c - 2.327105 \ T_c^2 + 1.360477 \times 10^{-2} \ T_c^3 \\
&\quad -5.155288 \times 10^{-5} \ T_c^4 \ .
\end{aligned}
$$

The value in seawater at 1 atm, i.e. at $P = 0$, is given by:

$$
\begin{aligned}
K(S, T_c, 0) \ &= \ K_{pw} + S(54.6746 - 0.603459 \ T_c + 1.09987 \times 10^{-2} \ T_c^2 \\
&\quad -6.1670 \times 10^{-5} \ T_c^3) + S^{3/2}(7.944 \times 10^{-2} \\
&\quad +1.6483 \times 10^{-2} \ T_c - 5.3009 \times 10^{-4} \ T_c^2) \ .
\end{aligned}
$$

The secant bulk modulus at S, T_c, and P is finally given by:

$$
\begin{aligned}
K(S, T_c, P) \ &= \ K(S, T_c, 0) + P(3.239908 + 1.43713 \times 10^{-3} \ T_c \\
&\quad +1.16092 \times 10^{-4} \ T_c^2 - 5.77905 \times 10^{-7} \ T_c^3) \\
&\quad +PS(2.2838 \times 10^{-3} - 1.0981 \times 10^{-5} \ T_c \\
&\quad -1.6078 \times 10^{-6} \ T_c^2) + 1.91075 \times 10^{-4} \ PS^{3/2} \\
&\quad +P^2(8.50935 \times 10^{-5} - 6.12293 \times 10^{-6} \ T_c \\
&\quad +5.2787 \times 10^{-8} \ T_c^2) + P^2 S(-9.9348 \times 10^{-7} \\
&\quad +2.0816 \times 10^{-8} \ T_c + 9.1697 \times 10^{-10} \ T_c^2) \ .
\end{aligned}
$$

Check values: $\rho(0, 5, 0) = 999.96675$, $\rho(35, 5, 0) = 1027.67547$, and $\rho(35, 25, 1000) = 1062.53817$ kg m^{-3}.

Appendix B

From two to six

In this appendix equations are derived that permit the calculation of carbonate system variables. We consider the carbonate system with the six variables (CO_2, HCO_3^-, CO_3^{2-}, H^+, DIC, TA)[1] (Table B.0.1). Given any two variables, one may calculate all other variables (Deffeyes, 1965, Park, 1969). A MATLAB file (csys.m) containing the corresponding numerical routines is available on our web-page: 'http://www.awi-bremerhaven.de/Carbon/co2book.html'.

Table B.0.1: The six variables of the carbonate system

	Variable	Remarks
1	$[H^+] =: h$	alternatively: $[OH^-]$
2	$[CO_2] =: s$	
3	$[HCO_3^-]$	
4	$[CO_3^{2-}]$	
5	TA	$= [HCO_3^-] + 2[CO_3^{2-}] + [B(OH)_4^-] + [OH^-] - [H^+]$
6	DIC	

With two out of six we have

$$N = \binom{6}{2} = \frac{6!}{2!(6-2)!} = 15$$

---

[1]Strictly speaking the carbonate system also includes OH^-, $B(OH)_3$, and $B(OH)_4^-$. The concentration of OH^- can be readily calculated for given pH ($[OH^-] = K_W^*/[H^+]$) and the boron components are derived from pH and total dissolved boron: $[B(OH)_3] = B_T/(1 + K_B^*/[H^+])$ and $[B(OH)_4^-] = B_T/(1 + [H^+]/K_B^*)$.

Table B.0.2: Estimated total errors in the calculated parameters of the carbonate system using various input measurements.[a]

Input	ΔpH	ΔTA ($\mu mol\ kg^{-1}$)	ΔDIC ($\mu mol\ kg^{-1}$)	ΔfCO_2 (μatm)
pH, TA	-	-	±3.8	±2.1
pH, DIC	-	±2.7	-	±1.8
pH, fCO_2	-	±18	±15	-
fCO_2, DIC	±0.0023	±3.0	-	-
fCO_2, TA	±0.0021	-	±3.4	-
TA, DIC	±0.0062	-	-	±5.8

[a] The following accuracies were assumed: DIC: ±2 $\mu mol\ kg^{-1}$, TA: ±4 $\mu mol\ kg^{-1}$, pH: 0.002, fCO_2: 2 μatm (compare Millero, 1995).

different combinations. We will discuss all of them below. It is desirable to measure more than two components to test the internal consistency of the measurements. With an accuracy in measuring of pH (±0.002), total alkalinity (±4 $\mu mol\ kg^{-1}$), dissolved inorganic carbon (±2 $\mu mol\ kg^{-1}$) and fugacity of CO_2 (±2 μatm), Millero et al. (1993b) estimated an internal consistency of ±0.003 − 0.006 in pH, ±5 − 7 $\mu mol\ kg^{-1}$ in total alkalinity, ±5 − 7 $\mu mol\ kg^{-1}$ in dissolved inorganic carbon and ±6 − 9 μatm in fugacity of CO_2.

(1) $[CO_2]$ and pH given

Eq. (1.1.9) $\Longrightarrow$

$$DIC = s\left(1 + \frac{K_1^*}{h} + \frac{K_1^* K_2^*}{h^2}\right)$$

$[HCO_3^-]$, $[CO_3^{2-}]$ and TA are given as functions of h, s and DIC by Eqs. (1.1.10), (1.1.11), and (1.5.80).

(2) $[CO_2]$ and $[HCO_3^-]$ given

Eqs. (1.1.9) and (1.1.10) $\Longrightarrow$

$$s\left(1 + \frac{K_1^*}{h} + \frac{K_1^* K_2^*}{h^2}\right) = [HCO_3^-]\left(1 + \frac{h}{K_1^*} + \frac{K_2^*}{h}\right)$$

$\Longrightarrow$ cubic equation for h

$$s\left(h^2 + K_1^* h + K_1^* K_2^*\right) = [\text{HCO}_3^-]\left(h^2 + \frac{h^3}{K_1^*} + K_2^* h\right)$$

Roots[2]: one positive, two negative. s and h known $\Longrightarrow$ proceed as in case 1.

(3) $[\text{CO}_2]$ and $[\text{CO}_3^{2-}]$ given

Eqs. (1.1.9) and (1.1.11) $\Longrightarrow$

$$s\left(1 + \frac{K_1^*}{h} + \frac{K_1^* K_2^*}{h^2}\right) = [\text{CO}_3^{2-}]\left(1 + \frac{h}{K_2^*} + \frac{h^2}{K_1^* K_2^*}\right)$$

$\Longrightarrow$ equation of fourth order for h

$$s\left(h^2 + K_1^* h + K_1^* K_2^*\right) = [\text{CO}_3^{2-}]\left(h^2 + \frac{h^3}{K_2^*} + \frac{h^4}{K_1^* K_2^*}\right)$$

Roots: one positive, three negative. s and h known $\Longrightarrow$ proceed as in case 1.

(4) $[\text{CO}_2]$ and TA given

Eq. (1.5.80) $\Longrightarrow$ equation of fourth order for h:

$$\begin{aligned}
\text{TA} \cdot h^2 \cdot (K_B^* + h) \;=\;& s\,(K_B^* + h)\,(K_1^* h + 2K_1^* K_2^*) \\
&+\; h^2 K_B^* B_T + (K_B^* + h)\left(K_W^* h - h^3\right)
\end{aligned}$$

Roots: one positive, one negative, two conjugate complex. s and h known $\Longrightarrow$ proceed as in case 1.

(5) $[\text{CO}_2]$ and DIC given

Eq. (1.5.79) $\Longrightarrow$ quadratic equation for h

$$\text{DIC} \cdot h^2 = s\left(h^2 + K_1^* h + K_1^* K_2^*\right)$$

Roots: one positive, one negative. s and h known $\Longrightarrow$ proceed as in case 1.

---

[2]Determined at DIC $= 2000\ \mu\text{mol kg}^{-1}$, TA $= 2200\ \mu\text{mol kg}^{-1}$, $T_c = 25°\text{C}$, $S = 35$.

(6) pH and $[HCO_3^-]$ given

DIC from Eq. (1.1.10)

$$\mathrm{DIC} = [\mathrm{HCO_3^-}] \left(1 + \frac{h}{K_1^*} + \frac{K_2^*}{h} \right)$$

s and $[\mathrm{CO_3^{2-}}]$: Eqs. (1.1.9) and (1.1.11); TA: Eq. (1.5.80).

(7) pH and $[CO_3^{2-}]$ given

DIC: Eq. (1.1.11)

$$\mathrm{DIC} = [\mathrm{CO_3^{2-}}] \left(1 + \frac{h}{K_2^*} + \frac{h^2}{K_1^* K_2^*} \right)$$

s and $[\mathrm{HCO_3^-}]$: Eqs. (1.1.9) and (1.1.10); TA: Eq. (1.5.80).

(8) pH and TA given

Eq. (1.5.80) $\Longrightarrow s$

$$s = \left(\mathrm{TA} - \frac{K_B^* B_T}{K_B^* + h} - \frac{K_W^*}{h} + h \right) \Big/ \left(\frac{K_1^*}{h} + 2\frac{K_1^* K_2^*}{h^2} \right)$$

s and h known $\Longrightarrow$ proceed as in case 1.

(9) pH and DIC given

Eq. (1.5.79) $\Longrightarrow s$

$$s = \mathrm{DIC} \Big/ \left(1 + \frac{K_1^*}{h} + \frac{K_1^* K_2^*}{h^2} \right)$$

s and h known $\Longrightarrow$ proceed as in case 1.

(10) $[HCO_3^-]$ and $[CO_3^{2-}]$ given

Eqs. (1.1.10) and (1.1.11) $\Longrightarrow$

$$[HCO_3^-]\left(1+\frac{h}{K_1^*}+\frac{K_2^*}{h}\right) = [CO_3^{2-}]\left(1+\frac{h}{K_2^*}+\frac{h^2}{K_1^*K_2^*}\right)$$

$\Longrightarrow$ cubic equation for h

$$[HCO_3^-]\left(h+\frac{h^2}{K_1^*}+K_2^*\right) = [CO_3^{2-}]\left(h+\frac{h^2}{K_2^*}+\frac{h^3}{K_1^*K_2^*}\right)$$

Roots: one positive, two negative. Eq. (1.1.10) $\Longrightarrow$ DIC

$$DIC = [HCO_3^-]\left(1+\frac{h}{K_1^*}+\frac{K_2^*}{h}\right)$$

Eq. (1.1.9) $\Longrightarrow$ s. Eq. (1.5.80) $\Longrightarrow$ TA.

(11) $[HCO_3^-]$ and TA given

Eqs. (1.1.9) and (1.1.10) $\Longrightarrow$

$$s\left(1+\frac{K_1^*}{h}+\frac{K_1^*K_2^*}{h^2}\right) = \left[HCO_3^-\right]\left(1+\frac{h}{K_1^*}+\frac{K_2^*}{h}\right).$$

Solve for s and insert into Eq. (1.5.80):

$$TA\cdot\left(1+\frac{K_1^*}{h}+\frac{K_1^*K_2^*}{h^2}\right) =$$

$$\left[HCO_3^-\right]\left(1+\frac{h}{K_1^*}+\frac{K_2^*}{h}\right)\cdot\left(\frac{K_1^*}{h}+2\frac{K_1^*K_2^*}{h^2}\right)$$

$$+\left(1+\frac{K_1^*}{h}+\frac{K_1^*K_2^*}{h^2}\right)\left(\frac{K_B^*B_T}{K_B^*+h}+\frac{K_W^*}{h}-h\right)$$

which gives (times $h^3\cdot(K_B^*+h)$) an equation of fifth order for h:

$$TA\cdot(K_B^*+h)\cdot\left(h^3+K_1^*h^2+K_1^*K_2^*h\right)$$

$$= [HCO_3^-]\left(h+\frac{h^2}{K_1^*}+K_2^*\right)$$

$$\cdot\left[(K_B^*+2K_2^*)K_1^*h+2K_B^*K_1^*K_2^*+K_1^*h^2\right]$$

$$+\left(h^2+K_1^*h+K_1^*K_2^*\right)\left(K_B^*B_Th+K_W^*K_B^*+K_W^*h-K_B^*h^2-h^3\right)$$

Roots: one positive, four negative. Now calculate s from Eq. (1.5.80)

$$s = \text{TA} / \left[\left(\frac{K_1^*}{h} + 2\frac{K_1^* K_2^*}{h^2} \right) + \frac{K_B^* B_T}{K_B^* + h} + \frac{K_W^*}{h} - h \right]$$

and proceed as in case 1.

(12) $[\text{HCO}_3^-]$ and DIC given

Eq. (1.1.10) is a quadratic equation for h

$$[\text{HCO}_3^-] \left(h + \frac{h^2}{K_1^*} + K_2^* \right) = h \cdot \text{DIC}$$

Roots: two positive (use the smaller one). Calculate s by Eq. (1.1.9) and proceed as in case 1.

(13) $[\text{CO}_3^{2-}]$ and TA given

This case is similar to case 11. Combine Eq. (1.1.9) with Eq. (1.1.11)

$$s \left(1 + \frac{K_1^*}{h} + \frac{K_1^* K_2^*}{h^2} \right) = \left[\text{CO}_3^{2-} \right] \left(1 + \frac{h}{K_2^*} + \frac{h^2}{K_1^* K_2^*} \right)$$

solve for s and insert into Eq. (1.5.80):

$$\text{TA} \left(1 + \frac{K_1^*}{h} + \frac{K_1^* K_2^*}{h^2} \right) =$$

$$\left[\text{CO}_3^{2-} \right] \left(1 + \frac{h}{K_2^*} + \frac{h^2}{K_1^* K_2^*} \right) \cdot \left(\frac{K_1^*}{h} + 2\frac{K_1^* K_2^*}{h^2} \right)$$

$$+ \left(1 + \frac{K_1^*}{h} + \frac{K_1^* K_2^*}{h^2} \right) \left(\frac{K_B^* B_T}{K_B^* + h} + \frac{K_W^*}{h} - h \right)$$

which gives (times $h^3 \cdot (K_B^* + h)$) an equation of fifth order for h:

$$\text{TA} \, (K_B^* + h) \left(h^3 + K_1^* h^2 + K_1^* K_2^* h \right)$$

$$= \left[\text{CO}_3^{2-} \right] \left(h + \frac{h^2}{K_2^*} + \frac{h^3}{K_1^* K_2^*} \right)$$

$$\cdot \left(K_1^* h^2 + K_1^* h \, (K_B^* + 2K_2^*) + 2K_B^* K_1^* K_2^* \right)$$

$$+ \left(h^2 + K_1^* h + K_1^* K_2^* \right) \left(K_B^* B_T h + K_W^* K_B^* + K_W^* h - K_B^* h^2 - h^3 \right)$$

Roots: two positive (use the larger one), three negative. Then calculate s from Eq. (1.5.80)

$$s = \left(\text{TA} - \frac{K_B^* B_T}{K_B^* + h} - \frac{K_W^*}{h} + h \right) \Big/ \left(\frac{K_1^*}{h} + 2\frac{K_1^* K_2^*}{h^2} \right)$$

and proceed as in case 1.

(14) $[CO_3^{2-}]$ and DIC given

This case is similar to case 12. Eq. (1.1.11) $\Longrightarrow$ quadratic equation for h

$$[CO_3^{2-}] \left(1 + \frac{h}{K_2^*} + \frac{h^2}{K_1^* K_2^*} \right) = \text{DIC}$$

Roots: one positive, one negative. Calculate s by Eq. (1.1.9) and proceed as in case 1.

(15) TA and DIC given

Eqs. (1.5.79) and (1.5.80) $\Longrightarrow$

$$\text{DIC} \left(\frac{K_1^*}{h} + 2\frac{K_1^* K_2^*}{h^2} \right)$$

$$= \left(\text{TA} - \frac{K_B^* B_T}{K_B^* + h} - \frac{K_W^*}{h} + h \right) \left(1 + \frac{K_1^*}{h} + \frac{K_1^* K_2^*}{h^2} \right)$$

which yields (times $h^3 \cdot (K_B^* + h)$) an equation of fifth order for h:

$$\text{DIC} \cdot (K_B^* + h) \cdot \left(K_1^* h^2 + 2K_1^* K_2^* h \right) =$$

$$\left[\text{TA} (K_B^* + h) \, h - K_B^* B_T h - K_W^* (K_B^* + h) + (K_B^* + h) h^2 \right]$$

$$\cdot \left(h^2 + K_1^* h + K_1^* K_2^* \right).$$

Roots: one positive, four negative. Once h has been determined, one can proceed as in case 9.

Appendix C

Details and Calculations

This appendix includes details and calculations from various sections of the book. The numbers of these sections are indicated at the beginning of each of the following subsections.

C.1 Total alkalinity and charge balance

In Section 1.2.5 an expression for alkalinity in terms of conservative ions was given (left-hand side of Eq. (1.2.40)). It was based on an approximation (PA) to the exact definition of total alkalinity (Eq. (1.2.33)) given by Dickson (1981). Here we will derive the explicit conservative alkalinity expression, $TA^{(ec)}$, corresponding to the full right-hand side of Eq. (1.2.33) given by Dickson. Effects of biogeochemical processes on total alkalinity such as algal uptake of ammonia, phosphate, or silicic acid are discussed.

We start with the charge balance in seawater. The sum of the charges of all ions present in seawater must equal zero:

$$[Na^+] + 2[Mg^{2+}] + 2[Ca^{2+}] + [K^+] + ... + [H^+]_F$$
$$-[Cl^-] - 2[SO_4^{2-}] - [NO_3^-] \qquad \qquad (C.1.1)$$
$$-[HCO_3^-] - 2[CO_3^{2-}] - [B(OH)_4^-] - [OH^-]$$
$$-[H_2PO_4^-] - 2[HPO_4^{2-}] - 3[PO_4^{3-}] + [NH_4^+] \quad \pm ... \; = 0 \; .$$

Now terms are added to both sides of Eq. (C.1.1) until Dickson's expression of total alkalinity (Eq. 1.2.33) appears on the right-hand side of the equation. The left-hand side then reads:

$$[Na^+] + 2[Mg^{2+}] + 2[Ca^{2+}] + [K^+] + ...$$
$$-[Cl^-] - [NO_3^-] - ...$$
$$\underbrace{-[H_3PO_4] - [H_2PO_4^-] - [HPO_4^{2-}] - [PO_4^{3-}]}_{= -P_T}$$

$$\underbrace{+[NH_3] + [NH_4^+]}_{= +NH_T} \underbrace{-2[SO_4^{2-}] - 2[HSO_4^-]}_{= -2\,S_T}$$

$$= \ TA$$

which can be written as:

$$[Na^+] + 2\,[Mg^{2+}] + 2\,[Ca^{2+}] + [K^+] + 2[Sr^{2+}] + ...$$
$$-[Cl^-] - [Br^-] - [NO_3^-] - ...$$
$$-P_T + NH_T - 2\,S_T$$

$$=: \ TA^{(ec)} \tag{C.1.2}$$

It is unusual that compounds such as total phosphate appear in the $TA^{(ec)}$ expression because it should contain conservative ions only. One reason for this is that the right-hand side of Eq. (1.2.33) contains species which are uncharged (refer to discussion in Section 1.2.3). Nevertheless, regardless whether P_T, NH_T, and S_T are included in $TA^{(ec)}$ or not, total alkalinity remains a conservative quantity - provided that the approach is based on electroneutrality. This is because P_T, NH_T, and S_T are *total concentrations* in units of mol kg^{-1} seawater that do not depend on temperature, pressure, and pH.

Changes of alkalinity due to biogeochemical processes such as precipitation or dissolution of $CaCO_3$ etc. have been discussed already in Section 1.2.7 on the basis of PA. The definition of $TA^{(ec)}$ (Eq. (C.1.2)) does not change any of the conclusions given in Section 1.2.7. The change of total alkalinity associated with various other processes derived from $TA^{(ec)}$ are as follows:

1. Uptake of one mole of ammonia (NH_3 or NH_4^+) by algae under the assumption that electroneutrality of algae is ensured by parallel uptake of OH^- or release of H^+: alkalinity decreases by one mole (term: $+NH_T$).

2. Uptake of one mole of phosphate (H_3PO_4, $H_2PO_4^-$, HPO_4^{2-}, or PO_4^{3-}) by algae under the assumption that electroneutrality of algae is ensured by parallel uptake of H^+ or release of OH^-: alkalinity increases by one mole (term: P_T).

3. Uptake of one mole of silicic acid ($Si(OH)_4$ or $SiO(OH)_3^-$) by algae under the assumption that electroneutrality of algae is ensured by parallel uptake of H^+ or release of OH^-: no change in alkalinity.

4. Remineralization of algal material has the reverse effects on alkalinity.

The conclusions 1 and 2 are consistent with results from laboratory experiments with algae grown on different nitrogen sources (Brewer and Goldman (1976); Goldman and Brewer (1980)). This supports the hypothesis that electroneutrality of

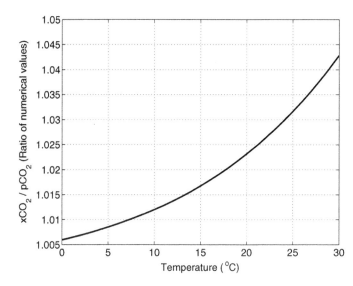

Figure C.2.1: The ratio of the numerical values of mole fraction and partial pressure, xCO_2/pCO_2, as a function of temperature at 100% humidity. At $xCO_2 = 360$ μmol mol^{-1}, the difference between the numerical values is $\sim 10 \times 10^{-6}$ ($T = 25°C$, $S = 35$).

algal cells which take up charged species is ensured by simultaneous uptake or release of H$^+$ or OH$^-$.[1] Please note that in this case the change of alkalinity does not depend on the species which is taken up, for example, H$_3$PO$_4$, H$_2$PO$_4^-$, HPO$_4^{2-}$, or PO$_4^{3-}$.

C.2   Saturation vapor pressure of water

When the pCO_2 of a seawater sample is calculated from its mole fraction measured in dry air, a correction has to be applied that accounts for the water vapor pressure (Section 1.4.1). The correction is quantified in the following.

The air equilibrated with the seawater is assumed to be at 100% humidity. Then the partial pressure of CO$_2$ is related to the mole fraction by:

$$pCO_2 = xCO_2 \cdot (p - pH_2O) \tag{C.2.3}$$

where p (atm) is total pressure and pH_2O (atm) is the saturation vapor pressure of water (Weiss and Price, 1980):

$$\ln pH_2O = 24.4543 - \frac{6745.09}{T} - 4.8489 \ln \frac{T}{100} - 0.000544\, S \ .$$

---

[1]The situation is different in transporter cells as, for example, in erythrocytes where HCO$_3^-$ is taken up and released at a different time by cotransport (antiport) with Cl$^-$: over long time spans the net transport of carbon as well as chlorine is zero. In growing algal cells, however, the net transport of carbon into the cell is positive. If this transport would be balanced by efflux of Cl$^-$, all chloride ions would be used up after a while.

Let us estimate the difference between the partial pressure and the mole fraction for a gas sample at 100% humidity and at standard pressure $p = 1$ atm. At $T = 25°C$ and $S = 35$, $pH_2O \simeq 0.03$ atm. Thus, the numerical value of the partial pressure is about 3% smaller than the mole fraction (compare Figure C.2.1). For example, if $xCO_2 = 360$ μmol mol^{-1}, then $pCO_2 = 349$ μatm. At the air-sea interface, where the air is assumed to be saturated with water vapor, the difference between the numerical values would therefore be about 10×10^{-6}.

In dry air and at 1 atm total pressure, the value of the mole fraction of CO_2 (in μmol mol^{-1}) is equal to the value of the partial pressure (μatm):

$$pCO_2\,(\mu\text{atm}) = xCO_2 \cdot \left(\underbrace{p}_{=1} - \underbrace{pH_2O}_{=0} \right) = xCO_2\,(\mu\text{mol mol}^{-1}). \qquad (\text{C.2.4})$$

C.3 The fugacity of a pure gas

In Section 1.4.2 an expression for the fugacity of CO_2 was given (Eq. (1.4.65)). In this section we will show how to calculate the fugacity of a pure gas when its equation of state is known.

The fugacity, f, of a pure gas can be calculated from its equation of state as follows (for derivation, see Klotz and Rosenberg, 2000):

$$\ln\left(\frac{f}{P}\right) = -\frac{1}{RT} \int_0^P \left[\frac{RT}{P'} - V_m(P') \right] dP' \qquad (\text{C.3.5})$$

where P is pressure, V the volume, R the gas constant ($= 8.314472$ J K^{-1} mol^{-1}), T the absolute temperature, and V_m is the molar volume. The task now is to find an expression for V_m (this constitutes the equation of state of the gas). Then integration of Eq. (C.3.5) yields the fugacity at a given pressure.

A gas is ideal if it obeys the equation of state

$$PV = nRT \quad \text{(ideal equation of state)} \qquad (\text{C.3.6})$$

or

$$PV_m = RT \qquad (\text{C.3.7})$$

where n the number of moles. The internal energy, U, of an ideal gas is a function of the temperature only. The ideal equation of state describes accurately the behavior of real gases at the limit of zero pressure. Inserting $V_m = RT/P$ into Eq. (C.3.5), it follows that the right-hand side equals zero and thus $f = p$. This is what we expect: the fugacity of an ideal gas is equal to its pressure.

Real gases can be approximated by semi-empirical representations such as the van der Waals equation: [2]

$$\left(P + \frac{a}{V_{\mathrm{m}}^2}\right)(V_{\mathrm{m}} - b) = RT \tag{C.3.8}$$

which contains two empirical constants a and b. For CO_2, $a \simeq 0.366$ m^6 mol^{-2} and $b \simeq 4.29 \cdot 10^{-5}$ m^3 mol^{-1}.

Another useful form of the equation of state for real gases is the virial equation (an empirical power series expansion):

$$PV_{\mathrm{m}} = A(T) + B(T)P + C(T)P^2 + ... \tag{C.3.9}$$

in which A, B, and C are functions of the temperature and are known as virial coefficients. The ideal equation of state can be recovered from the virial equation in the low pressure limit when $A(T) \rightarrow RT$.

Inserting an expression of the equation of state of a real gas into Eq. (C.3.5), one can calculate the fugacity of a real gas at given pressure. When the third and higher order virial terms in Eq. (C.3.9) are sufficiently small and $A(T) = RT$, Eq. (C.3.5) can be integrated analytically and one obtains

$$RT \ln\left(\frac{f}{P}\right) = BP$$

or

$$f = P \exp\left(\frac{BP}{RT}\right).$$

This equation corresponds to Eq. (1.4.65) given in Section 1.4. The calculation of the fugacity of a gas in a gas mixture is more involved and is therefore not discussed here (see, for example, Guggenheim, 1967).

Exercise C.1 (**))
Calculate the fugacity for a gas that obeys the van der Waals equation: a) Instead of solving a cubic equation for V_{m} apply a low pressure approximation for PV_{m}. b) Calculate the fugacity by numerical integration and compare the results for pressures up to 1 MPa with the approximate solution derived under a).

---

[2] The form of a semi-empirical relation has been suggested by theoretical considerations, while parameters are derived from observations.

C.4 Equilibrium at the air-sea interface and the chemical potential

In Section 1.4, the fugacity and partial pressure was discussed. Here it is demonstrated that the equilibrium at the air-sea interface can be characterized by equality of fugacities *or* partial pressures.

According to thermodynamics the flux of particles i from a subsystem s_1 to its neighboring one s_2 is proportional to the chemical potential $\mu_{i(s_1)}$ and equilibrium is reached when the chemical potentials of both subsystems are equal, i.e. $\mu_{i(s_1)} = \mu_{i(s_2)}$. The chemical potentials of CO_2 in water and air are related to activity and fugacity, respectively:

$$\mu_{CO_2(aq)} = \mu^0_{CO_2(aq)} + RT \ln \{CO_2\} \tag{C.4.10}$$

$$\mu_{CO_2(g)} = \mu^0_{CO_2(g)} + RT \ln fCO_2 \tag{C.4.11}$$

where $\mu^0_{CO_2(aq)}$ and $\mu^0_{CO_2(g)}$ are the chemical potentials at the respective standard state. Equilibrium at the air-sea interface is attained, when the chemical potentials of CO_2 are equal, i.e. $\mu_{CO_2(g)} = \mu_{CO_2(aq)}$. The latter equation can be solved for fCO_2

$$fCO_2 = \{CO_2\} \exp\left[\left(\mu^0_{CO_2(aq)} - \mu^0_{CO_2(g)}\right)/RT\right] \tag{C.4.12}$$

$$= [CO_2]\gamma_{CO_2(aq)} \exp\left[\left(\mu^0_{CO_2(aq)} - \mu^0_{CO_2(g)}\right)/RT\right] \tag{C.4.13}$$

$$= [CO_2]/K_0 \tag{C.4.14}$$

where we have expressed the activity of CO_2 as the product of molal concentration and activity coefficient. At equilibrium the fugacity is proportional to the concentration of CO_2. This relation is called Henry's law. Henry's constant, K_0, which is theoretically given by

$$K_0 = \frac{1}{\gamma_{CO_2(aq)} \exp\left[\left(\mu^0_{CO_2(aq)} - \mu^0_{CO_2(g)}\right)/RT\right]} \tag{C.4.15}$$

has been determined by Weiss (1974).

At a given temperature the chemical potential is a strictly monotonic function[3] of the fugacity. Therefore the inverse function of $\mu(f)$ yields a unique value of fugacity for any given chemical potential and thus air-sea equilibrium can be characterized also by equal fugacities of air and seawater. Because partial pressure is a strictly monotonic function of fugacity, the equilibrium can also be characterized by equality of partial pressures.

---

[3]This is a property of the logarithm function.

C.5   Change CO_2 concentration while keeping pH constant

Biologists like to culture plankton under various conditions such as light level, nutrient concentrations, temperature, pH etc. As an example consider the question whether the limitation of phytoplankton growth rates is due to carbon dioxide (Riebesell et al., 1993) or a pH-effect (Raven, 1993). For this and other investigations one would like to change the concentration of carbon dioxide while keeping pH constant. The following calculation is another worked out problem (cf. Section 1.6).

Increase $[CO_2]$

Method: increase (carbonate) alkalinity by adding sodium carbonate (Na_2CO_3) and sodium bicarbonate ($NaHCO_3$).

Theory: Addition of Na_2CO_3 and $NaHCO_3$ changes TA and DIC. The pH and $[CO_2]$ of the initial (i) and final (f) state are given. From these values one can calculate DIC and TA using Eqs. (1.5.79) and (1.5.80).

Example: increase CO_2 from 10 to 50 μmol kg^{-1} ($T_c = 25°C$, $S = 35$; $B_T = 4.106 \cdot 10^{-4}$ mol kg^{-1}):

- $pH_i = 8.1$; $[CO_2]_i = 10$ μmol kg^{-1}
  $\Longrightarrow$ DIC$_i = 2108$ μmol kg^{-1}; TA$_i = 2485$ μmol kg^{-1}

- $pH_f = 8.1$; $[CO_2]_f = 50$ μmol kg^{-1}
  $\Longrightarrow$ DIC$_f = 10541$ μmol kg^{-1}; TA$_f = 12004$ μmol kg^{-1}

Add Na_2CO_3 and $NaHCO_3$ in such portions that

$$[NaHCO_3] + \quad 2[Na_2CO_3] \quad = \quad TA_f - TA_i \quad = \quad 9520 \ \mu\text{mol kg}^{-1}$$
$$[NaHCO_3] + \quad [Na_2CO_3] \quad = \quad DIC_f - DIC_i \quad = \quad 8433 \ \mu\text{mol kg}^{-1}$$

which readily gives

$$[Na_2CO_3] \quad = \quad 1087 \ \mu\text{mol kg}^{-1}$$
$$[NaHCO_3] \quad = \quad 7346 \ \mu\text{mol kg}^{-1}$$

Remark: The salinity of the solution increases by less than 0.001 due to the addition of sodium ions.

Decrease $[CO_2]$

Method: One may proceed in two steps

1. Add a strong acid to the solution $\Longrightarrow$ $[CO_2]$ increases; then bubble with air (open system!) to speed up the outgassing of CO_2; DIC will decrease until the CO_2 concentration approaches equilibrium with the atmosphere.

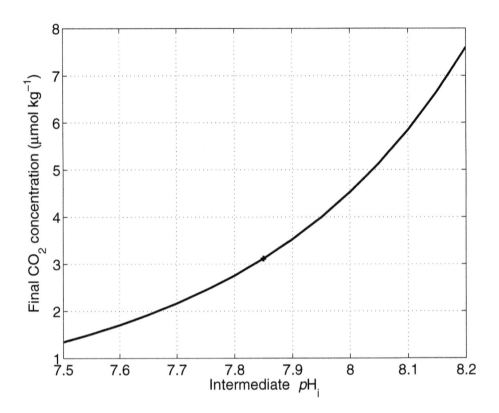

Figure C.5.2: Change $[CO_2]$ while keeping pH constant: For an initial and final pH 8.2 one has to acidify until an intermediate pH of 7.85 is reached in order to obtain a final CO_2 concentration of ≈ 3 μmol kg^{-1} ($T_c = 25°C$, $S = 35$, $DIC_i = 2000$ μmol kg^{-1}).

   2. Close the system and then add a strong base until the solution reaches the original pH value.

Theory: One can use the trial-and-error-method to get the desired CO_2 concentration. Here we calculate the intermediate (index m) pH to which the open system has to be acidified.

Step 1. Calculate the change of DIC due to the change (decrease) of pH while keeping $[CO_2]$ constant (equilibrium with atmosphere):

$$d\mathrm{DIC} = D_h\,dh + D_s\underbrace{\,ds\,}_{=\,0}$$

$$\Longrightarrow$$

$$\int d\mathrm{DIC} = \int D_h\,dh = -[CO_3^{2-}]_i \int \left(\frac{K_1^*}{h^2} + 2\frac{K_1^* K_2^*}{h^3} \right) dh.$$

Integration $\Longrightarrow$

$$\text{DIC}_m = \text{DIC}_i + [\text{CO}_3^{2-}]_i \left[\frac{K_1^*}{h_m} + \frac{K_1^* K_2^*}{h_m^2} - \frac{K_1^*}{h_i} - \frac{K_1^* K_2^*}{h_i^2} \right].$$

(please note that we currently do not know $h_m = 10^{-pH_m}$).

Step 2. Calculate the change of $s = [\text{CO}_2]$ due to the change (increase) of pH while keeping DIC constant (closed system):

$$\underbrace{d\text{DIC}}_{= 0} = D_h\, dh + D_s\, ds$$

$$\frac{ds}{dh} = -D_h/D_s = \frac{s \left(\dfrac{K_1^*}{h^2} + 2\dfrac{K_1^* K_2^*}{h^3} \right)}{1 + \dfrac{K_1^*}{h} + \dfrac{K_1^* K_2^*}{h^2}}. \tag{C.5.16}$$

The solution of (C.5.16)

$$s_f = s_i \frac{h_f^2 \left(h_m^2 + h_m K_1^* + K_1^* K_2^* \right)}{h_m^2 \left(h_f^2 + h_f K_1^* + K_1^* K_2^* \right)}$$

can be rewritten as a quadratic equation for h_m

$$h_m^2 \underbrace{\left[s_f \left(h_f^2 + h_f K_1^* + K_1^* K_2^* \right) - s_i h_f^2 \right]}_{=:\, a} - h_m \underbrace{s_i h_f^2 K_1^*}_{=:\, -b} \underbrace{-s_i h_f^2 K_1^* K_2^*}_{=:\, c} = 0.$$

The relevant solution of this equation reads

$$h_m = \frac{-b + \sqrt{b^2 - 4\, a\, c}}{2\, a}.$$

The results are plotted in Figure C.5.2 For an initial and final pH of 8.2 one has to acidify until an intermediate pH of 7.85 is reached in order to get a final CO_2 concentration of 3 μmol kg^{-1}.

C.6 The rate constant for the hydroxylation of CO_2, k_{+4}

The data for k_{+4} as obtained from the work of Johnson (1982) (Section 2.3.2, Figure 2.3.6b, diamonds) needs an explanation. It is demonstrated below, how the values for k_{+4} were determined from his reported data for $k_{+4} K_W$.

For the hydroxylation of CO_2:

$$\text{CO}_2 + \text{OH}^- \xrightarrow{k_{+4}} \text{HCO}_3^-$$

the rate law reads:

$$\frac{d[CO_2]}{dt} = - k_{+4}\,[OH^-]\,[CO_2] \tag{C.6.17}$$

where the brackets denote concentrations. Johnson used the activity of OH^-, $\{OH^-\}$, in his rate law (see his eq. (6); his rate constant k_{OH^-} is identical to k_{+4} used here):

$$\begin{aligned}
\frac{d[CO_2]}{dt} &= - k_{+4}\,\{OH^-\}\,[CO_2] \\
&= - \underbrace{(k_{+4}\,K_W)}_{x}\,[CO_2]/a_H
\end{aligned} \tag{C.6.18}$$

where he substituted the expression K_W/a_H for $\{OH^-\}$ ($K_W = a_H\{OH^-\}$ is the thermodynamic dissociation constant of water and a_H is the activity of H^+.) He then experimentally determined values for x which were assigned to $k_{+4}K_W$.

His values may be recalculated, this time using concentrations and the rate law (C.6.17). For the concentration of OH^-, we substitute $[OH^-] = K_W^*/[H^+]$ (where K_W^* is the ion product of water):

$$\begin{aligned}
\frac{d[CO_2]}{dt} &= - k_{+4}\,[OH^-]\,[CO_2] \\
&= - k_{+4}\,K_W^*\,[CO_2]/[H^+] \\
&= - \underbrace{(k_{+4}\,K_W^*\,\gamma_H)}_{x}\,[CO_2]/a_H
\end{aligned} \tag{C.6.19}$$

where the relationship between activity and concentration $a_H = \gamma_H[H^+]$ has been used. Comparing Eq. (C.6.18) and Eq. (C.6.19), it appears that Johnson actually determined values for $k_{+4}K_W^*\gamma_H$ rather than $k_{+4}K_W$. Using the values given in Table 1 of Johnson's paper and values for γ_H (Millero and Pierrot, 1998), we can calculate values for k_{+4} (Table C.6.1).

Table C.6.1: Values of the rate constant k_{+4} as determined from the data of Johnson (1982).[a]

T_c	$k_{+4}K_W^*\gamma_H$	K_W^* [b]	γ_H [c]	k_{+4}
(°C)	(mol kg^{-1} s^{-1})	(mol^2 kg^{-2})		(kg mol^{-1} s^{-1})
5	1.44×10^{-11}	8.40×10^{-15}	0.664	2.58×10^3
15	3.10×10^{-11}	2.33×10^{-14}	0.615	2.16×10^3
25	13.4×10^{-11}	5.93×10^{-14}	0.559	4.04×10^3
35	43.3×10^{-11}	1.39×10^{-13}	0.495	6.29×10^3

[a] At salinity $S = 33.77$.
[b] Values for K_W^* as given in DOE (1994).
[c] Millero and Pierrot (1998).

C.7 A formula for the equilibration time of CO_2

A formula is derived from which the CO_2 equilibration time in a closed seawater system on time scales longer ~ 1 s can be calculated as a function of pH (cf. Section 2.4.1).

The chemical reactions read:

$$CO_2 + H_2O \quad \underset{k_{-1}}{\overset{k_{+1}}{\rightleftharpoons}} \quad HCO_3^- + H^+ \tag{C.7.20}$$

$$CO_2 + OH^- \quad \underset{k_{-4}}{\overset{k_{+4}}{\rightleftharpoons}} \quad HCO_3^- \tag{C.7.21}$$

$$CO_3^{2-} + H^+ \quad \rightleftharpoons \quad HCO_3^-; \qquad K_2^* = [CO_3^{2-}][H^+]/[HCO_3^-] \tag{C.7.22}$$

$$H_2O \quad \rightleftharpoons \quad H^+ + OH^-; \quad K_W^* = [H^+][OH^-] \tag{C.7.23}$$

Using the conservation of $\Sigma CO_2 = [CO_2] + [HCO_3^-] + [CO_3^{2-}]$ and the alkalinity $TA = [HCO_3^-] + 2[CO_3^{2-}] + [OH^-] - [H^+]$ and the equilibrium relations (Eqs. (C.7.22) and (C.7.23)), four of the five variables can be eliminated. The linear differential equation for a perturbation in CO_2 $(= \delta CO_2)$ then reads:

$$\frac{d(\delta CO_2)}{dt} = -\frac{1}{\tau} \delta CO_2 \tag{C.7.24}$$

where the relaxation time is given by

$$\begin{aligned}
\frac{1}{\tau} = \quad &+ k_{+1} \quad &- \quad & k_{-1}(\partial_s b \, [H^+]^* + \partial_s h \, [HCO_3^-]^*) \\
&- k_{-4} \, \partial_s b \quad &+ \quad & k_{+4}([OH^-]^* + \partial_s oh \, [CO_2]^*)
\end{aligned} \tag{C.7.25}$$

with $[c_i]^*$ being the concentration of species i at equilibrium, and $\partial_s b, \partial_s h,$ and $\partial_s oh$ are the derivatives of $HCO_3^-, H^+,$ and OH^- with respect to CO_2 at equilibrium:

$$\begin{aligned}
\partial_s b &\equiv \left. \frac{\partial [HCO_3^-]}{\partial [CO_2]} \right|_* = -\frac{[H^+]^* + 2[CO_3^{2-}]^*/(1+\alpha)}{K_2^* + [H^+]^* + [CO_3^{2-}]^*/(1+\alpha)} \\
\partial_s h &\equiv \left. \frac{\partial [H^+]}{\partial [CO_2]} \right|_* = -\frac{\partial_s b + 2}{1+\alpha} \\
\partial_s oh &\equiv \left. \frac{\partial [OH^-]}{\partial [CO_2]} \right|_* = \alpha \, \frac{\partial_s b + 2}{1+\alpha}
\end{aligned} \tag{C.7.26}$$

with

$$\alpha = K_W^*/([H^+]^*)^2$$

and

$$[CO_3^{2-}]^* = \frac{\Sigma CO_2}{1 + [H^+]^*/K_2^* + ([H^+]^*)^2/K_1^*/K_2^*} \, .$$

C.8   Kinetic rate laws of the carbonate system

The set of coupled differential equations of the carbonate system reads (cf. Section 2.4.2):

$$
\frac{d[CO_2]}{dt} = \begin{aligned}[t] &+(k_{-1}[H^+] + k_{-4})[HCO_3^-] \\ &-(k_{+1} + k_{+4}[OH^-])[CO_2] \end{aligned}
\tag{C.8.27}
$$

$$
\frac{d[HCO_3^-]}{dt} = \begin{aligned}[t] &-(k_{-1}[H^+] + k_{-4})[HCO_3^-] \\ &+(k_{+1} + k_{+4}[OH^-])[CO_2] \\ &-(k_{-5}^{H^+} + k_{+5}^{OH^-}[OH^-])[HCO_3^-] \\ &+(k_{+5}^{H^+}[H^+] + k_{-5}^{OH^-})[CO_3^{2-}] \\ &+k_{+8}[B(OH)_3][CO_3^{2-}] - k_{-8}[B(OH)_4^-][HCO_3^-] \end{aligned}
\tag{C.8.28}
$$

$$
\frac{d[CO_3^{2-}]}{dt} = \begin{aligned}[t] &+(k_{-5}^{H^+} + k_{+5}^{OH^-}[OH^-])[HCO_3^-] \\ &-(k_{+5}^{H^+}[H^+] + k_{-5}^{OH^-})[CO_3^{2-}] \\ &-k_{+8}[B(OH)_3][CO_3^{2-}] + k_{-8}[B(OH)_4^-][HCO_3^-] \end{aligned}
\tag{C.8.29}
$$

$$
\frac{d[H^+]}{dt} = \begin{aligned}[t] &+k_{+1}[CO_2] - k_{-1}[H^+][HCO_3^-] \\ &+k_{-5}^{H^+}[HCO_3^-] - k_{+5}^{H^+}[H^+][CO_3^{2-}] \\ &+k_{+6} - k_{-6}[H^+][OH^-] \end{aligned}
\tag{C.8.30}
$$

$$
\frac{d[OH^-]}{dt} = \begin{aligned}[t] &+k_{-4}[HCO_3^-] - k_{+4}[CO_2][OH^-] \\ &-k_{+5}^{OH^-}[HCO_3^-][OH^-] + k_{-5}^{OH^-}[CO_3^{2-}] \\ &+k_{+6} - k_{-6}[H^+][OH^-] \\ &-k_{+7}[B(OH)_3][OH^-] + k_{-7}[B(OH)_4^-] \end{aligned}
\tag{C.8.31}
$$

$$
\frac{d[B(OH)_3]}{dt} = \begin{aligned}[t] &-k_{+7}[B(OH)_3][OH^-] + k_{-7}[B(OH)_4^-] \\ &-k_{+8}[B(OH)_3][CO_3^{2-}] + k_{-8}[B(OH)_4^-][HCO_3^-] \end{aligned}
\tag{C.8.32}
$$

$$
\frac{d[B(OH)_4^-]}{dt} = \begin{aligned}[t] &+k_{+7}[B(OH)_3][OH^-] - k_{-7}[B(OH)_4^-] \\ &+k_{+8}[B(OH)_3][CO_3^{2-}] - k_{-8}[B(OH)_4^-][HCO_3^-] \end{aligned}
\tag{C.8.33}
$$

C.9 Derivation of oxygen isotope partitioning in the system CO_2-H_2O

In the following, the oxygen isotope partitioning in the system $CO_2 - H_2O$ is examined in detail (cf. Section 3.3.5). This subject requires quite a bit of theoretical work. For further reading, see Usdowski et al. (1991) and Usdowski and Hoefs (1993). In contrast to the seawater system at $T = 25°C$ and $S = 35$ which is adopted as a standard system throughout the book, a fresh water system at $19°C$ is discussed in the following (see Section 3.3.5). Usdowski and co-workers calculations refer to ideal-dilute solutions, i.e. activity coefficients of ions approach 1, $\gamma_i \to 1$. In this approximation, stoichiometric equilibrium constants approach the value of thermodynamic equilibrium constants. Thus, equilibrium constants are simply denoted by K in the following.

Equilibrium constant and fractionation factor

Before getting into the details of oxygen isotope fractionation in the carbonate system the relation between the equilibrium constant K and the fractionation factor α is discussed. As an example, the oxygen isotope exchange between CO_2 and H_2O may be considered:

$$1/2 \; C^{16}O_2 + H_2^{18}O \quad \rightleftharpoons \quad 1/2 \; C^{18}O_2 + H_2^{16}O$$

(note that $C^{16}O_2 \equiv C^{16}O^{16}O$ and $C^{18}O_2 \equiv C^{18}O^{18}O$) for which the equilibrium constant is given by:

$$K = \frac{[C^{18}O_2]^{\frac{1}{2}}}{[C^{16}O_2]^{\frac{1}{2}}} \bigg/ \frac{[H_2^{18}O]}{[H_2^{16}O]} \quad .$$

On the other hand, the fractionation factor α is given by:

$$\alpha_{(CO_2-H_2O)} = \frac{2[C^{18}O^{18}O] + [C^{18}O^{16}O]}{2[C^{16}O^{16}O] + [C^{18}O^{16}O]} \bigg/ \frac{[H_2^{18}O]}{[H_2^{16}O]} \quad .$$

It can be shown that K is equal to α if the equilibrium constant for the reaction

$$2 \; C^{18}O^{16}O \quad \rightleftharpoons \quad C^{18}O^{18}O + C^{16}O^{16}O$$

is 4 which was done by Urey (1947). In general, the equilibrium constant is equal to the fractionation factor if the proportion of the isotopic molecules is determined from the symmetry numbers of the molecules alone (see Section C.10). However, the rule does not apply to hydrogen isotopes or to molecules in which atoms of the same element occupy nonequivalent positions (e.g. nitrogen in N_2O: $N\equiv N=O$).

Oxygen isotope equilibrium

The chemical reactions considered to describe oxygen isotope equilibrium in the system $CO_2 - H_2O$ are:

$$
\begin{aligned}
CO_2 + H_2O &\rightleftharpoons H_2CO_3 & K_{1'}; \\
H_2CO_3 &\rightleftharpoons H^+ + HCO_3^- & K_{H_2CO_3}; \\
CO_2 + OH^- &\rightleftharpoons HCO_3^- & K_1/K_W; \\
HCO_3^- &\rightleftharpoons CO_3^{2-} + H^+ & K_2; \\
H_2O &\rightleftharpoons H^+ + OH^- & K_W .
\end{aligned}
\qquad \text{(C.9.34)}
$$

where $CO_2 = CO_2(aq.)$ and H_2CO_3 is true carbonic acid. Note that for the discussion of the seawater carbonate system, the sum of these two species was denoted by CO_2. As a result, the set of constants used here differs from the set of constants used so far (cf. Eq. (3.2.21)). The constant $K_{1'} = 10^{-2.98}$ at $19°C$ (fresh water), describing the hydration equilibrium, and the dissociation constant of true carbonic acid $K_{H_2CO_3} = 10^{-3.41}$ at $19°C$ (fresh water), have been introduced. These constants are related to the dissociation constant K_1 by $K_{H_2CO_3} = K_1(1/K_{1'}+1)$. The values of K_2, and K_W at $19°C$ in fresh water are $K_2 = 10^{-10.39}$ and $K_W = 10^{-14.20}$ (Usdowski et al., 1991).

Including the possible combinations of ^{18}O and ^{16}O in the respective compounds, the first reaction of Eq. (C.9.34) reads in detail (Mills and Urey, 1940):

$$
\begin{aligned}
C^{18}O^{18}O + H_2^{18}O &\rightleftharpoons H_2C^{18}O^{18}O^{18}O & {}^OK_1; \\
C^{18}O^{16}O + H_2^{18}O &\rightleftharpoons H_2C^{18}O^{18}O^{16}O & {}^OK_2; \\
C^{16}O^{16}O + H_2^{18}O &\rightleftharpoons H_2C^{18}O^{16}O^{16}O & {}^OK_3; \\
C^{18}O^{18}O + H_2^{16}O &\rightleftharpoons H_2C^{18}O^{18}O^{16}O & {}^OK_4; \\
C^{18}O^{16}O + H_2^{16}O &\rightleftharpoons H_2C^{18}O^{16}O^{16}O & {}^OK_5; \\
C^{16}O^{16}O + H_2^{16}O &\rightleftharpoons H_2C^{16}O^{16}O^{16}O & {}^OK_6.
\end{aligned}
\qquad \text{(C.9.35)}
$$

Analogous equations can be written with respect to OH^-. Using explicit expressions for equilibrium constants, e.g.,

$$
{}^OK_1 = \frac{[H_2C^{18}O^{18}O^{18}O]}{[C^{18}O^{18}O][H_2^{18}O]} \,,
$$

Eq. (C.9.35) defines a set of coupled equations. Evaluating for instance the expression $({}^OK_1{}^OK_2{}^OK_3)/({}^OK_4{}^OK_5{}^OK_6)$ defines a new equilibrium constant; the overall isotope exchange reaction between carbonic acid and water may then be written as

$$
1/3\ H_2C^{16}O_3 + H_2^{18}O \rightleftharpoons 1/3\ H_2C^{18}O_3 + H_2^{16}O
$$

for which the fractionation factor is given by

$$
\alpha_{(H_2CO_3-H_2O)} = \frac{[H_2C^{18}O_3]^{\frac{1}{3}}}{[H_2C^{16}O_3]^{\frac{1}{3}}} \Bigg/ \frac{[H_2^{18}O]}{[H_2^{16}O]}
$$

$$
= \left(\frac{{}^OK_1{}^OK_2{}^OK_3}{{}^OK_4{}^OK_5{}^OK_6} \right)^{\frac{1}{3}} .
$$

Similarly, the isotope exchange reaction between bicarbonate and water may be written as

$$1/3\ HC^{16}O_3^- + H_2^{18}O \ \rightleftharpoons \ 1/3\ HC^{18}O_3^- + H_2^{16}O$$

with fractionation factor

$$\alpha_{(HCO_3^- - H_2O)} \ = \ \frac{[HC^{18}O_3^-]^{\frac{1}{3}}}{[HC^{16}O_3^-]^{\frac{1}{3}}} \Big/ \frac{[H_2^{18}O]}{[H_2^{16}O]}$$

$$= \ \alpha_{(H_2CO_3 - H_2O)} \left(\frac{^{18}K_{H_2CO_3}}{^{16}K_{H_2CO_3}} \right)^{\frac{1}{3}}$$

where $K_{H_2CO_3}$ is the dissociation constant of true carbonic acid, e.g:

$$^{18}K_{H_2CO_3} \ = \ \frac{[H^+][HC^{18}O_3^-]}{[H_2C^{18}O_3]} \ .$$

The isotope exchange reaction between carbonate ion and water may be written as

$$1/3\ C^{16}O_3^{2-} + H_2^{18}O \ \rightleftharpoons \ 1/3\ C^{18}O_3^{2-} + H_2^{16}O$$

with fractionation factor

$$\alpha_{(CO_3^{2-} - H_2O)} \ = \ \frac{[C^{18}O_3^{2-}]^{\frac{1}{3}}}{[C^{16}O_3^{2-}]^{\frac{1}{3}}} \Big/ \frac{[H_2^{18}O]}{[H_2^{16}O]}$$

$$= \ \alpha_{(HCO_3^- - H_2O)} \left(\frac{^{18}K_2}{^{16}K_2} \right)^{\frac{1}{3}}$$

and

$$^{18}K_2 \ = \ \frac{[H^+][C^{18}O_3^{2-}]}{[HC^{18}O_3^-]} \ .$$

Thus, the isotopic exchange between all the dissolved carbonate species and water at any pH may be expressed by the equation

$$1/3\ {}^{16}S + H_2^{18}O \ \rightleftharpoons \ 1/3\ {}^{18}S + H_2^{16}O \tag{C.9.36}$$

for which the fractionation factor is given by

$$\alpha_{(S-H_2O)} \ = \ \frac{^{18}S^{\frac{1}{3}}}{^{16}S^{\frac{1}{3}}} \Big/ \frac{[H_2^{18}O]}{[H_2^{16}O]} \ . \tag{C.9.37}$$

with

$$^{18}S \ = \ [H_2C^{18}O_3] + [HC^{18}O_3^-] + [C^{18}O_3^{2-}]$$
$$^{16}S \ = \ [H_2C^{16}O_3] + [HC^{16}O_3^-] + [C^{16}O_3^{2-}] \ .$$

The essence of the discussion so far is that the oxygen isotopic exchange between the sum of the dissolved carbonate species ($S = [H_2CO_3] + [HCO_3^-] + [CO_3^{2-}]$) and water can be expressed at any pH by the fractionation factor (Eq. (C.9.37)). As will be described later, this factor can actually be measured analytically. If a relationship between α, S, and the pH of the solution can be found, individual fractionation factors between water and H_2CO_3, HCO_3^-, and CO_3^{2-} can be calculated. This procedure is explained in the following.

Chemical equilibrium between CO_2 and $\mathcal{S}$

The chemical equilibrium between CO_2 and the sum of the dissolved carbonate species involves the following reactions:

$$
\begin{aligned}
CO_2 + H_2O &\rightleftharpoons H_2CO_3 & K_{1'}; \\
H_2CO_3 &\rightleftharpoons H^+ + HCO_3^- & K_{H_2CO_3}; \\
HCO_3^- &\rightleftharpoons CO_3^{2-} + H^+ & K_2;
\end{aligned} \qquad \text{(C.9.38)}
$$

The equilibrium constant relating $\mathcal{S} = [H_2CO_3] + [HCO_3^-] + [CO_3^{2-}]$ and CO_2 can be written as:

$$
\frac{\mathcal{S}}{[CO_2]} = \frac{[H_2CO_3]}{[CO_2]} + \frac{[HCO_3^-]}{[CO_2]} + \frac{[CO_3^{2-}]}{[CO_2]}
$$

Using Eqs. (C.9.38), the fractions such as $[H_2CO_3]/[CO_2]$ can be expressed in terms of the equilibrium constants and $[H^+]$. Thus, the equilibrium $CO_2 \rightleftharpoons \mathcal{S}$ at any pH is given by:

$$
\frac{\mathcal{S}}{[CO_2]} = K_{1'} + \frac{K_{1'} K_{H_2CO_3}}{[H^+]} + \frac{K_{1'} K_{H_2CO_3} K_2}{[H^+]^2} . \qquad \text{(C.9.39)}
$$

At very low pH, $\mathcal{S}$ is essentially H_2CO_3, and $\mathcal{S}/[CO_2]$ is equal to $K_{1'} = 10^{-2.98}$. On the other hand, at $pH = 14$, $\mathcal{S}$ is essentially CO_3^{2-}, $\mathcal{S}/[CO_2]$ being equal to $K_{1'} K_{H_2CO_3} K_2/[H^+]^2 = 10^{11.22}$. Since $\mathcal{S}/[CO_2]$ is constant at a given pH, describing the equilibrium between $\mathcal{S}$ and CO_2, the free energy $\Delta G^0 = -RT \ln(\mathcal{S}/[CO_2])$ may be calculated, ranging from 17.0 to -64.0 kJ mol^{-1} for $pH = 0$ to $pH = 14$ at 19°C. The quantity $\ln(\mathcal{S}/[CO_2])$ is displayed in Figure C.9.3 for the pH range from 0 to 14.

So far, a relationship between $\mathcal{S}/[CO_2]$ and pH has been obtained. Next, the relationship between $\alpha_{(\mathcal{S}-H_2O)}$ and pH is discussed which can be derived from precipitation experiments.

Precipitation experiments

The fractionation factor $\alpha_{(\mathcal{S}-H_2O)}$ can be determined from precipitation experiments of McCrea (1950) and Usdowski et al. (1991). Briefly, if calcium (barium) carbonate is *quantitatively* precipitated from a bicarbonate-carbonate solution (i.e., none of the carbonate in solution escapes precipitation) the oxygen isotopic composition of the precipitate is the average of the oxygen present in the bicarbonate and carbonate at the time of precipitation (McCrea, 1950). Thus, the calcium (barium) carbonate simply reflects the isotopic composition of $\mathcal{S}$. The results of McCrea (1950) (25°C) and of Usdowski et al. (1991) (19°C) are summarized in Table C.9.2.

As discussed at the beginning of Section C.9 the equilibrium constant of the isotope exchange reaction

$$
1/3 \ {}^{16}\mathcal{S} + H_2^{18}O \rightleftharpoons 1/3 \ {}^{18}\mathcal{S} + H_2^{16}O
$$

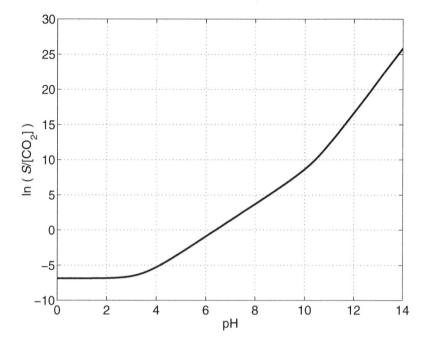

Figure C.9.3: The relationship between $\ln(\mathcal{S}/[CO_2])$ and pH calculated using Eq. (C.9.39). Note that $\ln(\mathcal{S}/[CO_2])$ which describes the equilibrium $CO_2 \rightleftharpoons \mathcal{S}$, is a function of pH only.

Table C.9.2: Fractionation factors $\alpha_{(\mathcal{S}-H_2O)}$ as a function of pH as derived from precipitation experiments.

McCrea (1950)	(25°C)	Usdowski et al. (1991)	(19°C)
pH	$\alpha_{(\mathcal{S}-H_2O)}$	pH	$\alpha_{(\mathcal{S}-H_2O)}$
8.34	1.03123	3.89	1.03863
9.03	1.03064	7.67	1.03316
9.38	1.03106	8.32	1.03231
10.69	1.02705	9.60	1.03023
11.65	1.02506	10.00	1.02986
		10.30	1.02907

is equal to the fractionation factor $\alpha_{(\mathcal{S}-H_2O)}$ which allows us to compare the free energy of isotopic equilibrium $\Delta G_\alpha^0 = -RT \ln \alpha_{(\mathcal{S}-H_2O)}$ to the free energy of the major component equilibrium $\Delta G^0 = -RT \ln(\mathcal{S}/[CO_2])$ (see above) at a given pH (Figure C.9.4). It appears that both quantities are linearly related. Linear regressions yield the following expressions:

$$\ln \alpha_{(\mathcal{S}-\mathrm{H_2O})} = -6.11 \times 10^{-4} \, \ln \frac{\mathcal{S}}{[\mathrm{CO_2}]} + 0.03402 \tag{C.9.40}$$

$$\ln \alpha_{(\mathcal{S}-\mathrm{H_2O})} = -6.10 \times 10^{-4} \, \ln \frac{\mathcal{S}}{[\mathrm{CO_2}]} + 0.03452 \tag{C.9.41}$$

with correlation coefficients of $r^2 = 0.957$ (Eq. (C.9.40), McCrea (1950), $25°\mathrm{C}$) and $r^2 = 0.998$ (Eq. (C.9.41), Usdowski et al. (1991), $19°\mathrm{C}$), respectively.

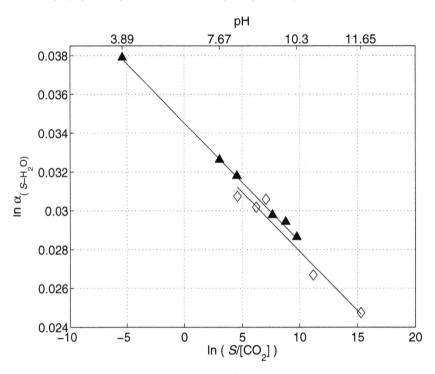

Figure C.9.4: The relationship between isotopic equilibrium ($\ln \alpha_{(\mathcal{S}-\mathrm{H_2O})}$) and major component equilibrium ($\ln(\mathcal{S}/[\mathrm{CO_2}])$) as derived from precipitation experiments by McCrea (1950) (open diamonds) and Usdowski et al. (1991) (closed triangles).

Individual fractionation factors

From the relation between $\alpha_{(\mathcal{S}-\mathrm{H_2O})}$ and $\mathcal{S}/[\mathrm{CO_2}]$ (Eq. (C.9.41)), individual fractionation factors for $\mathrm{H_2CO_3}$, $\mathrm{HCO_3^-}$, and $\mathrm{CO_3^{2-}}$ can be calculated. At a limiting $pH = 0$ the fractionation factor $\alpha_{(\mathcal{S}-\mathrm{H_2O})}$ is equal to the fractionation between $\mathrm{H_2CO_3}$ and $\mathrm{H_2O}$. All dissolved carbonate is essentially carbonic acid - the ratio of $[\mathrm{H_2CO_3}]/\mathcal{S}$ is 0.9996 here. Evaluating $\ln(\mathcal{S}/[\mathrm{CO_2}])$ at $pH = 0$, and inserting the result into Eq. (C.9.41) yields $\alpha_{(\mathrm{H_2CO_3}-\mathrm{H_2O})} = 1.0395$. At intermediate pH of ca.

6.9, $\alpha_{(\mathcal{S}-H_2O)}$ largely reflects the fractionation between HCO_3^- and H_2O. The pH value of 6.9 is obtained by calculating the maximum of $[HCO_3^-]/\mathcal{S}$ which gives a pH equal to $(pK_{H_2CO_3} + pK_2)/2$. A calculation analogous to that for carbonic acid yields $\alpha_{(HCO_3^- - H_2O)} = 1.0343$ ($[HCO_3^-]/\mathcal{S}$ is 0.9994 here). Finally, at a limiting $pH = 14$ the fractionation factor $\alpha_{(\mathcal{S}-H_2O)}$ is equal to the fractionation between CO_3^{2-} and H_2O; it follows $\alpha_{(CO_3^{2-} - H_2O)} = 1.0184$ ($[CO_3^{2-}]/\mathcal{S}$ is 0.9998 here). The individual fractionation factors obtained in this way are summarized in Table C.9.3 and in Figure 3.3.23 on page 204 (horizontal lines).

Table C.9.3: Oxygen isotope fractionation factors for the carbonate species with respect to water (19°C) after Usdowski et al. (1991).

pH	Value	$\alpha_{(\mathcal{S}-H_2O)}$	Value
0	0	$\alpha_{(H_2CO_3 - H_2O)}$	1.0395
$(pK_{H_2CO_3} + pK_2)/2$	6.9	$\alpha_{(HCO_3^- - H_2O)}$	1.0343[a]
pK_{H_2O}	14	$\alpha_{(CO_3^{2-} - H_2O)}$	1.0184

[a] Recalculated (Usdowski, pers. comm.).

Exercise C.2 (**)
Derive the pH value at which $[HCO_3^-]/\mathcal{S}$ is at maximum ($= (pK_{H_2CO_3} + pK_2)/2$). Tip: Calculate the minimum of $\mathcal{S}/[HCO_3^-]$.

C.10 Mathematical derivation of the partition function ratio

In this section a formula for the partition function is derived that permits the calculation of isotopic fractionation factors under equilibrium conditions (cf. Section 3.5). For further reading, see Urey (1947), Bigeleisen and Mayer (1947), Broecker and Oversby (1971), Andrews (1975), and Richet et al. (1977).

A relationship between the Gibbs free energy G, respectively Helmholtz free energy A, and the partition function $\mathbf{Q}$ of a thermodynamical system can be derived from statistical mechanics. The Helmholtz energy A is:

$$A = U - TS$$

with U being the internal energy of the system, and S being the entropy. Since the Gibbs free energy of the system is $G = U + pV - TS$ we have $G = A + pV$, where p is pressure and V is the volume of the system. The relationship between the partition function and the Helmholtz free energy is given by the fundamental equation (see e.g. Andrews, 1975):

$$A = -kT \ \ln(\mathbf{Q}) \ . \tag{C.10.42}$$

where $k = 1.38 \times 10^{-23}$ J K^{-1} is Boltzmann's constant, and T is the absolute temperature in Kelvin. Since all other thermodynamic functions (P, S, G, and U) may be found from A, thermodynamic considerations very often start with the calculation of the partition function (Eq. (C.10.42)).

The partition function $\mathbf{Q}$ which is a sum over states of the system is given by

$$\mathbf{Q} = \sum_i \exp(-E_i/kT) \tag{C.10.43}$$

where E_i is the energy of individual states. The partition function is the sum over all states of $\exp(-E_i/kT)$ of the system which can be interpreted as follows. For $T \to 0$ every term of the sum is zero, except for the one with $E_i = 0$ - only the zero-point energy is accessible to the system. On the other hand, for $T \to \infty$ every summand contributes a value of about 1 to the sum. Thus, the sum over all states of the system tends to infinity. The partition function measures the average number of states that are thermally accessible to a molecule at a given temperature.

For a mixture of ideal gases (A, B, C, D) the partition function (according to the reaction A + B $\rightleftharpoons$ C + D) can be written as:

$$\mathbf{Q} = \frac{Q_A^{N_A} Q_B^{N_B} Q_C^{N_C} Q_D^{N_D}}{N_A! \, N_B! \, N_C! \, N_D!}$$

where Q_i's are the molecular partition functions[4] and N_i's are the numbers of molecules. It can be shown that (Denbigh, 1971, Chapter 12.7):

$$\frac{N_C N_D}{N_A N_B} = \frac{Q_C Q_D}{Q_A Q_B} .$$

Dividing the number of particles of each chemical species (N) by the volume, it follows that the chemical equilibrium constant is given by:

$$K = \frac{[C][D]}{[A][B]} = \frac{Q_{C,V} \, Q_{D,V}}{Q_{A,V} \, Q_{B,V}} .$$

where $Q_{i,V}$'s are the molecular partition functions per unit volume. For the stoichiometry of the example given (2 reactants and 2 products) the ratio of the molecular partition functions is equal to the ratio of the molecular partition function per unit volume. This also holds for isotopic exchange reactions but is is not correct in the case of, e.g. 1 reactant and 2 products. The important result is that the chemical (and isotopic) equilibrium constant can be calculated from the molecular partition functions.

The equilibrium constant of the isotopic exchange reaction:

$$a\mathrm{A} + b\mathrm{B}' \quad \rightleftharpoons \quad a\mathrm{A}' + b\mathrm{B}$$

is given by:

$$K = \left(\frac{Q_{A'}}{Q_A}\right)^a \left(\frac{Q_{B'}}{Q_B}\right)^{-b} . \tag{C.10.44}$$

---

[4]The relationship between the molecular partition function Q and the partition function $\mathbf{Q}$ of N particles is $\mathbf{Q} = Q^N$ for distinguishable particles and $\mathbf{Q} = Q^N/(N!)$ for indistinguishable particles.

(Urey, 1947), where A and B are molecules which have one element as a common constituent and the primes label the heavy isotope. The Q_i's are the partition functions of the molecules. For example, the carbon isotope equilibrium between $CO_2(g)$ and aqueous CO_2 is described by:

$$^{12}CO_2(g) + {}^{13}CO_2 \quad \rightleftharpoons \quad {}^{13}CO_2(g) + {}^{12}CO_2$$

where the equilibrium constant (equal to the fractionation factor α in this particular case) is given by:

$$K = \alpha = \frac{[^{13}CO_2(g)][^{12}CO_2]}{[^{12}CO_2(g)][^{13}CO_2]}$$

which can be expressed by the ratio of the molecular partition functions:

$$K = \frac{(Q'_{CO_2(g)}/Q_{CO_2(g)})}{(Q'_{CO_2}/Q_{CO_2})} .$$

The essential step for the calculation of the equilibrium constant is therefore the calculation of the molecular partition function:

$$Q = \sum_i g_i \ \exp(-E_i/kT) \tag{C.10.45}$$

where g_i is the degeneracy of the state i (different states having the same energy).

The energy states of a molecule E_i are determined by solving the quantum-mechanical Schrödinger equation for the wavefunction Ψ:

$$-\frac{\hbar^2}{2M}(\nabla^2 + U)\Psi = E\Psi \tag{C.10.46}$$

where M is the mass of the molecule and $\hbar$ is Planck's constant divided by 2π (6.626×10^{-34} J s$/2\pi$); E and U are the total and potential energy, respectively. This equation has to be solved separately for translation, rotation and vibration. Provided the total energy is given by the sum of these energies (Ψ can be written as a product of translational, rotational and vibrational wavefunction), it follows

$$E = E_{tr} + E_{rot} + E_{vib} ,$$

and the partition function (Eq. (C.10.45)) is given by a product:

$$Q = Q_{tr} \ Q_{rot} \ Q_{vib}$$

Translation

The solution of Eq. (C.10.46) for a gas molecule in a cubic box with dimension a gives the energy of translation (see textbooks on quantum mechanics):

$$E_n = \frac{h^2}{8Ma^2}(n_x^2 + n_y^2 + n_z^2)$$

where n_i's are quantum numbers (positive integers) for translations in x, y, and z directions ($n_i = 0$ is forbidden due to Heisenberg's uncertainty principle) and M is the mass of the molecule. Note that n_i's have no upper bound, i.e., n_i's can tend to infinity. The one dimensional partition function (e.g. x-dimension) is the sum over all states:

$$Q_{\text{tr,x}} = \sum_{n=1}^{\infty} \exp\left(-\frac{n_x^2 h^2}{8Ma^2 kT}\right) \tag{C.10.47}$$

As mentioned in Section 3.5.1 the spacing between adjacent energy levels of translation can be shown to be very small compared to kT. The spacing is:

$$\Delta E_{n+1,n} = \frac{h^2}{8Ma^2}(2n+1) .$$

For example, the energy difference for $n = 1, n + 1 = 2$ of a $CO_2(g)$ molecule in a cubic box with $a = 10$ cm is 2.3×10^{-40} J. On the other hand, $kT = 4.1 \times 10^{-21}$ J at 25°C, giving a ratio of $\sim 10^{-20}$. The characteristic temperature of translation $\Theta_{tr,x}$ (where $\Delta E = kT$) is 1.6×10^{-17} K. For the parameter values discussed here the characteristic temperatures for translation, rotation, and vibration, and the quantum mechanical/classical partition functions of $CO_2(g)$ are summarized in Figure C.10.5. For translation, the sum in Eq. (C.10.47) can be very well approximated by an integral (classical value), yielding

$$Q_{\text{tr,x}} = \left(\frac{2\pi Ma^2 kT}{h^2}\right)^{\frac{1}{2}}$$

and finally for three dimensions:

$$Q_{\text{tr}} = Q_{\text{tr,x}}\, Q_{\text{tr,y}}\, Q_{\text{tr,z}} = \left(\frac{2\pi MkT}{h^2}\right)^{\frac{3}{2}} V$$

where V is the volume in which the molecule is enclosed.

Rotation

The energy levels of a linear, rigid rotator about a single axis are found by solving Schrödinger's equation. They are:

$$E_{\text{rot}} = j(j+1)\frac{h^2}{8\pi^2 I}$$

where j is zero or a positive integer and I is the moment of inertia of the molecule. The partition function of rotation is:

$$Q_{\text{rot}} = \sum_{j=0}^{\infty} (2j+1) \exp\left[-\frac{j(j+1)h^2}{8\pi^2 IkT}\right] \tag{C.10.48}$$

The spacing between adjacent energy levels in the case of rotation is:

$$\Delta E_{j+1,j} = \frac{h^2}{8\pi^2 I}2(j+1) .$$

Considering CO_2, for example, the moment of inertia is $I = 2m_O r^2 = 7.5 \times 10^{-46}$ kg m^2, where $m_O = 15.99 \times 1.66 \times 10^{-27}$ kg is the mass of the oxygen atom and $r = 1.189$ Å is the C−O bond length. The energy difference for $j = 0, j+1 = 1$ of a

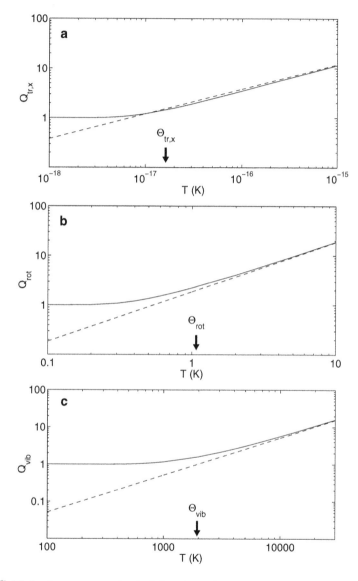

Figure C.10.5: Quantum mechanical (solid lines) and classical (dashed lines) partition functions of (a) translation, (b) rotation, and (c) vibration of $CO_2(g)$ as a function of temperature (note the different logarithmic temperature scales). For reasons of analogy the vibrational partition function $Q = 1/(1 - \exp(-u))$ is shown which refers to a measurement of energies with respect to the zero-point energy. The characteristic temperatures (Θ) calculated for parameter values as discussed in the text are indicated by arrows.

CO_2 molecule therefore is 1.5×10^{-23} J, whereas $kT = 4.1 \times 10^{-21}$ J at 25°C, giving a ratio of $\sim 10^{-3}$. The characteristic temperature of rotation (where $\Delta E = kT$) is

1.1 K. Thus, the rotational levels are closely spaced at room temperature (as is the case for translation) and the sum (Eq. (C.10.48)) can be replaced by an integral, yielding the classical partition function of rotation for diatomic molecules:

$$Q_{\rm rot} = \frac{1}{s}\frac{8\pi^2 IkT}{h^2} \tag{C.10.49}$$

and for polyatomic molecules:

$$Q_{\rm rot} = \frac{1}{s}\left(\frac{kT}{\hbar^2}\right)^{\frac{3}{2}}(8\pi I_1 I_2 I_3)^{\frac{1}{2}} \tag{C.10.50}$$

where s is the symmetry number of the molecule and I's are the principal moments of inertia. The symmetry number of a molecule is equal to the number of different values of the rotational coordinates which all correspond to one orientation of the molecule. For example, the oxygen atoms of the O_2 molecule are indistinguishable, thus a rotation by 180° about an axis perpendicular to the axis of the molecule transforms the molecule into the same orientation as before - the symmetry number s is 2. The linear molecule CO_2, has also symmetry number $s = 2$, whereas $s = 1$ for HCl (asymmetric diatomic molecule). The symmetry number of the planar CO_3^{2-} molecule (cf. Figure 3.2.11 on page 171) is 6: with either side of the plane facing upwards, three positions (differing by rotations of 120°) look identical. By a similar consideration one finds that s is 12 for the tetrahedral methane molecule (for detailed discussion, see Mayer and Mayer, 1966, Sec. 8e).

The symmetry number has to be introduced in Eq. (C.10.49) and Eq. (C.10.50) since quantum mechanics demands that like particles are truly indistinguishable. Thus, only $1/s$ of the classical value of $Q_{\rm rot}$ equals the value of the quantum mechanical partition function in the classical limit. This is in analogy to the introduction of the factor $1/N!$ in the expression of the partition function $\mathbf{Q} = Q^N/N!$ of a system of indistinguishable particles.

Vibration

The solution of Eq. (C.10.46) for a diatomic molecule, assuming a harmonic potential energy, gives vibrational energies of

$$E_n = (n + \tfrac{1}{2})hc\omega$$

where n is zero or a positive integer, $h = 6.626 \times 10^{-34}$ J s is Planck's constant, $c = 299792458$ m s^{-1} is the speed of light, $\omega = 1/\lambda$ is the wavenumber of vibration (often given in cm^{-1}), and λ is the wavelength. (The frequency of vibration is $\nu = c/\lambda$; the wavenumber ω and the frequency ν are often used equivalently). For polyatomic molecules there are a number of frequencies ν_i, with $i = 3N - 6$ and N the number of atoms in the molecule. If the potential energy is not sufficiently harmonic, corrections have to be applied (see Richet et al., 1977). In an harmonic approximation, however, the partition function reads:

$$Q_{\rm vib} = \sum_n \exp\left(-\frac{(n + \tfrac{1}{2})hc\omega}{kT}\right) \tag{C.10.51}$$

In this case the sum cannot be replaced by an integral because the spacing between adjacent vibrational levels is large compared to kT at room temperature. The spacing between adjacent energy levels of vibration is:

$$\Delta E_{n+1,n} = hc\omega \ .$$

For example, using a wavenumber of 1353.6 cm^{-1} for the symmetric stretch of CO_2 (see Figure 3.5.36, page 240), $\Delta E = 2.7 \times 10^{-20}$ J, whereas $kT = 4.1 \times 10^{-21}$ J at 25°C, giving a ratio of ~ 7. The characteristic temperature of vibration (where $\Delta E = kT$) is 1956 K - at room temperature most of the molecules are in their vibrational ground state. Thus, the sum (Eq. (C.10.51)) cannot be approximated by an integral. Fortunately, it can be calculated as:

$$Q_{\text{vib}} = \frac{\exp(-u/2)}{1 - \exp(-u)}$$

where $u = hc\omega/kT$.

Finally, the partition function for diatomic molecules reads:

$$Q = Q_{\text{tr}} Q_{\text{rot}} Q_{\text{vib}}$$

$$= \left(\frac{2\pi MkT}{h^2}\right)^{\frac{3}{2}} V \frac{1}{s} \frac{8\pi^2 IkT}{h^2} \frac{\exp(-u/2)}{1 - \exp(-u)}$$

For the calculation of the isotopic equilibrium constant (K), only the ratio of the partition functions of the molecule containing the light and the heavy isotope (denoted by a prime) is of interest (cf. Eq. (C.10.44)):

$$K = \left(\frac{Q_{\text{A}'}}{Q_{\text{A}}}\right)^a \left(\frac{Q_{\text{B}'}}{Q_{\text{B}}}\right)^{-b} \tag{C.10.52}$$

The ratio for diatomic molecules reads

$$\frac{Q'}{Q} = \frac{s}{s'} \frac{I'}{I} \left(\frac{M'}{M}\right)^{\frac{3}{2}} \frac{\exp(-u'/2)}{1 - \exp(-u')} \frac{1 - \exp(-u)}{\exp(-u/2)} \tag{C.10.53}$$

and for polyatomic molecules

$$\frac{Q'}{Q} = \frac{s}{s'} \left(\frac{I_1' I_2' I_3'}{I_1 I_2 I_3}\right)^{\frac{1}{2}} \left(\frac{M'}{M}\right)^{\frac{3}{2}} \prod_i \frac{\exp(-u_i'/2)}{1 - \exp(-u_i')} \frac{1 - \exp(-u_i)}{\exp(-u_i/2)} \tag{C.10.54}$$

where I_i's are the principal moments of inertia of the polyatomic molecules. Using the Redlich-Teller product rule (Redlich, 1935):

$$\frac{I'}{I} \left(\frac{M'}{M}\right)^{\frac{3}{2}} \left(\frac{m}{m'}\right)^{\frac{3}{2}n} \frac{u_i}{u_i'} = \left(\frac{I_1' I_2' I_3'}{I_1 I_2 I_3}\right)^{\frac{1}{2}} \left(\frac{M'}{M}\right)^{\frac{3}{2}} \left(\frac{m}{m'}\right)^{\frac{3}{2}n} \prod_i \frac{u_i}{u_i'} = 1$$

and introducing the reduced partition function ratio $(Q'/Q)_\text{r}$ which is simply $(Q'/Q)/(m'/m)^{3n/2}$ where n is the number and m the mass of isotopic atoms being exchanged, we have for diatomic molecules:

$$\left(\frac{Q'}{Q}\right)_\text{r} = \frac{s}{s'} \frac{u'}{u} \frac{\exp(-u'/2)}{1 - \exp(-u')} \frac{1 - \exp(-u)}{\exp(-u/2)} \tag{C.10.55}$$

and for polyatomic molecules

$$
\left(\frac{Q'}{Q}\right)_{\mathrm{r}} = \frac{s}{s'} \prod_i \frac{u'_i}{u_i} \frac{\exp(-u'_i/2)}{1 - \exp(-u'_i)} \frac{1 - \exp(-u_i)}{\exp(-u_i/2)}
\tag{C.10.56}
$$

which is exactly the formula given by Urey (1947). With Eq. (C.10.55) and Eq. (C.10.56) it is possible to calculate isotopic fractionation factors from a knowledge of the frequencies of the molecules only - the moments of inertia have been eliminated.

Appendix D

Answers to Exercises

1.1 HCO_3^-, in surface and deep ocean.

1.2 $[CO_2] = 7.65 \times 10^{-6}$ mol kg^{-1}, $[HCO_3^-] = 1.673 \times 10^{-3}$ mol kg^{-1}, $[CO_3^{2-}] = 3.19 \times 10^{-4}$ mol kg^{-1}.

1.3 The composition of the solution, i.e. the presence of e.g. Mg^{2+}, SO_4^{2-}, Ca^{2+}, and K^+ ions.

1.4 0.69734. 96%.

1.5 The equilibrium partial pressure of CO_2 decreases from 350.5 μatm at natural seasalt composition to 336.5 μatm at doubled $[Mg^{2+}]$ (decrease by 13.9 μatm).

1.6 The alkalinity is 34 μmol kg^{-1} higher.

1.7 $[CO_2]$ decreases by 6.1 μmol kg^{-1}. 0.4 μmol kg^{-1} (i.e. less than 10%) are due to the change in TA.

1.8 The carbonate alkalinity is not conserved. Consider ocean water initially with a total alkalinity (in the PA approximation PA $= [HCO_3^-] + 2[CO_3^{2-}] + [B(OH)_4^-] + [OH^-] - [H^+]$) of 2290 μmol kg^{-1} at $T_c = 25°C$ and $S = 35$ in equilibrium with an atmosphere with CO_2 partial pressure of 280 μatm. What happens when the atmospheric CO_2 content increases and the system is allowed to reach a new equilibrium at 360 μatm CO_2? CA increases from 2169 μmol kg^{-1} at 280 μatm to 2186 μmol kg^{-1} 360 μatm.

1.9 Volume is not conserved. $V_1 = 0.9808061$ m^3, $V_2 = 0.9735890$ m^3, $V_m = 0.9771894$ m^3, $V_{12} = (V_1 + V_2)/2 = 0.9771976$ m^3. The relative difference between V_m and V_{12} is $8 \cdot 10^{-4}$%.

1.10 Components are not conserved

$[CO_2]_1 = 9.00$ μmol kg^{-1}, $[CO_2]_2 = 13.47$ μmol kg^{-1}, $[CO_2]_m = 11.11$ μmol kg^{-1}, $[CO_2]_{12} := ([CO_2]_1 + [CO_2]_2)/2 = 11.24$ μmol kg^{-1}.

$[HCO_3^-]_1 = 1573.2 \ \mu mol \ kg^{-1}$, $[HCO_3^-]_2 = 1773.2 \ \mu mol \ kg^{-1}$, $[HCO_3^-]_m = 1673.6 \ \mu mol \ kg^{-1}$, $[HCO_3^-]_{12} = 1673.2 \ \mu mol \ kg^{-1}$.

$[CO_3^{2-}]_1 = 217.83 \ \mu mol \ kg^{-1}$, $[CO_3^{2-}]_2 = 213.29 \ \mu mol \ kg^{-1}$, $[CO_3^{2-}]_m = 215.25 \ \mu mol \ kg^{-1}$, $[CO_3^{2-}]_{12} = 215.56 \ \mu mol \ kg^{-1}$.

1.11 Titration of unbuffered water with a strong base. Charge conservation: $[B^-] + [H^+] = [OH^-]$ and $([B_0] \times V_B)/(V_A + V_B) + [H^+] = K_W^*/[H^+]$ yields a quadratic equation for $[H^+]$ as a function of V_B/V_A.

1.12 $[H^+] \approx [OH^-]$. No.

1.13 Use TA $= 2[CO_3^{2-}] + [HCO_3^-] + [B(OH)_4^-] + [OH^-] - [H^+]$ and express each term on the right-hand side as a function of equilibrium constants, B_T, s, and h.

1.14 Change of DIC is given by U: $\delta DIC = U$. $\delta CaCO_3 = \gamma U$; $\delta C_{org} = (1 - \gamma)U$; $\delta TA = -2\gamma U + (1 - \gamma)U/6.6$, where the 6.6 stems from the Redfield ratio $C : N = 6.6 : 1$. Thus to take the effect of nitrate assimilation into account, one has to substitute $-2\gamma U + (1 - \gamma)U/6.6$ for each $-2\gamma U$.

1.15 It decreases with increasing partial pressure of CO_2.

1.16 RF_0 is a measure for the carbonate concentration, $[CO_3^{2-}]$. The overall reaction may be written as: $CO_2 + CO_3^{2-} + H_2O = 2 HCO_3^-$.

2.1 For every molecule of B, two molecules of A are consumed. Thus the rate of change of [B] is half the rate of change of [A].

2.2 $[B](t) = [B]_0 + [A]_0 \{1 - \exp(-kt)\}$.

2.3 $A = 4.58 \times 10^{10} \ kg \ mol^{-1} \ s^{-1}$, $E_a = 20.8 \ kJ \ mol^{-1}$.

2.4 It is slightly higher than typical activation energies of diffusion-controlled reactions ($8 - 16 \ kJ \ mol^{-1}$).

2.5 $(+4, +2, -2, -4), (+2, +1, -1, -2), (-2, -1, +1, +2), (-4, -2, +2, +4)$.

2.6 pH $= 8.25, 8.21, 8.17$. Note that the result is independent of $[CO_2]$ or TA.

2.8 ~ 15 s (63%, Figure 2.4.9) and ~ 70 s (99%). Note that 1 mg $NaHCO_3$ per kg$= 1/84$ mmol kg$^{-1} \simeq 12 \ \mu mol \ kg^{-1}$, i.e. a small perturbation.

2.9 As ΣCO_2 goes to zero, so does $[CO_3^{2-}]$. Thus the right hand sides of Eq. (C.7.26) are independent of the concentrations of the carbonate species.

2.10 Using $A = \lambda N$ and $\lambda = \ln(2)/t_{1/2}$, it follows $N = A \times t_{1/2}/\ln(2)$. Since $[c] = N/N_A$ (in units of $\mu mol \ kg^{-1}$), and A is given in units of $\mu Ci \ kg^{-1}$, it follows: $[c] = A \times 3.7 \times 10^{10} \times t_{1/2}/(N_A \ \ln(2))$.

2.11 For example: c_2 and $(dc/dz)_{z_1}$. No.

2.12 $163 \ \mu mol \ kg^{-1}$.

2.13 At maximum uptake, $c(R_1) = 0$. Thus $F_1 = c_2 4\pi D R_1 = 140 \ nmol \ h^{-1}$.

2.14 Because the CO_2 concentration is small and chemical conversion of HCO_3^- to CO_2 might be too slow. Molecular CO_2 uptake (no HCO_3^- uptake) and small cell sizes ($R_1/a_k \ll 1$).

3.1 When a nucleus is formed from neutrons and protons, mass is converted into energy according to $E = \Delta m c^2$.

3.2 2.6×10^{10} kJ mol^{-1}. Nuclear energies are therefore much larger than e.g. energies of chemical bonds (~ 400 kJ mol^{-1}).

3.3 Using $\alpha_{(A-B)} = (\delta_A + 10^3)/(\delta_B + 10^3)$ yields $\varepsilon_{(A-B)} = (\delta_A - \delta_B)/(1 + \delta_B/10^3)$. Expansion of $\ln(1+x)$ for small x leads to $\ln(1+x) \simeq x$. Thus $10^3 \ln(\alpha) \simeq (\alpha - 1) \times 10^3 = \varepsilon$.

3.4 The distance would be fixed. This is forbidden because of Heisenberg's uncertainty principle.

3.5 $1.0209.$ $6.$

3.6 $x(t) = x_0 \cos(\omega t)$.

3.7 In equilibrium, the heavy isotope goes preferentially to the compound in which the element is bound most strongly.

3.8 This follows from $\alpha_{(A-B)} = (\delta_A + 10^3)/(\delta_B + 10^3)$.

3.9 As a first approximation, this feature may be explained by a Rayleigh process in which the water of a cloud traveling from low to high latitudes is successively depleted in ^{18}O because the rain forming from that cloud is enriched in ^{18}O relative to the remaining isotopic composition of the water vapor in the cloud.

3.10 Ansatz: $dn_2/dn_1 = (\alpha_1 x_1 + \alpha_2 x_2)N_2/N_1$. Note that the total amount removed is dn, thus $x_1 + x_2 = 1$.

3.11 General rule: the heavy isotope goes preferentially to the compound in which the element is bound most strongly. Thus the carbon atom is bound most strongly in HCO_3^-.

3.13 Parts of Antarctica are at higher latitude than Greenland. Thus fractionation effects due to Rayleigh processes and local temperature are different.

3.14 Respiration and photosynthesis of the biosphere + minor effects due to stratospheric diminution.

3.15 About 5°C.

3.16 About 3.5°C too low.

3.17 Because the dominant carbonate species in solution changes with increasing pH from $[H_2CO_3]$ to $[HCO_3^-]$ to $[CO_3^{2-}]$. As the oxygen isotopic composition of those species decrease in sequence from $[H_2CO_3]$ to $[HCO_3^-]$ to $[CO_3^{2-}]$, the oxygen isotope fractionation between H_2O and $\mathcal{S} = [H_2CO_3] + [HCO_3^-] + [CO_3^{2-}]$ decreases with pH.

3.18 Because $CO_2(aq)$ is much lower at high pH.

3.19 Abundances are equivalent to probabilities, p, hence multiply abundances. Example: $p(^{13}C^{18}O^{18}O) = 0.0111 \times 0.001995 \times 0.001995 = 4.4 \times 10^{-8}$. Note that e.g. $^{12}C^{16}O^{17}O$ and $^{12}C^{17}O^{16}O$ yield the same mass number. Thus the probability for a molecule containing ^{12}C, ^{16}O, and ^{17}O is $2 \times p(^{12}C^{16}O^{17}O)$.

3.20 $-39.9‰, 41.6‰$, and $-41.5‰$. In this case, it is therefore very important which quantity is used to express isotopic fractionation.

3.21 The number of protons/neutrons in the nucleus of ^{11}B and ^{10}B is odd/even and odd/odd, respectively. In ^{13}C and ^{12}C it is even/odd and even/even, respectively.

3.22 The intercept of the curves would move to higher pH. Lighter.

3.24 The pH was ~ 0.26 units higher in the glacial.

3.25 Carbonate ion concentration was higher in the glacial. Yes.

C.1 The van der Waals equation reads:

$$\left(P + \frac{a}{V_m^2}\right)(V_m - b) = RT = PV_m + \frac{a}{V_m} - bP - \frac{ab}{PV_m^2}$$

$\Rightarrow$ cubic equation for V_m:

$$PV_m^3 - (RT + bP)V_m^2 + aV_m - ab = 0$$

Integrand in Eq. (C.3.5)

$$\frac{RT}{P} - V_m = \frac{a}{PV_m} - b - \frac{ab}{PV_m^2}$$

Instead of solving the cubic equation for V_m insert the low pressure approximation $PV_m \approx RT \Rightarrow$

$$\frac{RT}{P} - V_m \approx \frac{a}{RT} - b - \frac{abP}{(RT)^2}.$$

Integrate:

$$
\begin{aligned}
RT \ln\left(\frac{f}{P}\right) &= -\int_0^P \left[\frac{a}{RT} - b - \frac{abP'}{(RT)^2}\right] dP' \\
&= -\frac{aP}{RT} + bP + \frac{abP^2}{2(RT)^2}
\end{aligned}
$$

C.2 Calculate minimum of $\mathcal{S}/[HCO_3^-] = [H_2CO_3]/[HCO_3^-] + 1 + [CO_3^{2-}]/[HCO_3^-]$. Insert equilibrium constants and differentiate with respect to $[H^+]$. Since this expression is equal to zero at minimum, it can be solved for $[H^+]$.

Appendix E

Notation and Symbols

Table E.0.1: Notation (Greek letters)

Symbol	Meaning
$^{13}\alpha_{(A-B)}$	Fractionation factor. Ratio of $^{13}C/^{12}C$ in samples A and B; Eq. (3.1.3)
γ	Rain rate parameter; Eq. (1.6.93)
γ_i	Ion activity coefficient
δ	Cross virial coefficient; Section 1.4
$\delta^n X_A$	δ value of sample A with respect to element X; Eq. (3.1.5)
δD	δ value (Hydrogen)
$\delta^{11}B$	δ value (Boron)
$\delta^{13}C$	δ value (Carbon)
$\delta^{18}O$	δ value (Oxygen)
ϵ	Dielectric constant of water (F m^{-1}), $\epsilon = \epsilon_0 \times \epsilon_r$; Eq. (2.3.26)
ϵ_0	Vacuum dielectric constant, $\epsilon_0 = 8.8542 \times 10^{-12}$ F m^{-1}
ϵ_r	Relative dielectric constant of water, $\epsilon_r = 80$
$^{13}\varepsilon_{(A-B)}$	Isotopic fractionation ($^{13}C/^{12}C$; samples A and B); Eq. (3.1.4)
μ_i	Chemical potential (J mol^{-1})
τ	Relaxation time (s)

Table E.0.2: Notation (Latin letters)

Symbol	Meaning
A^-	Anion of a weak acid; Eq. (1.5.69)
A_h	Partial derivative of TA; Eq. (1.5.86)
A_s	Partial derivative of TA; Eq. (1.5.85)
a	First constant in van der Waals equation; Eq. (C.3.8)
B	First virial coefficient of CO_2; Eq. (1.4.66)
B^+	Cation of a weak base; Eq. (1.5.69)
B_T	Total boron concentration; Eq. (A.7.14)
b	Second constant in van der Waals equation; Eq. (C.3.8)
CA	$CA = [HCO_3^-] + 2[CO_3^{2-}]$. Carbonate alkalinity; Eq. (1.2.29)
D	Diffusion coefficient ($m^2\ s^{-1}$)
D_a	Damköhler number; Eq. (2.6.73)
D_h	Partial derivative of DIC with respect to h; Eq. (1.5.84)
D_s	Partial derivative of DIC with respect to s; Eq. (1.5.83)
DIC	$DIC \equiv \Sigma CO_2 = [CO_2] + [HCO_3^-] + [CO_3^{2-}]$; Eq. (1.1.7)
d_{ML}	Mixed layer thickness (m); Eq. (1.6.101)
e_0	Elementary charge, $e_0 = 1.602 \times 10^{-19}$ C
E_a	Activation energy ($kJ\ mol^{-1}$)
F_T	Total fluoride concentration; Eq. (A.6.11)
fCO_2	Fugacity of CO_2
h	$h = [H^+]$
I	Ionic strength; Eq. (1.1.16)
K	Thermodynamic equilibrium constant
K^*	Stoichiometric equilibrium constant
K'	Hybrid equilibrium constant
K_1^*	First dissociation constant of carbonic acid ($mol\ kg^{-1}$); Eq. (1.1.5)
K_2^*	Second dissociation constant of carbonic acid ($mol\ kg^{-1}$); Eq. (1.1.6)

Table E.0.3: Notation (Latin letters; continuation)

Symbol	Meaning
K_B^*	Dissociation constant of boric acid (mol kg^{-1}); Eq. (1.1.14)
K_D	Distribution coefficient (dimensionless); Eq. (3.4.54)
K_F^*	Stability constant of hydrogen fluoride (mol kg^{-1}); Eq. (A.6.10)
K_0	Henry's constant (mol kg^{-1} atm^{-1}); Eq. (1.5.68)
$K_{H_2CO_3}^*$	Dissociation constant of true carbonic acid (mol kg^{-1}); Eq. (A.2.5)
K_S^*	Stability constant of hydrogen sulfate (mol kg^{-1}); Eq. (A.5.8)
K_w^*	Ion product of water (mol^2 kg^{-2}); Eq. (1.1.13)
K_{sp}^*	Solubility product (calcite/aragonite); Table 3.2.4
k	Kinetic rate constant; Eq. (2.1.3)
k_B	Boltzmann constant, $k_B = 1.38 \times 10^{-23}$ J K^{-1}
k_{ge}	Gas exchange coefficient (m d^{-1}); Eq. (1.6.101)
N_A	Avogadro constant, $N_A = 6.022 \times 10^{23}$ mol^{-1}
P	Pressure (bars, atm)
PA	PA $= [HCO_3^-] + 2[CO_3^{2-}] + [B(OH)_4^-] + [OH^-] - [H^+]$. Alkalinity for most practical purposes; Eq. (1.2.31)
PCO_2	Partial pressure of CO$_2$ in the sea (atm)
pCO_2	Partial pressure of CO$_2$ in the atmosphere (atm)
pK	$-\log(K)$
$\mathbf{Q}$	Partition function; Eq. (C.10.43)
Q	Molecular partition function; Eq. (C.10.45)
q_i	Charge concentration; Table 1.2.7
R	Gas constant, $R = 8.3145$ J mol^{-1} K^{-1}
$^{13}R_A$	Isotopic ratio (^{13}C/^{12}C) in sample A; Eq. (3.1.1)
RF	Revelle factor; Eq. (1.6.91)
RF_0	Revelle factor at constant TA; Eq. (1.5.89)
r	Reaction rate (mol kg^{-1} s^{-1}); Eq. (2.1.2)
$^{13}r_A$	Fractional abundance (^{13}C/^{12}C) in sample A; Eq. (3.1.2)
S	Salinity

Table E.0.4: Notation (Latin letters; continuation)

Symbol	Meaning
$\mathcal{S}$	$\mathcal{S} = [H_2CO_3] + [HCO_3^-] + [CO_3^{2-}]$; Eq. (3.3.40)
S_T	Total sulfate concentration; Eq. (A.5.9)
s	$s = [CO_2]$
s_{K*}	Sensitivity parameter; Subsection 1.1.5
T	Absolute temperature (in K)
T_c	Temperature (in °C)
TA	Total alkalinity; Eq. (1.2.33)
U	Export rate of carbon; Eq. (1.6.95)
V	Volume
V_m	Molar volume; Subsection C.3
xCO_2	Mole fraction of CO_2 (μmol mol^{-1}); Eq. (C.2.3)
z_i	(Charge of ion i)/e_0; Eq. 1.1.16

Table E.0.5: Notation (miscellaneous symbols)

Symbol	Meaning
∂	partial derivative
$a_{H+} = \{H^+\}$	activity of H^+
$[H^+]$	concentration of H^+ (mol kg^{-1})
m_{H+}	molality of H^+ (mol l^{-1})

Table E.0.6: Abbreviations

Acronym	Meaning
DIC $\equiv \Sigma CO_2$	Total dissolved inorganic carbon
SST	Sea surface temperature
POC	Particulate organic carbon
TA	Total alkalinity
SMOW	Standard Mean Ocean Water

References

Alonso, M. and E. J. Finn. *Physics*. Addison-Wesley Publishing Company, Inc., Massachusetts, pp. 1138, 1992.

Anderson, L. G., D. R. Turner, M. Wedborg, and D. Dyrssen. Determination of total alkalinity and total dissolved inorganic carbon. In Grasshoff, K., K. Kremling, and M. Ehrhardt, editor, *Methods of Seawater Analysis*, pages 127–148. Wiley-VCH, Weinheim, 600 pp., 1999.

Andrews, F. C. *Equilibrium Statistical Mechanics*. Wiley & Sons, New York, pp. 255, 1975.

Archer, D., H. Kheshgi and E. Maier-Reimer. Dynamics of fossil fuel CO_2 neutralization by marine $CaCO_3$. *Global Biogeochem. Cycles*, 12:259–276, 1998.

Atkins, P. W. *Physical Chemistry (6th ed.)*. Oxford Univ. Press, Oxford, pp. 1014, 1998.

Atkins, P. W. *Physical Chemistry (4th ed.)*. Oxford Univ. Press, Oxford, pp. 995, 1990.

Bacastow, R. B., C. D. Keeling, T. J. Lueker, M. Wahlen and W. G. Mook. The ^{13}C Suess effect in the world surface oceans and its implications for oceanic uptake of CO_2: Analysis of observations at Bermuda. *Global Biogeochem. Cycles*, 10(2):335–346, 1996.

Bakker, D. C. E. *Process studies of the air-sea exchange of carbon dioxide in the Atlantic Ocean*. PhD thesis, Rijksuniversiteit Groningen, 1998.

Bard, E. Ice age temperatures and geochemistry. *Science*, 284:1133–1134, 1999.

Barron, J. L., D. Dyrssen, E. P. Jones, and M. Wedborg. A comparison of computer methods for seawater alkalinity titrations. *Deep-Sea Res.*, 30:441–448, 1983.

Battle, M., M. L. Bender, P. P. Tans, J. W. C. White, J. T. Ellis, T. Conway, and R. J. Francey. Global carbon sinks and their variability inferred from atmospheric O_2 and $\delta^{13}C$. *Science*, 287:2467–2470, 2000.

Beck, W. S., K. F. Liem, and G. G. Simpson. *Life: An Introduction to Biology (3rd ed.)*. HarperCollins Publishers Inc., New York, pp. 1361, 1991.

Bellerby, R. G. J., D. R. Turner, and J .E. Robertson. Surface pH and pCO_2 distributions in the Bellingshausen Sea, Southern Ocean, during the early austral summer. *Deep-Sea Res. II*, 42(4-5):1093–1107, 1995.

Bemis, B. E., H. J. Spero, J. Bijma, and D. W. Lea. Reevaluation of the oxygen isotopic composition of planktonic foraminifera: Experimental results and revised paleotemperature equations. *Paleoceanography*, 13:150–160, 1998.

Ben-Yaakov, S. and M .B. Goldhaber. The influence of sea water composition on the apparent constants of the carbonate system. *Deep-Sea Res.*, 20:87–99, 1973.

Bender, M., T. Sowers, and L. Labeyrie. The Dole effect and its variations during the last 130,000 years as measured in the Vostok ice core. *Global Biogeochem. Cycles*, 8:363–376, 1994.

Berger, W. H. Planktonic foraminifera: Selective solution and paleoclimatic interpretation. *Deep-Sea Res.*, 15:31–43, 1968.

Berger, W. H. and R. S. Keir. Glacial-Holocene changes in atmospheric CO_2 and the deep-sea record. In Hansen, J. E. and T. Takahashi, editor, *Climate Processes and Climate Sensitivity, Geophys. Monogr. Ser. Vol. 29*, pages 337–351. AGU, Washington, D.C., 1984.

Berger, W. H., C. G. Adelseck and L. A. Mayer. Distribution of carbonate in surface sediments of the Pacific Ocean. *J. Geophys. Res.*, 81(15):2617–2627, 1976.

Berner, R. A. *Principles of Chemical Sedimentology*. McGraw-Hill, New York, pp. 240, 1971.

Bigeleisen, J. The effects of isotopic substitution on the rates of chemical reactions. *J. Phys. Chem.*, 56:823–828, 1952.

Bigeleisen, J. Chemistry of Isotopes. *Science*, 147:463–471, 1965.

Bigeleisen, J. and M. G. Mayer. Calculation of equilibrium constants for isotopic exchange reactions. *J. Chem. Phys.*, 15:261–267, 1947.

Bigeleisen, J. and Wolfsberg M. Theoretical and experimental aspects of isotope effects in chemical kinetics. *Adv. Chem. Phys.*, 1:15–76, 1958.

Bjerrum, N. Die Theorie der alkalischen und azidemetrischen Titrierung. *Sammlung chemischer und chem-technischer Vorträge*, 21:1–128, 1914.

Bjerrum, N. Ionic association. I. Influence of ionic association on the activity of ion at moderate degree of association. *Kgl. Danske Videnskab. Mat.-Fys. Medd.*, 7(9):1–48, 1926.

Bottinga, Y. Calculation of fractionation factors for carbon and oxygen isotopic exchange in the system calcite-carbon dioxide-water. *J. Phys. Chem.*, 72:800–808, 1968.

Boucher, D. F. and G. E. Alves. Dimensionless numbers. *Chem. Eng. Progr.*, 55:55–64, 1959.

Bouvier-Soumagnac Y. and J.-C. Duplessy. Carbon and oxygen isotopic composition of planktonic foraminifera from laboratory culture, plankton tows and recent sediment: implications for the reconstruction of paleoclimatic conditions and of the global carbon cycle. *J. Foram. Res.*, 15:302–320, 1985.

Bowen, R. *Isotopes in the Earth Sciences*. Elsevier Applied Science, London, pp. 647, 1988.

Boyle, E. A. The role of vertical chemical fractionation in controlling late Quaternary atmospheric carbon dioxide. *J. Geophys. Res.*, 93:15,701–15,714, 1988.

Bradshaw, A. L. and P. G. Brewer. High precision measurements of alkalinity and total carbon dioxide in seawater by potentiometric titration: 2. Measurements on standard solutions. *Marine Chemistry*, 24:155–162, 1988b.

Bradshaw, A.L. and P.G. Brewer. High precision measurements of alkalinity and total carbon dioxide in seawater by potentiometric titration: 1. Presence of unknown protolyte(s)? *Marine Chemistry*, 23:69–86, 1988a.

Brenninkmeijer, C. A. M., P. Kraft, and W. G. Mook. Oxygen isotope fractionation between CO_2 and H_2O. *Isotope Geoscience*, 1:181–190, 1983.

Brewer, P. G. and J. C. Goldman. Alkalinity changes generated by phytoplankton growth. *Limnol. Oceanogr.*, 21:108–117, 1976.

Broecker, W. S. Ocean chemistry during glacial times. *Geochim. Cosmochim. Acta*, 46:1689–1705, 1982.

Broecker, W. S. *Chemical Oceanography*. Harcourt Brace Jovanovich, Inc., New York, pp. 214, 1974.

Broecker, W. S. *How to Build a Habitable Planet*. Eldigio Press, Lamont-Doherty Geological Observatory, Palisades, New York, pp. 291, 1985.

Broecker, W. S. and T.-H. Peng. Gas exchange rates between air and sea. *Tellus*, 26:21–35, 1974.

Broecker, W. S. and T.-H. Peng. The role of $CaCO_3$ compensation in the glacial to interglacial atmospheric CO_2 change. *Global Biogeochem. Cycles*, 1:15–29, 1987.

Broecker, W. S. and T.-H. Peng. The cause of the glacial to interglacial atmospheric CO_2 change: a polar alkalinity hypothesis. *Global Biogeochemical Cycles*, 3:215–239, 1989.

Broecker, W. S. and T.-H. Peng. Greenhouse Puzzles: Keelings's World, Martin's World, Walker's World (2nd ed.). Eldigio Press, Palisades, New York, 1998.

Broecker, W. S. and T.-H. Peng. *Tracers in the Sea*. Eldigio Press, Lamont-Doherty Geological Observatory, Palisades, New York, pp. 690, 1982.

Broecker, W. S. and T. Takahashi. Calcium carbonate precipitation on the Bahama Banks. *J. Geophys. Res.*, 71:1575–1602, 1966.

Broecker, W. S. and T. Takahashi. Neutralization of fossil fuel CO_2 by marine calcium carbonate. In Anderson, N. R. and A. Malahoff, editor, *The Fate of Fossil Fuel CO_2 in the Oceans*, pages 213–241. Plenum Press, New York, 1977.

Broecker, W. S. and T. Takahashi. The relationship between lysocline depth and *in situ* carbonate ion concentration. *Deep-Sea Res.*, 25:65–95, 1978.

Broecker, W. S. and V. M. Oversby. *Chemical Equilibria in the Earth.* McGraw-Hill, New York, pp. 318, 1971.

Broecker, W. S., T. Takahashi, H. J. Simpson, and T.-H. Peng. Fate of fossil fuel carbon dioxide and the global carbon budget. *Science*, 206:409–418, 1979.

Carslaw, H. S., and J. C. Jaeger. *Conduction of Heat in Solids.* Oxford University Press, 1959.

Chacko, T., T. K. Mayeda, R. N. Clayton, and J. R. Goldsmith. Oxygen and carbon isotope fractionations between CO_2 and calcite. *Geochim. Cosmochim. Acta*, 55:2867–2882, 1991.

Chadwick, J. The existence of a neutron. *Proc. Roy. Soc. London*, ser. A, 136:692–708, 1932.

Ciais, P., P. P. Tans, A. S. Denning, R. J. Francey, M. Trolier, H. A. J. Meijer, J. W. C. White, J. A. Berry, D. R. Randall, J. G. Collatz, P. J. Sellers, P. Monfray, and M. Heimann. A three-dimensional synthesis study of $\delta^{18}O$ in atmospheric CO_2, 2. Simulations with the TM2 transport model. *J. Geophys. Res.*, 102:5873–5883, 1997.

Ciais, P., P. P. Tans, J. W. C. White, M. Trolier, R. J. Francey, J. A. Berry, D. R. Randall, P. J. Sellers, J. G. Collatz, and D. S. Schimel. Partitioning of ocean and land uptake of CO_2 as inferred by $\delta^{13}C$ measurements from the NOAA Climate Monitoring and Diagnostics Laboratory Global Air Sampling Network. *J. Geophys. Res.*, 100:5051–5070, 1995.

Clark, I. D. and P. Fritz. *Environmental Isotopes in Hydrogeology.* Lewis Publishers, Boca Raton, New York, pp. 328, 1997.

CLIMAP Project Members. The surface of the ice-age earth. *Science*, 191:1131–1137, 1976.

CLIMAP Project Members. Seasonal reconstruction of the earth's surface at the last glacial maximum. Geol. Soc. Am. Chart Ser. MC-36, Boulder, Colorado, 1981.

Copin-Montegut, C. A new formula for the effect of temperature on the partial pressure of CO_2 in seawater. *Mar. Chem.*, 25:29–37, 1988.

Coplen, T. B., C. Kendall, and J. Hopple. Comparison of stable isotope reference samples. *Nature*, 302:236–238, 1983.

Craig, H. Carbon 13 in plants and the relationships between carbon 13 and carbon 14 variations in nature. *J. Geol.*, 62:115–149, 1954.

Craig, H. Isotopic variations in meteoric waters. *Science*, 133:1702–1703, 1961.

Craig H. and Gordon L. I. Deuterium and oxygen 18 variations in the ocean and the marine atmosphere. Symposium on Marine Geochemistry. University of Rhode Island Publication, Vol. 3, pages 277–374, 1965.

Criss, R. E. *Principles of Stable Isotope Distribution*. Oxford University Press, 1999.

Cussler, E. L. *Diffusion. Mass Transfer in Fluid Systems*. Cambridge University Press, New York, pp. 525, 1984.

Davis, A. R. and B. G. Oliver. A vibrational-spectroscopic study of the species present in the $CO_2 - H_2O$ system. *J. Sol. Chem.*, 1(4):329–339, 1972.

De La Rocha, C. L., M. A. Brzezinski, and M. J. DeNiro. Fractionation of silicon isotopes by marine diatoms during biogenic silica formation. *Geochim. Cosmochim. Acta*, 61:5051–5056, 1997.

De La Rocha, C. L., M. A. Brzezinski, M. J. DeNiro, and A. Shemesh. Silicon-isotope composition of diatoms as an indicator of past oceanic changes. *Nature*, 395:680–683, 1998.

Debye, P. Reaction rates in ionic solutions. *Trans. Electrochem. Soc.*, 82:265–272, 1942.

Deffeyes, K. S. Carbonate equilibria: a graphic and algebraic approach. *Limnol. Oceanogr.*, 10:412–426, 1965.

Denbigh, K. *The Principles of Chemical Equilibrium*. Cambridge Univ. Press, London, pp. 377-383, 1971.

Dickson, A. G. An exact definition of total alkalinity and a procedure for the estimation of alkalinity and total inorganic carbon from titration data. *Deep-Sea Res.*, 28A:609–623, 1981.

Dickson, A. G. pH scales and proton-transfer reactions in saline media such as sea water. *Geochim. Cosmochim. Acta*, 48:2299–2308, 1984.

Dickson, A. G. Thermodynamics of the dissociation of boric acid in synthetic seawater from 273.15 to 318.15 K. *Deep-Sea Res.*, 37:755–766, 1990.

Dickson, A. G. The development of the alkalinity concept in marine chemistry. *Marine Chemistry*, 40(1-2):49–63, 1992.

Dickson, A. G. pH buffers for sea water media based on the total hydrogen ion concentration scale. *Deep-Sea Res.*, 40:107–118, 1993a.

Dickson, A. G. The measurement of sea water pH. *Marine Chemistry*, 44:131–142, 1993b.

Dickson, A. G. and F. J. Millero. A comparison of the equilibrium constants for the dissociation of carbonic acid in seawater media. *Deep-Sea Res.*, 34:1733–1743, 1987.

Dickson, A. G. and J. P. Riley. The estimation of acid dissociation constants in seawater media from potentiometric titrations with strong base. I. The ionic product of Water (K_W). *Marine Chemistry*, 7:89–99, 1979.

DOE. *Handbook of methods for the analysis of the various parameters of the carbon dioxide system in sea water; version 2.* Dickson, A. G. and Goyet, C., editors. ORNL/CDIAC-74, 1994.

Dole, M., G. A. Lane, D. P. Rudd, and D. A. Zaukelies. Isotopic composition of atmospheric oxygen and nitrogen. *Geochim. Cosmochim. Acta*, 6:65–78, 1954.

Drever, J. I. *The Geochemistry of Natural Waters*. Prentice Hall, Englewood Cliffs, 1982.

Dreybrodt, W., L. Eisenlohr, B. Madry, and R. Ringer. Precipitation kinetics of calcite in the system $CaCO_3$-H_2O-CO_2: The conversion to CO_2 by the slow process $H^+ + HCO_3^- \rightarrow CO_2 + H_2O$ as a rate limiting step. *Geochim. Cosmochim. Acta*, 61:3897–3904, 1997.

Dyrssen, D., and L. G. Sillén. Alkalinity and total carbonate in sea water. A plea for *P-T*-independent data. *Tellus*, 19:113–121, 1967.

Edsall, J. T. Carbon dioxide, carbonic acid, and bicarbonate ion: Physical properties and kinetics of interconversion. In Forster, R. E. et al., editor, *CO_2: Chemical, Biochemical, and Physiological Aspects*, pages 15–27. NASA SP-188. U.S. GPO, Washington, D.C., 1969.

Eigen, M. Protron transfer, acid-base catalysis, and enzymatic hydrolysis. *Angew. Chem. Int. Ed. Eng.*, 3:1–19, 1964.

Eigen, M., and G. Hammes. Elementary steps in enzyme reactions. In Nord, F. F., editor, *Advances in Enzymology*, pages 1–38. Wiley & Sons, New York, 1963.

Eigen, M. and L. De Maeyer. Relaxation methods. In Friess, S. L., E. S. Lewis, and A. Weissberger, editor, *Technique of Organic Chemistry (2nd ed.)*, volume VIII, No. 2, pages 895–1054. Wiley-Interscience, New York, 1963.

Eigen, M., K. Kustin, and G. Maass. Die Geschwindigkeit der Hydratation von SO_2 in wäßriger Lösung. *Z. Phys. Chem. NF*, 30:130–136, 1961.

Eisma, D., W. G. Mook, and H. A. Das. Shell characteristics, isotopic composition and trace-element contents of some euryhaline molluscs as indicators of salinity. *Palaeogeogr., Palaeoclimatol., Palaeoecol.*, 19:39–62, 1976.

Emerson, S. Chemically enhanced CO_2 gas exchange in a eutrophic lake: A general model. *Limnol. Oceanogr.*, 20:743–753, 1975.

Emerson, S. Enhanced transport of carbon dioxide during gas exchange. In B. Jähne and E. C. Monahan, editor, *Air-Water Gas Transfer. Selected papers from the Third International Symposium on Air-Water Gas Transfer July 24 - 27, 1995, Heidelberg University*, pages 23–36. AEON Verlag & Studio, Hanau, Germany, 1995.

Emiliani, C. Pleistocene temperatures. *J. Geol.*, 63:538–578, 1955.

Emiliani, C. Isotopic paleotemperatures. *Science*, 154:851–857, 1966.

Emiliani, C. and N. J. Shackleton. The Brunhes epoch: Isotopic paleotemperatures and geochronology. *Science*, 183:511–514, 1974.

Emrich, K., D. H. Ehhalt, and J. C. Vogel. Carbon isotope fractionation during the precipitation of calcium carbonate. *Earth Planet. Sci. Lett.*, 8:363–371, 1970.

Epstein, S. and T. Mayeda. Variation of O^{18} content of waters from natural sources. *Geochim. Cosmochim. Acta*, 4:213–224, 1953.

Epstein, S., H. A. Buchsbaum, H. A. Lowenstam, and H. C. Urey. Revised carbonate-water isotopic temperature scale. *Bull. Geol. Soc. Am.*, 64:1315–1326, 1953.

Epstein, S., P. Thompson, and C. J. Yapp. Oxygen and hydrogen isotopic ratios in plant cellulose. *Science*, 198:1209–1215, 1977.

Erez, J. and B. Luz. Experimental paleotemperature equation for planktonic foraminifera. *Geochim. Cosmochim. Acta*, 47:1025–1031, 1983.

Fairbanks, R. G. A 17,000-year glacio-eustatic sea level record: Influence of glacial melting rates on the Younger Dryas event and deep-ocean circulation. *Nature*, 342:637–642, 1989.

Farquhar, G. D., J. Lloyd, J. A. Taylor, L. B. Flanagan, J. P. Syvertsen, K. T. Hubick, S. C. Wong, and J. R. Ehleringer. Vegetation effects on the isotope composition of oxygen in atmospheric CO_2. *Nature*, 363:439–443, 1993.

Faure, G. *Principles of Isotope Geology*. Wiley & Sons, New York, pp. 589, 1986.

Forster, R. E., J. T. Edsall, A. B. Otis, and F. J. W. Roughton (editors). CO_2: *Chemical, Biochemical, and Physiological Aspects*. NASA SP-188. U.S. GPO, Washington, D.C., pp. 291, 1969.

Fourier, J. B. J. *Théorie Analytique de la Chaleur*. F. Didot, Paris, 1822.

Fowler, W. A. The quest for the origin of the elements. *Science*, 226:922–935, 1984.

Francey, R. J. and P. P. Tans. Latitudinal variation in oxygen-18 of atmospheric CO_2. *Nature*, 327:495–497, 1987.

Francey, R. J., C. E. Allison, D. M. Etheridge, C. M. Trudinger, I. G. Enting, M. Leuenberger, R. L. Langenfelds, E. Michel, and L. P. Steele. A 1000-year high precision record of $\delta^{13}C$ in atmospheric CO_2. *Tellus*, 51B:170–193, 1999.

Freeman, K. H. and J. M. Hayes. Fractionation of carbon isotopes by phytoplankton and estimates of ancient CO_2 levels. *Global Biogeochem. Cycles*, 6:185–198, 1992.

Friedman, I. and J. R. O'Neil. Compilation of stable isotope fractionation factors of geochemical interest. In M. Fleischer, editor, *Data of
Geochemisty (6th ed.)*. Chap. KK, U.S. Geol. Surv. Prof. Pap. 440-KK,
1977.

Galimov, E. M. *The Biological Fractionation of Isotopes*. Academic Press,
New York, pp. 261, 1985.

Garrels, R. M. and M. E. Thompson. A chemical model for seawater at 25°C
and one atmosphere total pressure. *Amer. J. Sci.*, 260:57–66, 1962.

Gattuso, J. P, D. Allemand, and M. Frankignoulle. Photosynthesis and
calcification at cellular, organismal and community levels in coral reefs:
A review on interactions and control by carbonate chemistry. *Am. Zoo.*,
39(1):160–183, 1999.

Gavis, J. and J. F. Ferguson. Kinetics of carbon dioxide uptake by phytoplankton at high pH. *Limnol. Oceanogr.*, 20:211–221, 1975.

Gill, A. E. *Atmosphere-Ocean Dynamics*. Academic Press, San Diego,
pp. 662, 1982.

Goldman, J. C. and P. G. Brewer. Effect of nitrogen source and growth rate
on phytoplankton mediated changes in alkalinity. *Limnol. Oceanogr.*,
25:352–357, 1980.

González, L. A. and K. C. Lohmann. Carbon and oxygen isotopic composition of Holocene reefal carbonates. *Geology*, 13:811–814, 1985.

Goyet, C. and A. Poisson. New determination of carbonic acid dissociation
constants in seawater as a function of temperature and salinity. *Deep-Sea
Res.*, 36:1635–1654, 1989.

Goyet, C., F. J. Millero, A. Poisson and D. K. Shafer. Temperature dependence of CO_2 fugacity in seawater. *Mar. Chem.*, 44:205–219, 1993.

Gran, G. Determination of the equivalence point in potentiometric titrations, II. *Analyst*, 77:661–671, 1952.

Grossman, E. L. Carbon isotopic fractionation in live benthic foraminifera
- comparison with inorganic precipitate studies. *Geochim. Cosmochim.
Acta*, 48:1505–1512, 1984.

Grossman, E. L. and T.-L. Ku. Oxygen and carbon isotope fractionation in
biogenic aragonite: Temperature effects. *Chem. Geol.*, 59:59–74, 1986.

Guggenheim, E. A. *Thermodynamics: An Advanced Treatment for Chemists and Physicists (5th ed.).* North-Holland, Amsterdam, pp. 390, 1967.

Guilderson, T. P., R. G. Fairbanks, and J. L. Rubenstone. Tropical temperature variations since 20,000 years ago: Modulating interhemispheric climate change. *Science*, 263:663–665, 1994.

Guillard, R. R. L. and B. H. Ryther. Studies of marine planktonic diatoms. *Can. J. Microbiol.*, 8:229–239, 1962.

Guy, R. D., M. L. Fogel, and J. A. Berry. Photosynthetic fractionation of the stable isotopes of oxygen and carbon. *Plant Physiol.*, 101:37–47, 1993.

Hagemann, R., G. Nief, and E. Roth. Absolute isotopic scale for deuterium analysis of natural waters. Absolute D/H ratio for SMOW. *Tellus*, 22:712–715, 1970.

Halas, S., J. Szaran, and H. Niezgoda. Experimental determination of carbon isotope equilibrium fractionation between dissolved carbonate and carbon dioxide. *Geochim. Cosmochim. Acta*, 61:2691–2695, 1997.

Hansson, I. A new set of pH-scales and standard buffers for sea water. *Deep-Sea Res.*, 20:479–491, 1973a.

Hansson, I. A new set of acidity constants for carbonic acid and boric acid in sea water. *Deep-Sea Res.*, 20:461–478, 1973b.

Hastings, D. W., A. D. Russell, and S. R. Emerson. Foraminiferal magnesium in *Globigerinoides sacculifer* as a paleotemperature proxy. *Paleoceanography*, 13:161–169, 1998.

Hayes, J. M. Fractionation et al.: An introduction to isotopic measurements and terminology. *Spectra*, 8:3–8, 1982.

Hayes, J. M. Factors controlling ^{13}C contents of sedimentary organic compounds: Principles and evidence. *Marine Geology*, 113:111–125, 1993.

Hayes, J. M., H. Strauss and A. J. Kaufman. The abundance of ^{13}C in marine organic matter and isotopic fractionation in the global biogeochemical cycle of carbon during the past 800 Ma. *Chem. Geol.*, 161:103–125, 1999.

Heinze, C., E. Maier-Reimer, and K. Winn. Glacial pCO_2 reduction by the world ocean: Experiments with the Hamburg Carbon Cycle Model. *Paleoceanography*, 6:395–430, 1991.

Hemming, N. G. and G. N. Hanson. Boron isotopic composition and concentration in modern marine carbonates. *Geochim. Cosmochim. Acta*, 56:537–543, 1992.

Hemming, N. G., R. J. Reeder, and G. N. Hanson. Mineral-fluid partitioning and isotopic fractionation of boron in synthetic calcium carbonate. *Geochim. Cosmochim. Acta*, 59:371–379, 1995.

Hemming, N. G., R. J. Reeder, and S. R. Hart. Growth-step-selective incorporation of boron on the calcite surface. *Geochim. Cosmochim. Acta*, 62:2915–2922, 1998a.

Hemming, N. G., T. P. Guilderson and R. G. Fairbanks. Seasonal variations in the boron isotopic composition of coral: A productivity signal? *Global Biogeochem. Cycles*, 12:581–586, 1998b.

Hershey, J. P., M. Fernandez, P. J. Milne, and F. J. Millero. The ionization of boric acid in NaCl, Na-Ca-Cl and Na-Mg-Cl solutions at 25°C. *Geochim. Cosmochim. Acta*, 50:143–148, 1986.

Hoefs, J. *Stable Isotope Geochemistry (4th ed.).* Springer Verlag, Berlin, Heidelberg, Germany, pp. 201, 1997.

Holland, H. D. *The chemical evolution of the atmosphere and oceans.* Princeton University Press, Princeton, NJ, pp. 598, 1984.

Holligan, P. M. and J. E. Robertson. Significance of ocean carbonate budgets for the global carbon cycle. *Global Change Biology*, 2:85–95, 1996.

Hostetler, S. W. and A. C. Mix. Reassessment of ice-age cooling of the tropical ocean and atmosphere. *Nature*, 399:673–676, 1999.

Hurlbut, C. S. *Dana's Manual of Mineralogy (18rd ed.).* Wiley & Sons, New York, pp. 579, 1971.

Hut, G. Consultants' group meeting on stable isotope reference samples for geochemical and hydrological investigations, Rep. to Dir. Gen., Int. At. Energy Agency, Vienna, pp. 42, 1987.

Hynes, J. T. The protean proton in water. *Nature*, 397:565–567, 1999.

Imbrie, J. and N. G. Kipp. A new micropaleontological method for quantitative paleoclimatology: Application to a late Pleistocene Caribbean core. In K. K. Turekian, editor, *The Late Cenozoic Glacial Ages*, pages 71–179. Yale Univ. Press, New Haven, Conn., 1971.

Indermühle, A., T. F. Stocker, F. Joos, H. Fischer, H. J. Smith, M. Wahlen, B. Deck, D. Mastroianni, J. Tschumi, T. Blunier, R. Meyer, and B. Stauffer. Holocene carbon-cycle dynamics based on CO_2 trapped in ice at Taylor Dome, Antarctica. *Nature*, 398:121–126, 1999.

Inoue, H. and Y. Sugimura. Carbon isotopic fractionation during the CO_2 exchange process between air and sea water under equilibrium and kinetic conditions. *Geochim. Cosmochim. Acta*, 49:2453–2460, 1985.

IPCC. *Climate Change 1995 - The Science of Climate Change.* Houghton, J. T., L. G. M. Filho, B. A. Callander, N. Harris, A. Kattenberg, and K. Maskell, editors, Cambridge University Press, 1996.

Ishikawa, T. and E. Nakamura. Boron isotope systematics of marine sediments. *Earth Planet. Sci. Lett.*, 117:567–580, 1993.

IUPAC. Isotopic compositions of the elements 1997. *Pure Appl. Chem.*, 70:217–235, 1998.

IUPAC. Atomic weights of the elements 1997. *Pure Appl. Chem.*, 71:1593–1607, 1999.

Jähne, B., G. Heinz, and W. Dietrich. Measurement of the diffusion coefficients of sparingly soluble gases in water. *J. Geophys Res.*, 92:10,767–10,776, 1987.

Johansson, O., and M. Wedborg. On the evaluation of potentiometric titrations of seawater with hydrochloric acid. *Oceanol. Acta*, 5:209–218, 1982.

Johnson, K. M., A. Körtzinger, L. Mintrop, J. C. Duinker, and D. W. R. Wallace. Coulometric total carbon dioxide analysis for marine studies: measurement and internal consistency of underway TCO_2 concentrations. *Marine Chemistry*, 67:123–144, 1999.

Johnson, K. S. Carbon dioxide hydration and dehydration kinetics in seawater. *Limnol. Oceanogr.*, 27:849–855, 1982.

Kakihana, H. and M. Kotaka. Equilibrium constants for boron isotope-exchange reactions. *Bull. of Research Laboratory for Nuclear Reactors*, 2:1–12, 1977.

Kakihana, H., M. Kotaka, S. Satoh, M. Nomura, and M. Okamoto. Fundamental studies on the ion-exchange separation of boron isotopes. *Bull. Chem. Soc. Japan*, 50:158–163, 1977.

Keeling, C. D., R. B. Bacastow, and P. P. Tans. Predicted shift in the $^{13}C/^{12}C$ ratio of atmospheric carbon dioxide. *Geophys. Res. Lett.*, 7:505–508, 1980.

Keeling, C. D., T. P. Whorf, M. Wahlen, and J. van der Plicht. Interannual extremes in the rate of rise of atmospheric carbon dioxide since 1980. *Nature*, 375:666–670, 1995.

Kern, D. M. The hydration of carbon dioxide. *J. Chem. Education*, 37:14–23, 1960.

Kiddon, J., M. L. Bender, J. Orchardo, D. A. Caron, J. C. Goldman, and M. Dennett. Isotopic fractionation of oxygen by respiring marine organisms. *Global Biogeochem. Cycles*, 7:679–694, 1993.

Killingley, J. K. Effects of diagenetic recrystallization on $^{18}O/^{16}O$ values of deep-sea sediments. *Nature*, 301:594–597, 1983.

Kim, S.-T. and J. R. O'Neil. Equilibrium and nonequilibrium oxygen isotope effects in synthetic carbonates. *Geochim. Cosmochim. Acta*, 61:3461–3475, 1997.

Kleypas, J. A., R. W. Buddemeier, D. Archer, J.-P. Gattuso, C. Langdon, and B. N. Opdyke. Geochemical consequences of increased atmospheric carbon dioxide on coral reefs. *Science*, 284:118–120, 1999.

Klotz, I. M. and R. M. Rosenberg. *Chemical Thermodynamics - Basic Theory and Methods*. Wiley-Interscience, New York, pp. 576, 2000.

Knoche, W. Chemical reactions of CO_2 in water. In Bauer, C., G. Gros, and H. Bartels, editor, *Biophysics and physiology of carbon dioxide*, pages 3–11. Springer-Verlag, Berlin, 1980.

Knox, F. and M. B. McElroy. Changes in atmospheric CO_2: influence of the marine biota at high latitude. *J. Geophys. Res.*, 89:4629–4637, 1984.

Korb, R. E., P. J. Saville, A. M. Johnston, and J. A. Raven. Sources of inorganic carbon for photosynthesis by three species of marine diatom. *J. Phycol.*, 33:433–440, 1997.

Körtzinger, A. Determination of carbon dioxide partial pressure ($p(CO_2)$). In Grasshoff, K., K. Kremling, and M. Ehrhardt, editor, *Methods of Seawater Analysis*, pages 149–158. Wiley-VCH, Weinheim, 600 pp., 1999.

Kotaka, M. and H. Kakihana. Thermodynamic isotope effect of trigonal planar and tetrahedral molecules. *Bull. of Research Laboratory for Nuclear Reactors*, 2:13–29, 1977.

Kroopnick, P. and H. Craig. Atmospheric oxygen: Isotopic composition and solubility fractionation. *Science*, 175:54–55, 1972.

Lane, G. and M. Dole. Fractionation of oxygen isotopes during respiration. *Science*, 123:574–576, 1956.

Langdon, Ch., T. Takahashi, C. Sweeny, D. Chipman, J. Goddard, F. Marubini, H. Aceves, H. Barnett, and M. J. Atkinson. Effect of calcium carbonate saturation state on the calcification rate of an experimental coral reef. *Global Biogeochem. Cycles*, 14:639–654, 2000.

Lasaga, A. C. Rate laws of chemical reactions. In Lasaga, A. C. and R. J. Kirkpatrick, editor, *Kinetics of Geochemical Processes*, pages 1–68. Mineral. Soc. Am., Rev. Miner., Vol.8, Washington, D.C., 1981.

Lee, K. E. and N. C. Slowey. Cool surface waters of the subtropical North Pacific Ocean during the last glacial. *Nature*, 397:512–514, 1999.

Lee, K., F. J. Millero, R. H. Byrne, R. A. Feely and R. Wanninkhof. The recommended dissociation constants for carbonic acid in seawater. *Geophys. Res. Let.*, 27:229–232, 2000.

Lehman, J. T. Enhanced transport of inorganic carbon into algal cells and its implication for biological fixation of carbon. *J. Phycol.*, 14:33–42, 1978.

Lemarchand, D., J. Gaillardet, É. Lewin, and C. J. Allègre. The influence of rivers on marine boron isotopes and implications for reconstructing past ocean pH. *Nature*, 408:951–954, 2000.

Leśniak, P. M. and H. Sakai. Carbon isotope fractionation between dissolved carbonate (CO_3^{2-}) and $CO_2(g)$ at 25° and 40°. *Earth Planet. Sci. Lett.*, 95:297–301, 1989.

Lueker, T. J., A. G. Dickson, and Ch. D. Keeling. Ocean pCO_2 calculated from dissolved inorganic carbon, alkalinity, and equations for K_1 and

K_2: validation based on laboratory measurements of CO_2 in gas and seawater at equilibrium. *Mar. Chem.*, 70:105–119, 2000.

Luz, B., E. Barkan, M. L. Bender, M. H. Thiemens, and K. A. Boering. Triple-isotope composition of atmospheric oxygen as a tracer of biosphere productivity. *Nature*, 400:547–550, 1999.

Lynch-Stieglitz, J., T. F. Stocker, W. S. Broecker, and R. G. Fairbanks. The influence of air-sea exchange on the isotopic composition of oceanic carbon: Observations and modeling. *Global Biogeochem. Cycles*, 9:653–665, 1995.

Mackenzie, F. T. and J. A. Mackenzie. *Our Changing Planet*. Prentice Hall, Englewood Cliffs, NJ, pp. 387, 1995.

Majoube, M. Fractionnement en oxygèn 18 et en deutérium entre l'eau et sa vapeur. *J. Chim. Phys.*, 68:1423–1436, 1971.

Mallo P., G. Waton, and S. J. Candau. Temperature-jump rate study of chemical relaxations in sea-water. *Nouv. J. Chim.*, 8:373–379, 1984.

Marx, D., M. E. Tuckerman, J. Hutter, and M. Parrinello. The nature of the hydrated excess proton in water. *Nature*, 397:601–604, 1999.

Mashiotta, T. A., D. W. Lea, and H. J. Spero. Glacial-interglacial changes in Subantarctic sea surface temperature and $\delta^{18}O$-water using foraminiferal Mg. *Earth Planet. Sci. Lett.*, 170:417–432, 1999.

Mayer, J. E. and M. G. Mayer. *Statistical Mechanics*. Wiley & Sons, New York, pp. 495, 1966.

McClendon, J. F., G. C. Gault, and S. Mulholland. The hydrogen ion concentration, CO_2 tension, and CO_2 content of seawater. Carnegie Inst. Washington Publ., 51: 21-69, 1917.

McConnaughey, T. ^{13}C and ^{18}O isotopic disequilibrium in biological carbonates: I. Patterns. *Geochim. Cosmochim. Acta*, 53:151–162, 1989a.

McConnaughey, T. ^{13}C and ^{18}O isotopic disequilibrium in biological carbonates: II. *In vitro* simulation of kinetic isotope effects. *Geochim. Cosmochim. Acta*, 53:163–171, 1989b.

McCrea. On the isotopic chemistry of carbonates and a paleotemperature scale. *J. Chem. Phys.*, 18(6):849–857, 1950.

McCulloch M. T., A. W. Tudhope, T. M. Esat, G. E. Mortimer, J. Chappell, B. Pillans, A. R. Chivas, and A. Omura. Coral record of equatorial sea-surface temperatures during the penultimate deglaciation at Huon Peninsula. *Science*, 283:202–204, 1999.

Mehrbach, C, C. H. Culberson, J. E. Hawley, and R. M. Pytkowicz. Measurement of the apparent dissociation constant of carbonic acid in seawater at atmospheric pressure. *Limnol. Oceanogr.*, 18:897–907, 1973.

Melander, L. and Saunders, W. H. *Reaction Rates of Isotopic Molecules.* Wiley & Sons, New York, pp. 331, 1980.

Mellen R. H, D. G. Browning, and V. P. Simmons. Investigation of chemical sound absorption in seawater: Part III. *J. Acoust. Soc. Am.*, 70:143–148, 1981.

Mellen R. H, D. G. Browning, and V. P. Simmons. Investigation of chemical sound absorption in seawater: Part IV. *J. Acoust. Soc. Am.*, 74:987–993, 1983.

Miller, K. G., R. G. Fairbanks, and G. S. Mountain. Tertiary oxygen isotope synthesis, sea level history, and continental margin erosion. *Paleoceanography*, 2:1–19, 1987.

Miller, R. F., D. C. Berkshire, J. J. Kelley, and D. W. Hood. Method for determination of reaction rates of carbon dioxide with water and hydroxyl ion in seawater. *Environ. Sci. Technol.*, 5:127–133, 1971.

Millero, F. J. The thermodynamics of the carbonate system in seawater. *Geochim. Cosmochim. Acta*, 43:1651–1661, 1979.

Millero, F. J. The thermodynamics of seawater at one atmosphere. *Ocean Sci. Eng.*, 7:403–460, 1982.

Millero, F. J. Thermodynamics of the carbon dioxide system in the oceans. *Geochim. Cosmochim. Acta*, 59:661–677, 1995.

Millero, F. J. *The Physical Chemistry of Natural Waters.* John Wiley & Sons, 880 pp., 2001.

Millero, F. J. *Chemical Oceanography (2nd ed.).* CRC Press, Boca Raton, pp. 469, 1996.

Millero, F. J., and A. Poisson. International one-atmosphere equation of state of seawater. *Deep-Sea Res.*, 28A:625–629, 1981.

Millero, F. J. and D. Pierrot. A chemical equilibrium model for natural waters. *Aquatic Geochemistry*, 4:153–199, 1998.

Millero, F. J. and D .R. Schreiber. Use of the ion pairing model to estimate activity coefficients of the ionic components of natural waters. *Am. J. Sci.*, 282:1508–1540, 1982.

Millero, F. J. and R. Roy. A chemical model for the carbonate system in natural waters. *Croatica Chemica Acta*, 70:1–38, 1997.

Millero, F. J., J. Z. Zhang, S. Fiol, S. Sotolongo, R. N. Roy, K. Lee, and S. Mane. The use of buffers to measure the pH of seawater. *Marine Chemistry*, 44:143–152, 1993a.

Millero, F. J., R. H. Byrne, R. Wanninkhof, R. Feely, T. Clayton, P. Murphy, and M. F. Lamb. The internal consistency of CO_2 measurements in the equatorial Pacific. *Marine Chemistry*, 44:269–280, 1993b.

Milliman, J. D. Production and accumulation of calcium carbonate in the ocean: Budget of a nonsteady state. *Global Biogeochem. Cycles*, 4:927–957, 1993.

Milliman, J. D. and A. W. Droxler. Neritic and pelagic carbonate sedimentation in the marine environment: Ignorance is not bliss. *Geol. Rundsch.*, 85:496–504, 1996.

Mills, G. A. and H. C. Urey. The kinetics of isotopic exchange between carbon dioxide, bicarbonate ion, carbonate ion and water. *J. Am. Chem. Soc.*, 62:1019–1026, 1940.

Mook, W. G. Carbon-14 in hydrogeological studies. In Fritz, P. and J. Ch. Fontes, editor, *Handbook of Environmental Isotope Geochemistry, 1*, pages 49–74. Elsevier, Amsterdam, 1980.

Mook, W. G. ^{13}C in atmospheric CO_2. *Netherlands Journal of Sea Research*, 20(2/3):211–223, 1986.

Mook, W. G. Principles of Isotope Hydrology. Free University of Amsterdam, Amsterdam, pp. 153, 1994.

Mook, W. G., J. C. Bommerson, and W. H. Staverman. Carbon isotope fractionation between dissolved bicarbonate and gaseous carbon dioxide. *Earth Planet. Sci. Lett.*, 22:169–176, 1974.

Moore, J. W. and R. G. Pearson. *Kinetics and Mechanism (3rd ed.).* Wiley-Interscience, New York, pp. 455, 1981.

Morel, F. M. M. and J. G. Hering. *Principles and Applications of Aquatic Chemistry.* Wiley & Sons, New York, 1993.

Morse, J. W. and F. T. Mackenzie. *Geochemistry of Sedimentary Carbonates.* Developments in sedimentology, 48, Elsevier, Amsterdam, 1990.

Mucci, A. The solubility of calcite and aragonite in seawater at various salinities, temperatures, and one atmosphere total pressure. *Am. J. Sci.,* 283:780–799, 1983.

Mucci, A., R. Canuel, and S. Zhong. The solubility of calcite and aragonite in sulfate-free seawater and seeded growth kinetics and composition of precipitates at 25°C. *Chem. Geol.,* 74:309–329, 1989.

Müller, T. J. Determination of salinity. In Grasshoff, K., K. Kremling, and M. Ehrhardt, editor, *Methods of Seawater Analysis.* Wiley-VCH, Weinheim, 600 pp., 1999.

Nguyen, M. T. and T.-K. Ha. A theoretical study of the formation of carbonic acid from the hydration of carbon dioxide: A case of active solvent catalysis. *J. Am. Chem. Soc.,* 106:599–602, 1984.

Nürnberg, D., J. Bijma, and C. Hemleben. Assessing the reliability of magnesium in foraminiferal calcite as a proxy for water mass temperatures. *Geochim. Cosmochim. Acta,* 60:803–814, Erratum: pp. 2483–2484, 1996.

O'Leary, M. H. Measurement of the isotope fractionation associated with diffusion of carbon dioxide in aqueous solution. *J. Phys. Chem.,* 88:823–825, 1984.

O'Leary, M. H., S. Madhaven, and P. Paneth. Physical and chemical basis of carbon isotope fractionation in plants. *Plant, Cell Environ.,* 15:1099–1104, 1992.

O'Neil, J. R. Terminology and standards. In Valley, J. W, J. R. O'Neil, and H. P. Taylor, editor, *Stable isotopes in high temperature geological processes*, pages 561–570. Mineral. Soc. Am., Rev. Miner., Vol.16, Washington, D.C., 1986.

O'Neil, J. R., L. H. Adami, and S. Epstein. Revised value for the O^{18} fractionation between CO_2 and H_2O at 25°C. *J. Res. US Geol. Surv.,* 3:623–624, 1975.

Ortiz, J. D. and A. C. Mix. Comparison of Imbrie-Kipp transfer function and modern analog temperature estimates using sediment trap and core top foraminiferal faunas. *Paleoceanography*, 12:175–190, 1997.

Palmer, M. R., A. J. Spivack, and J. M. Edmond. Temperature and pH controls over isotopic fractionation during adsorption of boron on marine clay. *Geochim. Cosmochim. Acta*, 51:2319–2323, 1987.

Palmer, M. R. and G. H. Swihart. Boron isotope geochemistry: An overview. In Grew, E. S. and L. M. Anovitz, editor, *Boron: Mineralogy, Petrology, and Geochemistry*, pages 709–744. Mineral. Soc. Am., Rev. Miner., Vol.33, Washington, D.C., 1996.

Palmer, M. R., P. N. Pearson, and S. J. Cobb. Reconstructing past ocean pH-depth profiles. *Science*, 282:1468–1471, 1998.

Paneth, P. and M. H. O'Leary. Carbon isotope effect on dehydration of bicarbonate ion catalyzed by carbonic anhydrase. *Biochemistry*, 24:5134–5147, 1985.

Park, K. Oceanic CO_2 system: an evaluation of ten methods of investigation. *Limnol. Oceanogr.*, 14:179–186, 1969.

Pitzer, K. S. Thermodynamics of electrolytes. I. Theoretical basis and general equations. *J. Phys. Chem.*, 77:268–277, 1973.

Plummer, L. N. and E. Busenberg. The solubilities of calcite, aragonite, and vaterite in CO_2-H_2O-solutions between 0 and 90°C and an evaluation of the aqueous model for the system $CaCO_3$-CO_2-H_2O. *Geochim. Cosmochim. Acta*, 46:1011–1040, 1982.

Popp, B. N., R. Takigiku, J. M. Hayes, J. W. Louda and E. W. Baker. The post-Paleozoic chronology and mechanism of ^{13}C depletion in primary marine organic matter. *Amer. J. Sci.*, 289:436–454, 1989.

Quay, P. D., B. Tilbrook, and C. S.Wong. Oceanic uptake of fossil fuel CO_2: Carbon-13 evidence. *Science*, 256:74–79, 1992.

Rakestraw, N. W. The conception of alkalinity or excess base of sea water. *J. Mar. Res.*, 8:14–20, 1949.

Raven, J. A. Limits on growth rates. *Nature*, 361:209–210, 1993.

Redlich, O. Eine allgemeine Beziehung zwischen den Schwingungsfrequenzen isotoper Molekeln. *Z. physikal. Chem.*, B, 28:371–382, 1935.

Reeder, R. J. Crystal chemistry of the rhombohedral carbonates. In Reeder, R. J., editor, *Carbonates: Mineralogy and chemistry*, pages 1–47. Mineral. Soc. Am., Washington, D.C., 1983.

Reinfelder, J. R., A. M. L. Kraepiel, and F. M. M. Morel. Unicellular C_4 photosynthesis in a marine diatom. *Nature*, 407:996–999, 2000.

Richet, P., Y. Bottinga, and M. Javoy. A review of hydrogen, carbon, nitrogen, oxygen, sulphur, and chlorine stable isotope fractionation among gaseous molecules. *Ann. Rev. Earth Planet. Sci.*, 5:65–110, 1977.

Riebesell, U., D. Wolf-Gladrow, and V. Smetacek. Carbon dioxide limitation of marine phytoplankton growth rates. *Nature*, 361:249–251, 1993.

Riebesell U., I. Zondervan, B. Rost, P. D. Tortell, R. E. Zeebe, and F. M. M. Morel. Reduced calcification of marine plankton in response to increased atmospheric CO_2. *Nature*, 407:364–367, 2000.

Rohling, E. J., M. Fenton, F. J. Jorissen, P. Bertrand, G. Ganssen, and J. P. Caulet. Magnitudes of sea-level lowstands of the past 500,000 years. *Nature*, 394:162–165, 1998.

Romanek, C. S., E. L. Grossman, and J. W. Morse. Carbon isotope fractionation in synthetic aragonite and calcite: Effects of temperature and precipitation rate. *Geochim. Cosmochim. Acta*, 56:419–430, 1992.

Rosell-Melé, A., E. Bard, K.-C. Emeis, P. Farrimond, J. Grimalt, P. J. Müller, and R. R. Schneider. Project takes a new look at past sea surface temperatures. *EOS, AGU*, 79:393–394, 1998.

Roy, R. N., L. N. Roy, K. M. Vogel, C. Porter-Moore, T. Pearson, C. E. Good, F. J. Millero, and D. M. Campbell. The dissociation constants of carbonic acid in seawater at salinities 5 to 45 and temperatures 0 to 45°C. *Marine Chemistry*, 44:249–267, 1993a.

Roy, R. N., L. N. Roy, M. Lawson, K. M. Vogel, C. Porter-Moore, W. Davis, and F. J. Millero. Thermodynamics of the dissociation of boric acid in seawater at S = 35 from 0 to 55°C. *Marine Chemistry*, 44:243–248, 1993b.

Rubinson, M. and R. N. Clayton. Carbon-13 fractionation between aragonite and calcite. *Geochim. Cosmochim. Acta*, 33:997–1002, 1969.

Sackett, W. M. A history of the $\delta^{13}C$ composition of oceanic plankton. *Marine Chemistry*, 34:153–156, 1991.

Sanyal, A., M. Nugent, R. J. Reeder, J. Bijma. Seawater pH control on the boron isotopic composition of calcite: Evidence from inorganic calcite precipitation experiments. *Geochim. Cosmochim. Acta*, 64:1551–1555, 2000.

Sanyal, A., N. G. Hemming, G. N. Hanson, and W. S. Broecker. Evidence for a higher *p*H in the glacial ocean from boron isotopes in foraminifera. *Nature*, 373:234–236, 1995.

Sanyal, A., N. G. Hemming, W. S. Broecker, D. W. Lea, H. J. Spero, and G. N. Hanson. Oceanic *p*H control on the boron isotopic composition of foraminifera: Evidence from culture experiments. *Paleoceanography*, 11:513–517, 1996.

Sarmiento, J. L. and J. R. Toggweiler. A new mode of the role of the oceans in determining atmospheric P_{CO_2}. *Nature*, 308:621–624, 1984.

Sarnthein and the Working Group GLAMAP-2000. Atlantic sea-surface temperatures during the LGM - The "GLAMAP-2000" record. 1. EPI-LOG Workshop on Global Ocean and Land Surface Temperatures during the Last Ice Age, Delmenhorst, Germany 3-6 May 1999.

Schrag, D. P., G. Hampt, and D. W. Murray. Pore fluid constraints on the temperature and oxygen isotopic composition of the glacial ocean. *Science*, 272:1930–1932, 1996.

Schwarcz, H. P., E. K. Agyei, and C. C. McMullen. Boron isotopic fractionation during clay adsorption from sea-water. *Earth Planet. Sci. Lett.*, 6:1–5, 1969.

Shackleton, N. J. Oxygen isotope evidence for Cenozoic climatic change. In P. Brenchley, editor, *Fossils and Climate*, pages 27–34. Wiley & Sons, New York, 1984.

Shackleton, N. J. Oceanic carbon isotope constraints on oxygen and carbon dioxide in the Cenozoic atmosphere. In Sundquist, E. T. and W. S. Broecker, editor, *The Carbon Cycle and Atmospheric CO_2: Natural Variations Archean to Present, Geophys. Monogr. Ser., Vol. 32*, pages 412–417. AGU, Washington D.C., 1985.

Shackleton, N. J. and N. D. Opdyke. Oxygen isotope and palaeomagnetic stratigraphy of equatorial Pacific core V28-238: Oxygen isotope temperatures and ice volume on a 10^5 and 10^6 year scale. *Quat. Res.*, 3:39–55, 1973.

Siegenthaler, U. Carbon dioxide: Its natural cycle and anthropogenic perturbation. In Buat-Ménard, P., editor, *The Role of Air-Sea Exchange in Geochemical Cycling*, pages 209–248. Reidel, Dordrecht, 1986.

Siegenthaler, U. and K. O. Münnich. $^{13}C/^{12}C$ fractionation during CO_2 transfer from air to sea. In Bolin, B., editor, *SCOPE 16 - The Global Carbon Cycle*, pages 249–257. Wiley & Sons, New York, 1981.

Siegenthaler, U. and Th. Wenk. Rapid atmospheric CO_2 variations and ocean circulation. *Nature*, 308:624–626, 1984.

Sigman, D. M. and Boyle, E. A. Glacial/interglacial variations in atmospheric carbon dioxide. *Nature*, 407:859–869, 2000.

Sillén, L. G. The physical chemistry of seawater. In M. Sears, editor, *Oceanography*, pages 549–581. Amer. Assoc. Adv. Sci., Publ. 67, Washington, D.C., 1961.

Sillén, L. G. The ocean as a chemical system. *Science*, 156:1189–1197, 1967.

Skirrow, G. The dissolved gases - carbon dioxide. In J. P. Riley and G. Skirrow, editor, *Chemical Oceanography, Vol. 1*, pages 227–322. Academic Press, 1965.

Skirrow, G. The dissolved gases - carbon dioxide. In J. P. Riley and G. Skirrow, editor, *Chemical Oceanography, Vol. 2 (2nd ed.)*, pages 1–192. Academic Press, New York, 1975.

Soddy, F. Intra-atomic charge. *Nature*, 92:399–400, 1913-14.

Sørensen, S. P. L. Enzymstudien. II. Mitteilung. Über die Messung und die Bedeutung der Wasserstoffionenkonzentration bei enzymatischen Prozessen. *Biochem. Z.*, 21:131–304, 1909.

Spero H. J. and D. F. Williams. Extracting environmental information from planktonic foraminiferal $\delta^{13}C$ data. *Nature*, 335:717–719, 1988.

Spero, H. J., J. Bijma, D. W. Lea, and A. D. Russell. Deconvolving glacial ocean carbonate chemistry from the planktonic foraminifera carbon isotope record. In Abrantes, F. and A. C. Mix, editor, *Reconstructing Ocean History: A Window into the Future*, pages 329–342. Kluwer Academic/Plenum Pub., New York, 1999.

Spero H. J., J. Bijma, D. W. Lea, and B. E. Bemis. Effect of seawater carbonate concentration on foraminiferal carbon and oxygen isotopes. *Nature*, 390:497–500, 1997.

Spivack, A. J. and J. M. Edmond. Boron isotope exchange between seawater and the oceanic crust. *Geochim. Cosmochim. Acta*, 51:1033–1043, 1987.

Spivack, A. J., C.-F. You, and H. J. Smith. Foraminiferal boron isotope ratios as a proxy for surface ocean pH over the past 21 Myr. *Nature*, 363:149–151, 1993.

Steemann Nielsen, E. The use of radio-active carbon ^{14}C for measuring organic production in the sea. *J. Con. Perm. int. Explor. Mer.*, 18(2):117–140, 1952.

Steinfeld, J. I., J. S. Francisco, and W. L. Hase. *Chemical Kinetics and Dynamics (2nd ed.)*. Prentice Hall, Upper Saddle River, NJ, pp. 518, 1999.

Stoll, M. H. C. *Inorganic carbon behaviour in the North Atlantic Ocean*. PhD thesis, Rijksuniversiteit Groningen, 1994.

Stumm, W. and J. J. Morgan. *Aquatic Chemistry (3rd ed.)*. Wiley & Sons, New York, pp. 1022, 1996.

Su, C. and D. L. Suarez. Coordination of adsorbed boron: a FTIR spectroscopic study. *Environ. Sci. Technol.*, 29:302–311, 1995.

Sundquist, E. T. Geologic Analogs: Their value and limitations in carbon dioxide research. In J. R. Trabalka and D. E. Reichle, editor, *The Changing Carbon cycle: A Global Analysis*, pages 371–402. Springer-Verlag, New York, 1986.

Szaran, J. Carbon isotope fractionation between dissolved and gaseous carbon dioxide. *Chem. Geol.*, 150:331–337, 1998.

Takahashi, T., R. A. Feely, R. F. Weiss, R. H. Wanninkhof, D. W. Chipman, S. C. Sutherland, and T. T. Takahashi. Global air-sea flux of CO_2: An estimate based on measurements of sea-air pCO_2 difference. *Proc. Natl. Acad. Sci.*, 94:8292–8299, 1997.

Tans, P. P., J. A. Berry and R. F. Keeling. Oceanic $^{13}C/^{12}C$ observations: A new window on oceanic CO_2 uptake. *Global Biogeochem. Cycles*, 7(2):353–368, 1993.

Tarutani, T., R. N. Clayton, and T. K. Mayeda. The effect of polymorphism and magnesium substitution on oxygen isotope fractionation between calcium carbonate and water. *Geochim. Cosmochim. Acta*, 33:987–996, 1969.

Thode, H. G., M. Shima, C. E. Rees, and K. V. Krishnamurty. Carbon-13 isotope effects in systems containing carbon dioxide, bicarbonate, carbonate, and metal ions. *Canad. J. Chem.*, 43:582–595, 1965.

Thomson, J. J. Rays of positive electricity. *Proc. Roy. Soc. London*, ser. A, 89:1–20, 1914.

Turner, J. V. Kinetic fractionation of carbon-13 during calcium carbonate precipitation. *Geochim. Cosmochim. Acta*, 46:1183–1191, 1982.

UNESCO. Algorithms for computation of fundamental properties of seawater. Endorsed by UNESCO/SCOR/ICES/IAPSO Joint Panel on Oceanographic Tables and Standards and SCOR Working Group 51. no. 44, 1983.

UNESCO. Thermodynamics of the carbon dioxide system in seawater. *Unesco technical papers in marine science*, 51:1–55, 1987.

Urey, H. C. The thermodynamic properties of isotopic substances. *J. Chem. Soc.*, pages 562–581, 1947.

Urey, H. C., F. G. Brickwedde, and G. M. Murphy. A hydrogen isotope of mass 2 and its concentration. *Phys. Rev.*, 40:1–15, 1932.

Usdowski, E. Reactions and equilibria in the systems CO_2-H_2O and $CaCO_3$-CO_2-H_2O (0° - 50°C). A review. *N. Jahrb. Miner. Abh.*, 144:148–171, 1982.

Usdowski E. and J. Hoefs. Oxygen isotope exchange between carbonic acid, bicarbonate, carbonate, and water: A re-examination of the data of McCrea (1950) and an expression for the overall partitioning of oxygen isotopes between the carbonate species and water. *Geochim. Cosmochim. Acta*, 57:3815–3818, 1993.

Usdowski, E., G. Menschel, and J. Hoefs. Kinetically controlled partitioning and isotopic equilibrium of ^{13}C and ^{12}C in the system $CO_2 - NH_3 - H_2O$. *Z. Phys. Chem.*, 130:13–21, 1982.

Usdowski, E., J. Michaelis, M. E. Böttcher, and J. Hoefs. Factors for the oxygen isotope equilibrium fractionation between aqueous and gaseous CO_2, carbonic acid, bicarbonate, carbonate, and water (19°C). *Z. Phys. Chem.*, 170:237–249, 1991.

Vengosh, A., Y. Kolodny, A. Starinsky, A. R. Chivas, and M. T. McCulloch. Coprecipitation and isotopic fractionation of boron in modern biogenic carbonates. *Geochim. Cosmochim. Acta*, 55:2901–2910, 1991.

Vogel, J. C., P. M. Grootes, and W. G. Mook. Isotopic fractionation between gaseous and dissolved carbon dioxide. *Z. Phys.*, 230:225–238, 1970.

Wachter, E. A. and J. M. Hayes. Exchange of oxygen isotopes in carbon dioxide-phosphoric acid systems. *Chem. Geol.*, 52:365–374, 1985.

Waldmann, L. Zur Theorie der Isotopentrennung durch Austauschreaktionen. *Naturwiss.*, 31:205–206, 1943.

Wanninkhof, R. Kinetic fractionation of the carbon isotopes ^{13}C and ^{12}C during transfer of CO_2 from air to seawater. *Tellus*, 37B:128–135, 1985.

Wanninkhof, R. and M. Knox. Chemical enhancement of CO_2 exchange in natural waters. *Limnol. Oceanogr.*, 41:689–697, 1996.

Wanninkhof, R., E. Lewis, R. A. Feely, and F. J. Millero. The optimal carbonate dissociation constants for determining surface water pCO_2 from alkalinity and total inorganic carbon. *Marine Chemistry*, 65:291–301, 1999.

Waton, G., P. Mallo, and S. J. Candau. Temperature-jump rate study of the chemical relaxation of aqueous boric acid solutions. *J. Phys. Chem.*, 88:3301–3305, 1984.

Wedborg, M., D. R. Turner, L. G. Anderson and D. Dyrssen. Determination of pH. In Grasshoff, K., K. Kremling, and M. Ehrhardt, editor, *Methods of Seawater Analysis*, pages 109–125. Wiley-VCH, Weinheim, 600 pp., 1999.

Wefer, G. and W. H. Berger. Isotope paleontology: growth and composition of extant calcareous species. *Mar. Geol.*, 100:207–248, 1991.

Weiss, R. F. Carbon dioxide in water and seawater: The solubility of a non-ideal gas. *Mar. Chem.*, 2:203–215, 1974.

Weiss, R. F. and B. A. Price. Nitrous oxide solubility in water and seawater. *Mar. Chem.*, 8:347–359, 1980.

Williams, W. S. C. *Nuclear and Particle Physics.* Oxford University Press, pp. 385, 1991.

Wolf-Gladrow, D. A., and U. Riebesell. Diffusion and reactions in the vicinity of plankton: A refined model for inorganic carbon transport. *Mar. Chem.*, 59:17–34, 1997.

Wolf-Gladrow, D. A., J. Bijma, and R. E. Zeebe. Model simulation of the carbonate system in the microenvironment of symbiont bearing foraminifera. *Mar. Chem.*, 64:181–198, 1999a.

Wolf-Gladrow, D. A., U. Riebesell, S. Burkhardt, and J. Bijma. Direct effects of CO_2 concentration on growth and isotopic composition of marine plankton. *Tellus*, B 51:461–476, 1999b.

Wollast, R. Rate and mechanism of dissolution of carbonates in the system $CaCO_3$-$MgCO_3$. In W. Stumm, editor, *Aquatic chemical kinetics*, pages 431–445. Wiley & Sons, New York, 1990.

Zachos, J. C., L. D. Stott, and K. C. Lohmann. Evolution of early Cenozoic marine temperatures. *Paleoceanography*, 9:353–387, 1994.

Zeebe, R. E. An explanation of the effect of seawater carbonate concentration on foraminiferal oxygen isotopes. *Geochim. Cosmochim. Acta*, 63:2001–2007, 1999.

Zeebe, R. E. Modelling carbon and oxygen isotope fractionation in foraminifera. In J. Erez and T. McConnaughey, editor, *Biomineralization in Marine Symbiotic Associations and its Implications for Paleoceanographic Studies*, page . In press, 2001a.

Zeebe, R. E. Seawater pH and Isotopic Paleotemperatures of Cretaceous Oceans. *Palaeogeogr., Palaeoclimatol., Palaeoecol.*, 170:49–57, 2001b.

Zeebe R. E., A. Sanyal, J. D. Ortiz, and D. A. Wolf-Gladrow. A theoretical study of the kinetics of the boric acid - borate equilibrium in seawater. *Mar. Chem.*, 74(2):113–124, 2001.

Zeebe, R. E., H. Jansen, and D. A. Wolf-Gladrow. On the time required to establish chemical and isotopic equilibrium in the carbon dioxide system in seawater. *Mar. Chem.*, 65:135–153, 1999b.

Zeebe, R. E., J. Bijma, and D. A. Wolf-Gladrow. A diffusion-reaction model of carbon isotope fractionation in foraminifera. *Mar. Chem.*, 64:199–227, 1999a.

Zhang, J., P. D. Quay, and D. O. Wilbur. Carbon isotope fractionation during gas-water exchange and dissolution of CO_2. *Geochim. Cosmochim. Acta*, 59:107–114, 1995.

Zondervan, I., R. E. Zeebe, B. Rost, and U. Riebesell. Decreasing marine biogenic calcification: A negative feedback on rising atmospheric pCO_2. *Global Biogeochem. Cycles*, 15(2):507–516, 2001.

Zuddas, P. and A. Mucci. Kinetics of calcite precipitation from seawater: II. The influence of the ionic strength. *Geochim. Cosmochim. Acta*, 62:757–766, 1998.

Index